W9-BAS-579

Aircraft Safety

Aircraft Safety
Accident Investigations, Analyses, and Applications

Shari Stamford Krause, Ph.D.

McGraw-Hill

New York San Francisco Washington, D.C. Auckland Bogotá
Caracas Lisbon London Madrid Mexico City Milan
Montreal New Delhi San Juan Singapore
Sydney Tokyo Toronto

McGraw-Hill

A Division of The **McGraw·Hill** *Companies*

©1996 by The McGraw-Hill Companies, Inc.

Printed in the United States of America. All rights reserved. The publisher takes no responsibility for the use of any materials or methods described in this book, nor for the products thereof.

pbk   2 3 4 5 6 7 8 9 FGR/FGR 9 0 0 9 8 7 6
hc   1 2 3 4 5 6 7 8 9 FGR/FGR 9 0 0 9 8 7 6

Product or brand names used in this book may be trade names or trademarks. Where we believe that there may be proprietary claims to such trade names or trademarks, the name has been used with an initial capital or it has been capitalized in the style used by the name claimant. Regardless of the capitalization used, all such names have been used in an editorial manner without any intent to convey endorsement of or other affiliation with the name claimant. Neither the author nor the publisher intends to express any judgment as to the validity or legal status of any such proprietary claims.

Library of Congress Cataloging-in-Publication Data
Krause, Shari Stamford.
    Aircraft safety : accident investigations, analyses & applications
/ by Shari Stamford Krause.
       p.   cm.
    Includes bibliographical references and index.
    ISBN 0-07-036026-X (hardcover).   ISBN 0-07-036027-8 (pbk.)
    1. Aeronautics—Safety measures. 2. Aircraft accidents-
-Investigation. I. Title.
    TL553.5.K73 1996
    363.12'41—dc20   95-47931
                                                           CIP

McGraw-Hill books are available at special quantity discounts to use as premiums and sales promotions, or for use in corporate training programs. For more informa-tion, please write to the Director of Special Sales, McGraw-Hill, 11 West 19th Street, New York, NY 10011. Or contact your local bookstore.

Acquisitions editor: Shelley IC. Chevalier
Editorial team: Charles Spence, Book Editor
           Susan W. Kagey, Managing Editor
           Joanne Slike, Executive Editor
           Jennifer M. Secula, Indexer
Production team: Katherine G. Brown, Director
           Rose McFarland, Desktop Operator
           Lorie L. White, Proofreading
           Jeffrey Miles Hall, Computer Artist
Design team: Jaclyn J. Boone, Designer   0360278
           Katherine Lukaszewicz, Associate Designer   PFS

Disclaimer

The author used multiple sources to derive the facts and opinions expressed in *Aircraft Safety: Accident Investigation, Analyses, and Applications*. Although reasonable care was used to verify facts and instructions, neither the author nor publisher assume liability for errors in details which inevitably result from interpretation of evidence and details.

Contents

PART II: METEOROLOGY
AND ATMOSPHERIC PHENOMENA

6 Cloud formation   125

7 Thunderstorms   133

8 Microbursts and low-level windshear   139

9 Icing conditions   149

PART III: COLLISION AVOIDANCE

PART IV: MECHANICAL DEFICIENCIES AND MAINTENTANCE OVERSIGHTS

Acknowledgments

I WOULD LIKE TO EXTEND MY HEARTFELT APPRECIATION TO MY wonderful husband, Merrick, who has always supported my research and writing endeavors. I especially thank him for the hundreds of hours and sleepless nights that he devoted to this project as my extremely talented graphic illustrator. All of the graphic artwork in this text was either created or enhanced by him. I am truly grateful for all of his tireless efforts.

I could not have produced such a book without the many aviation professionals who helped in my quest for material. My sincerest appreciation to:

Gerry Brown, National Aeronautics and Space Administration

Cynthia Redding, National Transportation Safety Board

Pablo Santa Maria, Air Line Pilots Association

Dan Simonsen, aviation photographer

Loretta Stailey, National Technical Information Services

I also want to thank Shelley Chevalier, McGraw-Hill Aviation Acquisitions Editor, and John Cellini for their invaluable guidance and unrelenting support.

And a special thank you to Charlie Spence for his superb editing and gracious insight.

Introduction

IN THE WAKE OF THE JANUARY 1995 AVIATION SUMMIT, TRANSPORTATION Secretary Federico Pena announced to the world his commitment to a zero-accident rate for the U.S. airline industry. This bold and innovative move challenged government and airline officials to "elevate margins of safety, and anticipate rather than react" to safety problems before they occur. According to the National Transportation Safety Board (NTSB), the airline fatality rate for 1994 had climbed to a five-year high. The 264 deaths, the majority of which resulted from four horrific accidents in a five-month period, have spawned these most recent and aggressive measures.

To successfully achieve a zero-accident rate is indeed a lofty goal. However, we in the flying community must wholeheartedly and unequivocally strive towards obtaining that mindset. One way to begin is by studying the material in this book.

Although there has always been a genuine concern for safety throughout the airline industry, it appears that the lessons learned from past tragedies often go unnoticed, or worse yet, disregarded altogether. You may balk at the notion, but the evidence is quite clear. The causes of accidents are frequently repeated, simply due to the common thread that entwines all accidents—human intervention. But why does there seem to be this "disconnect" between our sincere desire to be safe and our penchant for continuously committing identical errors? One reason is that we often do not see ourselves when we pick up an accident report or a safety article in a magazine. We become passive readers and, therefore, we fail to apply those valuable insights and important lessons to our own flying routine.

The contents of this book present a unique blend of research material and instructional guidelines against the backdrop of actual aircraft accident cases. Additional references are used to expound on the NTSB's analyses and to fill in any gaps. The purpose is to provide pilots and aviation professionals, regardless of experience levels, the clear and realistic lessons that have evolved from these accidents. In turn, the reader will learn ways to customize those lessons into practical techniques suitable for any flight environment.

The book is divided into four comprehensive parts: Human Factors, Weather, Mid-Air Collisions, and Mechanical Failure. At the beginning of each of these sections is a complete study of issues that relate to a particular area of concentration. Each chapter is devoted to a detailed analysis of a specific accident that falls under one of these categories.

Part I, "Human factors," includes accidents associated with crew resource management (CRM): distraction in the cockpit, communication errors, cockpit discipline, pilot judgment, aeronautical decision-making (ADM), and substance abuse. Part II, "Meteorology and atmospheric phenomena," includes accidents associated with severe thunderstorm activity, microbursts, windshear, turbulence, and icing conditions. Part III, "Collision avoidance," includes accidents associated with the physical limitations, equipment shortcomings, and air traffic control constraints in a collision-avoidance environment. Part IV, "Mechanical deficiencies and maintenance oversights," includes accidents associated with aircraft maintenance and mechanical problems.

For the benefit of both self-study readers and classroom students, each area begins with learning objections and concludes with detailed, quick-reference summaries. Every chapter also includes lessons learned and practical applications.

Striving toward a zero-accident rate can only be accomplished if everyone associated with aviation accepts that challenge with a personal goal to "elevate the margins of safety, and anticipate rather than react."

PART I

Human factors

IN THE PAST DECADE, THE DISCIPLINE OF HUMAN FACTORS HAS BECOME instrumental in determining the causes of aircraft accidents. Advancements have been remarkable, and we in the aviation community are fortunate to have access to much valuable research. Yet, the practicality of this information is often lost somewhere between the laboratory and the cockpit. As a result, we tend to repeat the same *avoidable* mistakes over and over. But why? We have the tools to prevent an accident: experience, skill, knowledge, training, the desire to not die—so what stops us? Part of the answer is that many pilots and controllers have a misconception or lack of understanding about what can affect human performance to the point of disaster.

Data collected by NASA (National Aeronautics and Space Administration) and studied by the Ohio State University Aviation Psychology Laboratory in 1995, revealed that four out of five pilot errors that cause an aircraft incident occur before the flight leaves the ground. A recent study by the National Transportation Safety Board (NTSB) on the causes of major U.S. air carrier accidents revealed similar conclusions. Of 302 specific errors identified in 37 crashes analyzed between 1978 and 1990, more than three-fourths—78 percent— of the factors were contributed by other persons, including air traffic controllers, maintenance personnel, airline management, airport staff, and pilots of other aircraft. The results of these two expansive studies indicate that everything and everybody that we come in contact with from the time we pull out of the garage to head for the airport to when we secure the aircraft at the end of the day has the potential of interfering with the safety of that flight.

I know these ideas have been expressed in other literature and might be viewed with skepticism, but rest assured that this book will not address such important issues in those typically vague and broad concepts. In my professional opinion, these concepts are partly to blame for the information disconnect that pilots experience when studying human-error-related accidents. The collection of chapters in this section, therefore, was written with the sole intent to give you the insights and specific guidance that might help in your pursuit of becoming the safest and best pilot possible. They include crew resource management (CRM), pilot judgment and aeronautical decision-making (ADM), distractions in the cockpit, and cockpit discipline.

A QUICK LESSON IN HUMAN FACTORS

Why does every action, or inaction, we do in the air seem to have a domino effect; and why can the cause sometimes be traced back to something we did, or did not do, before we ever walked out to the airplane? This is one of those hotly-debatable questions that every pilot has asked, particularly after a flight when everything that could have gone wrong did. Although I've touched upon the answer a little earlier, the best place to start is by referring to Fig. I-1.

In human factors terms, a SHEL model represents the interrelationships between three types of resources and their environment. Every resource falls into one of the following categories: software (S), hardware (H), and liveware (L). These elements are then surrounded by an environment (E). Hardware is the aircraft itself, and its systems. Software includes all the regulations, SOPs, and policies. Manuals, checklists, maps, performance charts, tables, and graphs are also resources contained in the software element. And lastly, there is liveware. This includes you, and all the people you deal with on the ground and in the air. All three of these elements are then encompassed in a flight environment. This pertains directly to the actual physical environment and weather in which the aircraft operates. We obviously have no control over the environment but we do have some control over the factors.

No resource is totally isolated from the others. If one element is inferior, it will have a negative impact on all the others. In case you're still a little hazy on how this applies to the real world, following are a few examples of these interrelationships.

Liveware-Liveware Interface: This represents you (liveware) and any other person (liveware) who comes in contact with the operation of the flight. You were complacent about a checklist. The second-in-command (SIC) was preoccupied with ATC. The controller was stressed from a high workload, and therefore delayed your clearance. The mechanic was fatigued due to lack of sleep and didn't install a new component to the

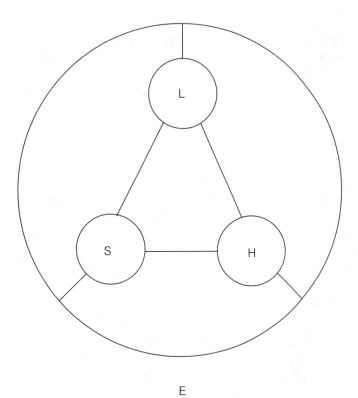

Fig. I-1. *SHEL Model. The SHEL model illustrates the interrelationships of software, liveware, hardware, and environment.*

VOR correctly, which subsequently fails in-flight. The meteorologist, in a hurry to get home, misidentified the intensity of some building thunderstorms, which meant that you had a busted forecast before you left the ground. You and the SIC never really clicked, so when an abnormal situation arose in-flight, due to the poorly installed part, there was a lot of miscommunication. Because you hadn't stayed current with your pubs and system knowledge, you displayed poor judgment and made a bad decision.

Liveware-Hardware Interface: The airplane (hardware) was broken. You (liveware) didn't catch it because you weren't paying attention. Your SIC (liveware) didn't catch it because he lacked the experience and knowledge. Maintenance (liveware) didn't catch it, either. Perhaps due to poor quality control, the mechanic being distracted, or the mechanic's lack of knowledge in certain areas.

Liveware-Software Interface: Neither one of you looked at the maintenance logs (software) for previous write-ups. You said that you would take care of getting the weather forecast (software), but you wasted so much time chit-chatting with an old classmate (L-L) that you bumped into on the ramp that by the time you got to the weather shop you were running late and didn't have the extra minutes to get a thorough briefing.

Liveware-Environment Interface: You approach unexpected bad weather (environment). It was unexpected only because both you and the meteorologist succumbed to "get-home-itis." Although the presence of weather is uncontrollable, what we do near and in the weather is totally up to us.

Hardware-Environment Interface: VOR broken in bad weather, marginal VFR rapidly deteriorating to IFR.

Software-Environment Interface: The regulations for marginal VFR, VFR partial panel, and VFR clearance in IMC.

OTHER FACTORS TO CONSIDER

Each person involved with a particular flight has a "personal" SHEL model. Therefore, a pilot or controller does not have just an individual L-H, L-S, and L-E interface to be concerned with. There are, in addition, L-L (pilot-pilot/pilot-controller) and H-H (ATC equipment-onboard equipment/TCAS II-TCAS II) interrelationships that can interfere with any given flight.

REFERENCES

Elwyn Edwards. 1988. "Introductory Overview." *Human Factors in Aviation.* Ed. Wiener, Earl L. and David C. Nagel. San Diego: Academic Press: 3-25.

1
Crew resource management

A<small>FTER COMPLETING THIS CHAPTER, YOU SHOULD BE ABLE TO:</small>

1. Discuss the basic concepts and goals of CRM.
2. Describe the main points of a CRM program.
3. Discuss how airline safety has improved since the introduction of CRM training.
4. Discuss the ways in which CRM can be translated to general aviation.

The principles of crew resource management (CRM) are based on one simple premise: the effective management of a pilot's available resources. Every pilot can benefit from such a concept, as studies have shown that most safe and accident-free flights are the direct result of CRM.

Whether you go to work with four stripes on your sleeve or fly Cherokees every other weekend, there are an abundance of resources to manage. Yourself, your fellow crewmembers, your airplane, air traffic control, maintenance, dispatchers, meteorologists, paperwork, checklists, approach plates, SOPs, and so on. Therefore, CRM probably could be considered the biggest chunk of a SHEL model. A breakdown in one area of CRM most likely will spread into other parts of a flight, and as you will see in the case studies presented at the end of each section, can ultimately lead to a serious accident.

THE GOALS OF CRM

This concept has proven so successful that many of the world's airlines have developed at least some form of CRM training. The core of the program is to create teamwork. The various techniques are designed to enhance management and leadership methods, emphasize decision-making and judgment skills, improve effective communication with others, and provide an overall productive work environment. The training also develops a keen insight into a person's behavior pattern during normal and emergency situations.

A STUDY IN TEAMWORK

As early as 1976, NASA researchers studied the value of total crew involvement and teamwork. One simulated exercise involved 18 experienced Boeing 747 pilots. The test included a progressive emergency coupled with bad weather and numerous cockpit distractions. Several flight engineers abandoned their critical fuel-burn calculations to oblige flight attendants who had interrupted their work with trivial matters. As a result, NASA noted a number of gross operational errors, including one miscalculation of 100,000 pounds of fuel. Although some captains resumed a leadership role and regained the flightcrew's attention, others either never heard what the flight attendant asked for or did nothing to stop the second officer from discontinuing his duties.

A SKEWED SENSE OF LEADERSHIP

Although most of the professional flying community has accepted CRM training as a viable asset to one's leadership style, the National Transportation Safety Board (NSTB) continues to document cases in which the cockpit is more like a demilitarized zone than a positive work environment. Usually being the more experienced airline pilot, the captain is most likely to dictate the objectives and priorities of the flight. Therefore, an underlying tone might be created that does not allow other crewmembers to speak freely, or worse, to do their jobs. In those cases, CRM training might be the only way crewmembers can form a bridge between conflicting viewpoints or personalities and achieve the mutual goal of ending the day without an incident.

CRM AND OVERALL SAFETY

It is important to note that a breakdown in CRM can cause problems other than accidents. Researchers have found that the relationship between poor communication and crew coordination contributes to even the simplest operating errors. Incidents associated with the mishandling of engine settings, hydraulics, and fuel systems, and the misreading or missetting of instruments are higher among crews that do not practice CRM techniques.

Pilots who have been involved in accidents or incidents have the opportunity to report their opinions and observations through NASA's Aviation Safety Reporting System (ASRS). Interestingly enough, 35 percent mentioned weak crew coordination, 16 percent named a relaxed cockpit environment with too much extraneous conversation, 15 percent spoke of misunderstanding other crewmembers, 10 percent specified complacency, and 5 percent cited a lack of confidence in subordinates, which in turn caused them to do other crewmembers' duties.

THE INTEGRATION OF CRM IN AIR CARRIER OPERATIONS

United Airlines, with the assistance of NASA's human factors division, pioneered one of the most respected CRM programs in the world. Although other airlines have outstanding CRM training in their organizations, this chapter focuses primarily on the CRM concepts that United has formally recognized.

United's program begins with an intensive three-and-a-half day seminar at a suburban Denver hotel. The airline purposely wants the participants to be in a location far removed from the surroundings of a flight training center. Each seminar is limited to 24 participants, who are assigned to four six-person teams. The groups are equally mixed with captains, first officers, and second officers. Oddly enough, the airline has since found that putting together teams of people who probably will disagree on some matters, like nonsmokers and smokers, works best. Once the teams are set, they stay that way for the duration of the seminar. That philosophy comes from the belief that if crews are forced to eat, work, and socialize together, the necessary bonding needed to form a cohesive group is then able to take place.

The seminars have very little structure and are conducted by airline pilots who do not instruct, but rather guide, the participants in the right direction. They have found this kind of environment to be conducive to the learning process that is so critical to the goals of CRM.

THE THREE-PHASE PROCESS

A specialized team of psychologists, researchers, and pilots determined that effective CRM training must incorporate three distinctive phases: awareness, practice and feedback, and continued reinforcement. Each phase is composed of several key elements, which when applied collectively, have been known to enhance the crew's overall work performance.

AWARENESS PHASE

The awareness phase focuses on the functioning of crews as teams during routine and emergency situations. This stage also helps pilots learn how best to develop a positive leadership style while understanding their own behavioral traits.

Team effort

Prior to attending the seminar, pilots receive eight booklets that describe a series of nontechnical in-flight situations, such as weather-related scenarios. The questions that follow ask what the pilot would do if in that position. Responses are graded and an individual score is later given to the pilot. While attending the CRM seminar, the pilots take a similar questionnaire, only this time they must answer as a team. Surprisingly, most crews try to work on these problems individually. Only when time is running out do these crews start working together; and they soon realize that if they had worked as a team from the start, they would have completed the assignment.

Each exercise presents a series of questions designed to measure leadership and behavioral attitudes in five separate areas:

1. Inquiry—whether the person takes an active or passive role in compiling facts.
2. Advocacy—whether the person aggressively states his or her feelings.
3. Conflict resolution—how the pilot typically tries to find a solution when working with others.
4. Critique—how well the participant is able to evaluate his or her performance and that of others.
5. Decision-making—assertiveness of the pilot.

In each area, five choices are given and the participant is asked to circle the one believed to be the best personal description. For instance, the choices for conflict resolution are:

A. When conflict arises, I try to remain neutral or stay out of it.
B. I try to avoid generating conflict, but when it does appear, I try to soothe feelings and to keep my crewmembers together.
C. I try to find a position that is acceptable.
D. I try to cut off conflict or win my position.
E. I seek out reasons for it in order to resolve underlying causes.

The results of this phase reveal how decisions made individually are much poorer than those made from a team effort. Most importantly, crewmembers are able to analyze how they react to various leadership styles in the cockpit and how their own behavior can affect operational outcomes.

PRACTICE AND FEEDBACK PHASE

Pilots are exposed to actual team-building as they go through problem-solving exercises. The focus is split between simulated flight decisions and scenarios that have nothing to do with flying. This split puts the pilots into situations in which they must solve problems without drawing on their technical experience. Once again the motive is to show the crews the importance of communicating and sticking together as a team.

The first part of this phase, which usually takes place on the second day of the seminar, is called Line-Oriented Flight Training (LOFT). The six-person teams are divided into three-person crews (captain, first officer, and second officer) to begin the flight simulator portion, which is videotaped. Scenarios from actual accident reports are used to introduce a wide spectrum of in-flight problems, including windshear and other weather phenomena, mechanical failures, crewmember incapacitation, and sick passengers. It is important that the LOFT scenario itself does not overload the crew, but presents believable and relevant situations. After two simulated flights, the tape is played back only to that specific crew and a CRM seminar guide. The tape is later erased. For the first time, these individuals see themselves functioning in a third-person perspective. Pilots have noted this can be the most powerful stimulus for attitude and behavioral change.

LOFT studies

From years of data collection, researchers have developed valuable insight as to how pilots interact. Among the discoveries, the following three points are of particular interest with regards to accident prevention.

First, any level of stress can produce a narrowing of perceptual attention. Therefore, a crew's overall performance will be weakened if the captain does not delegate duties. For instance, if a captain uses poor judgment and attempts to fly *and* direct crewmember activities in an emergency, the decision-making process will be negatively affected. When LOFT experiments involved high workloads and multiple aircraft emergencies, the crews noticed immediate benefits when the captain transferred aircraft control to the first officer while concentrating on effective decision-making and problem-solving. In these cases, the captain had more time to analyze the situation and develop effective plans.

This arrangement also provided a greater sense of responsibility to other crewmembers and promoted a high degree of vigilance and participation. Very simply, if treated and accepted as a valuable part of the team, a crewmember is going to respond accordingly.

The second finding is a problem with overfamiliarity of standing operating procedures and checklists. Pilots who have flown the same airplane or route for years have a tendency to hear and see what they've come to expect. Even familiarity between crewmembers can have the same results.

And lastly, the LOFT exercises have proven the need for the captain to include the coordination of the cabin crew's activities. The margin of safety for any particular flight dramatically increases when the captain involves the flight attendants during the initial preflight brief. Discussing specific emergencies with the entire crew has proven to enhance the overall safety of the flight. In addition, by setting the stage for open communication with the cabin crew, potentially serious malfunctions in the aircraft, such as pressurization, electrical, and air-conditioning problems have been detected and corrected early.

The culmination of the LOFT segment of the seminar is a dramatic videotaped reenactment of the last hour of United Flight 173 (refer to case study I-8). Once pilots have been exposed to the concepts of CRM and have viewed the tape, more than 70 situations have been observed in which the crew of Flight 173 could have reacted more effectively.

CONTINUOUS REINFORCEMENT PHASE

This stage is to be carried out of the seminar and into the actual cockpit environment. After each phase of CRM training, a poll is taken to determine how many pilots in the course believe they already use CRM principles in the cockpit. At the beginning, typically 95 percent of the participants feel they fall into that category. By the second day, that figure has dropped to 75 percent; and at the end of the program, it has dropped even further to only 15 percent. The reinforcement phase, therefore, is considered the most critical step for the ongoing success of CRM training.

This is where the true test of CRM takes place. The feeling among many seminar graduates is that since every United pilot has gone through CRM, they all know what is expected from each other. As soon as a pilot notices another person falling into the old way of conducting cockpit business, immediate action should take place. Some pilots say that even being as blunt as saying "You're being a jerk, let's practice some CRM," has been known to have a great impact. The philosophy at the airline is that the issue is too serious to overlook a potentially dangerous cockpit environment.

Therefore, peer pressure is viewed as a key element during the reinforcement phase. But encouraging crewmembers to stick together as a team is just as important. After all, no one wants to be known as the weak link of a crew.

CRM IN ACTION

An example of how these CRM principles can translate into the real world became evident when a United Airlines 727 encountered windshear while taking off from Denver's Stapleton Airport. In this 1984 incident, the jet struck a localizer shack, ripping a 15-foot gash in the belly of the airplane. The crew managed to land safely and was later praised by the NTSB for their outstanding performance.

The flightcrew reacted in an exemplary manner, according to the board's report, which included the comment that United's CRM training played a positive role in preventing a more serious accident from occurring in Denver. During the preflight briefing the crew discussed the threat of windshear, and the captain elected to make a full-power takeoff as a precaution. Each pilot was allowed to freely enter into the decision-making process. The flight attendants were also included in the briefing, allowing them the opportunity to prepare themselves with their own procedures should the aircraft encounter a windshear or subsequent emergency situation.

CRM FOR THE ENTIRE CREW

A recent study conducted by NASA researchers Chute and Wiener revealed an area in flight operations that is often overlooked in CRM training: pilot-flight attendant communication. As the aforementioned incident attests, the overall safety of a flight is dependent on the inclusion of flight attendants as a valuable part of the crew.

Historically, flight attendants have been cited as contributing factors to accidents and incidents. These instances have often been needless and avoidable and were usually attributed to a lack of communication between all crewmembers. For instance, in the case of a fatal 1989 Air Ontario F-28 accident, investigators discovered that flight attendants failed to inform the pilots that there was wet snow on the wings. Among the reasons given were professional respect, an assumption that the pilots were aware of all pertinent information, and a reluctance to second-guess the pilots.

According to the study, this reluctance was evident in another fatal accident in 1989. When the captain of a British Midlands 737-400 notified passengers over the PA system that there was a problem with the right engine, flight attendants could see that the left engine was on fire. The cabin crew never advised the pilots, and the cockpit crew proceeded to shut down the only good engine.

Cultural differences

The researchers noted that hundreds of communication and coordination problems between crews have been documented. In addition to the thorough examination of accident and incident reports, they also visited joint training classes of pilots and flight attendants. Based upon the cumulative information that they obtained, Chute and Wiener suggested that five factors have influenced the perpetual division of these crewmembers. Philosophical and corporate cultural differences appear to be at the core of these factors.

1. Historical background. Pilots were the persons who flew the airplane. The first flight attendants were called "skygirls" and were instructed to give a sharp salute to the pilots when they entered or exited the aircraft. Even today, there can be an impression that flight attendants are considered as "glorified waitresses" or referred to as "the girls in the back."

2. Physical separation. A closed cockpit door can serve as a point of division in the same manner as when flight attendants are huddled in the aft galley. A territorial attitude of "You take care of your part of the airplane, I'll take care of mine," is another problem indicative to the physical separation in an airplane.

3. Psychological isolation. Pilot skepticism; a misunderstanding of each other's duties, responsibilities, and problems; and flight-attendant ambivalence about chain of command are issues that crews have raised.

4. Regulatory factors. It's been noted by pilots that many flight attendants either disregard or don't understand that the sterile cockpit concept is a federal regu-

lation. When a flight attendant violates this rule and disrupts the cockpit activity during a critical phase of flight, pilots get angry. That hostility and frustration can roll over to subsequent flights and can be unfairly directed to all flight attendants.

5. Organizational separation. Often this division is brought on by training and scheduling differences and separation of departments. At some airlines, pilots work for flight operations whereas flight attendants fall under the marketing department.

The importance of communication

As part of the study, 302 crew surveys were analyzed. Of the 177 pilots who participated, 99 percent were male and had been pilots with their current airlines for a mean number of slightly more than seven years. The breakdown was 91 captains, 81 first officers, and 5 second officers. The mean number of total flight hours was 10,658; and 53 percent received their initial training in the military, while the remaining 47 percent were civilian-trained.

Of the 125 flight attendants who participated, 84 percent were female and had been flight attendants with their current airlines between seven and 20 years. About 81 percent considered their jobs as a long-term career, not a temporary fun thing to do. Most of the flight attendants were qualified on three to five aircraft types.

Numerous questions on the survey pertained to pilot-flight attendant interaction and communication. The results are quite interesting.

Question: Did you notice any work-related differences when paired with the same crew for several legs as opposed to one or two legs a trip?

The majority of crewmembers—77 percent of flight attendants and 71 percent of pilots—said they noticed a difference in the quality of interactions when paired for greater lengths of time. Some of the comments from flight attendants included: "Increased level of confidence and support." "You know how the flight-deck crew flies, what is normal and what is not." "You can depend on them." A sampling of pilot comments included: "Cockpit and cabin crews learn what to expect from one another, sets routines, etc." "Carry-over procedures and problems (delays, mechanicals) are understood." "Better communication and working relationship. More openness between crews."

The researchers believed that these responses corroborated other related data that indicated that crewmembers preferred to have the opportunity to foster good working relationships. Chute and Wiener also suggested that safety was enhanced through increased contact.

Question: How often do flight attendants receive formal briefings from the pilots?

More than one-third of the pilots said that they briefed the cabin crew frequently. However, more than 50 percent of the flight attendants noted the briefings occurred infrequently. In a further breakdown of the survey, it was revealed that 65 percent of pilots who routinely provide briefings do so only with the lead flight attendant. The results of the study strongly suggested, therefore, that lead flight attendants, for the most part, do not relay the information to the rest of the cabin crew.

Question: What are the benefits derived from briefings?

Flight attendants and pilots believed that setting the tone for crew communication was the most important result. Flight attendants ranked emergency procedures as a close second, whereas pilots chose weather-related problems.

Question: What is the frequency of flight-attendant introductions and pilot introductions?

According to the results, each crewmember usually waits for the other to introduce themselves. But both agreed that introductions were extremely important.

Question for flight attendants: I like it when pilots . . .

The participants responded by completing the question with: "Hold briefings, or at least introduce themselves and establish communication." "Introduce themselves and give a short briefing regarding communication, it shows respect." "Introduce themselves, give us a briefing on what they like to do in emergencies," and "Let us know about problems that might arise, including weather and delays."

The researchers concluded that, in many ways, pilots and flight attendants want the same level of communication, but for various reasons fail to take the extra steps to do so. They also noted that pilots are often unaware of the importance of certain pieces of information to flight attendants (i.e., weather delays, expected turbulence, and the timing of the pilots' meal service).

Recommendations

Chute and Wiener made several valid recommendations concerning pilot and flight attendant communication. The following are a few of those suggestions.

1. CRM training, whether done jointly or separately, should address cockpit/cabin crew communication.
2. Airlines should thoroughly investigate reorganizing their pilots and flight attendants under the same administrative structure.
3. Cockpit crews and cabin crews should be scheduled for flight sequences as a team.
4. Recurrent training should emphasize the importance of crew briefings. In particular, these briefings should be incorporated in captain upgrade training.
5. Crews should observe the courtesy of an introduction.

Remember, the term CRM is defined as, "using all available resources, information, equipment, and people to achieve safe and efficient flight operations." Don't forget the other half of your crew.

CRM AND AIR-CARRIER SAFETY

According to documented statistics, since United began the CRM training program, the airline's accident rate went from one hull loss per one million operations to one hull loss per 4.8 million operations. The average for all U.S. carriers is still one hull loss per one million operations. Nevertheless, the progress of these managerial concepts and techniques can be measured only during the aftermath of an incident or accident. That is the true test whether CRM training is a success or not.

EVERYDAY EXAMPLES

Let's look at a few examples that are relevant to professional and general aviation scenarios.

Pilot-pilot

Two issues are associated with this factor: One: Are you operating the aircraft as a crew? And two: Is your nonflying passenger also a licensed pilot?

If you fall into the first category, ask yourself the following questions:

1. Do I have a personal (nonprofessional) problem with this pilot?
   If, "Yes," then make the decision to leave it at home.

2. Do I think this person is an incompetent pilot?
   If, "Yes," then monitor and cross-check or take a bus.

3. Am I overly familiar with this person's flying techniques and habits?
   If, "Yes," then monitor and cross-check. Don't expect or assume anything. Make absolutely sure that checklists, ATC clearances, frequency changes, etc., have been completed correctly.

4. Is my communication clear, direct, and concise?
   If, "No," then make sure to think before you talk. Speak up. Say exactly what you mean. And watch extraneous chit-chat that might cloud the real message.

5. Do I easily get intimidated by this person?
   If, "Yes," then be confident with your knowledge and skills. More flight time and a few gray hairs don't automatically make another person a better pilot. But if you are a bit rusty in certain areas, then crack open a book. Flying is a constant learning process.

6. Does this other pilot have a tendency to get us into a lot of unsafe situations?
   If, "Yes," then speak up. I guarantee that suffering through a few dirty looks is better than dying.

What if you're the pilot and your passenger is one too?

1. My friend always wants to help out, but I'd rather fly and keep him as the passenger.
   First of all, a passenger who is also a pilot can be useful, especially if you insist that he or she help with cross-checks and scan for traffic. But you are still pilot-in-command. Therefore, fly whatever way makes you most comfortable.

2. Do I communicate in a clear, direct, and concise manner?

If "No," and if you want your pilot friend to help, then you must say exactly what the passenger-pilot's limits are—in a clear, direct, and concise manner. If you want that person to stop touching the radios, say so in a clear, direct, and concise manner. Just remember to set the limits before you're shooting a night ILS approach at minimums.

3. Do I let my friend talk me into things that I know are either unsafe or beyond my ability?
   If "Yes," remember: You are pilot-in-command. Good advice should always be well-received. That's part of good judgment. But don't allow the talking to continue if you know the information is wrong or you get that uneasy feeling in the pit of your stomach. You are the pilot-in-command.

PILOT-CONTROLLER

1. Am I clear, direct, and concise when I talk to ATC?
   If "No," then speak up. Think before you talk. Know exactly what you're going to say before you pick up the microphone. Be short and to the point.

2. I'm the pilot-in-command, controllers work for me.
   Pilots and controllers make up a very unique team. Each sees things that the other can't. An uncoordinated effort between pilot and controller can often lead to disaster.

3. Do I ask ATC for flight-following services when I'm flying VFR?
   If "Yes," then you're catching on.

CASE STUDY REFERENCES

I-1, I-2, I-3, I-4, I-5, I-6, I-7, I-8, II-2, II-3, III-1, III-2, III-3, III-5, III-6, IV-1, IV-3, IV-4

CHAPTER REVIEW

- The principles of crew resource management (CRM) are based on the effective management of a pilot's available resources.
- The goals of CRM are to create teamwork, enhance management skills, improve effective communication skills, and provide a productive work environment.
- Pilots of all experience levels can apply CRM.

A breakdown of CRM
The lack of CRM can cause:

- Poor communication
- Poor crew coordination
- Mishandling of engine and systems' settings
- Misreading instruments

CRM and overall safety

Pilots involved in accidents or incidents report the following as causal factors:

- Weak crew coordination
- Too-relaxed environment in the cockpit
- Misunderstanding crewmembers
- Complacency
- Lack of confidence in other crewmembers

CRM concept in air-carrier operations

- Airline CRM training programs are often developed after an accident or a series of incidents.

There are three primary phases of CRM:

- Awareness phase
- Practice and feedback phase
- Continuous reinforcement phase

Awareness phase deals with a pilot's personal leadership style and behavioral traits. Leadership and behavioral attitudes can be measured by the following:

- Inquiry
- Advocacy
- Conflict resolution
- Critique
- Decision-making

Practice and feedback phase focuses on team-building and problem solving exercises through:

- Line-oriented flight training (LOFT)
- Continuous reinforcement phase applied on the job

Flying the line with CRM

The lessons learned from the incident involving a United Airlines 727 are:

- Cockpit crew coordinated emergency procedures before they left the ground.
- Each crewmember had input to the decision-making process.
- Captain included flight attendants in pretakeoff briefing.

CRM for the entire crew

- Historical background
- Physical separation
- Psychological isolation
- Regulatory factors
- Organizational separation
- Importance of pilot-flight attendant communication

Air carrier safety
- Industry average: one hull loss per one million operations.
- United Airlines average since CRM: one hull loss per 4.8 million operations.

Lessons for every day
Pilot-pilot:

- Personal problems
- Professional concerns
- Technical skills
- Communication skills
- Intimidation
- Pilot-in-command

Pilot-controller communication skills:

- Knowledge of regulations and responsibilities
- Intimidation
- Pilot-in-command

CHAPTER REFERENCES

Baron, Sheldon. "Pilot Control." *Human Factors in Aviation*. Ed. Wiener, Earl L. and David C. Nagel. San Diego: Academic Press, 1988: 347–385.

Burrows, William E. "Cockpit Encounters." *Psychology Today*. November 1982: 43–7.

Caro, Paul W. "Flight Training and Simulation." *Human Factors in Aviation*. Ed. Wiener, Earl L. and David C. Nagel. San Diego: Academic Press, 1988: 258–9.

Chute, Rebecca D. and Earl L. Wiener. March 1995. "On a Collision Course?" *Air Line Pilot*: 20–25.

Driskeal, James, Richard Adams. 1992. *Crew Resource Management: An Introductory Handbook*. U.S. Department of Transportation. Federal Aviation Administration.

Edwards, Elwyn. "Introductory Overview." *Human Factors in Aviation*. Ed. Wiener, Earl L. and David C. Nagel. San Diego: Academic Press, 1988: 3–25.

Englade, Kenneth F. "Better Managers in the Friendly Skies." *Across the Board*. June 1988: 36–45.

Foushee, H. Clayton and Robert L. Helmreich. "Group Interaction and Flight Crew Performance." *Human Factors in Aviation*. Ed. Wiener, Earl L. and David C. Nagel. San Diego: Academic Press, 1988: 189–227.

Helmreich, Robert L., Ph.D. "Does CRM Training Work?" *Air Line Pilot*. May 1991: 17–20.

Kantowitz, Barry H. and Patricia A. Casper. "Human Workload in Aviation." *Human Factors in Aviation*. Ed. Wiener, Earl L. and David C. Nagel. San Diego: Academic Press, 1988: 157–187.

Lederer, Jerome. May 1992. "Communication is More Than Words." *Air Line Pilot*: 34–38.

Merritt, Ashleigh. May 1995. "Cross-Cultural Issues in CRM Training." *Air Line Pilot*: 32–35.

Nagel, David C. "Human Error in Aviation Operations." *Human Factors in Aviation*. Ed. Wiener, Earl L. and David C. Nagel. San Diego: Academic Press, 1988: 263–301.

Nance, John. Blind Trust. New York: Quill, 1986.

Negrette, Arthur J. "Cockpit Communication Could Save Your Life." *Rotor and Wing International*. February 1987: 22–23.

"Personality in the Cockpit." *Air Transport*. May 1988: 14–16.

Steenblik, Jan W. "Two Pilots, One Team—Part One." *Air Line Pilot*. August 1988: 10–5.

Steenblik, Jan W. "Two Pilots, One Team—Part Two." *Air Line Pilot*. September 1988: 10–4, 49.

Stone, Richard B. and Gary L. Babcock. "Airline Pilots' Perspective." *Human Factors in Aviation*. Ed. Wiener, Earl L. and David C. Nagel. San Diego: Academic Press, 1988: 529–560.

Viets, Jack. "Cockpit Resource Management and Airline Safety." *San Francisco Chronicle*. 15 January 1986: A9.

Instructions

1. Read over each of the situations and the choices. Decide which one is the most likely reason why you might make the choice that is described. Place a numeral 1 in the space provided on the answer sheet.

2. Continue by placing a 2 by the next most probable reason, and so on until you have filled in all five blanks with 1, 2, 3, 4 and 5.

3. Do all 10 situations and fill in each blank even though you might disagree with all of the choices listed. Remember, there are no correct or "best" answers.
   Example:
   _5_a. (your least likely response)
   _3_b.
   _1_c. (your most likely response)
   _2_d.
   _4_e.

Attitude inventory

Situation 1. Nearing the end of a long flight, your destination airport is reporting a ceiling of 600 feet and ½ mile visibility, fog and haze. You have just heard another aircraft miss the approach (ILS minimums are 200 and ½). You decide to attempt the ILS approach. Why do you make the attempt?

____a. Ceiling and visibility estimates are often not accurate.

____b. You are a better pilot than the one who just missed the approach.

____c. You might as well try, you can't change the weather.

____d. You are tired and just want to land.

____e. You've always been able to complete approaches under these circumstances in the past.

Situation 2. You plan an important business flight under instrument conditions in an aircraft with no deicing equipment. You'll be flying through an area in which light to moderate rime or mixed icing in clouds, and precipitation above the freezing level has been forecast. You decide to make the trip, thinking:

____a. You believe that your altitudes en route can be adjusted to avoid ice accumulation.

____b. You've been in this situation many times and nothing has happened.

____c. You must get to the business meeting in two hours and can't wait.

____d. You do not allow an icing forecast to stop you; weather briefers are usually overly cautious.

____e. There's nothing you can do about atmospheric conditions.

Situation 3. You arrive at the airport for a flight with a friend and plan to meet his friend who is arriving on a commercial airplane at your destination. The airplane you scheduled has been grounded for avionics repairs. You are offered another airplane equipped with unfamiliar avionics. You depart on an instrument flight without a briefing on the unfamiliar equipment. Why?

____a. If the avionics are so difficult to operate, the FBO would not have offered the plane as a substitute.

____b. You are in a hurry to make the scheduled arrival.

____c. Avionics checkouts are not usually necessary.

____d. You do not want to admit that you are not familiar with the avionics.

____e. You probably won't need to use those radios anyway.

Situation 4. You arrive at your destination airport to pick up a passenger after the fuel pumps have closed. Your calculations before departing determined that there would be enough fuel to complete the trip with the required reserves. The winds on the trip were stronger than anticipated, and you are not certain of the exact fuel consumption. You decide to return home without refueling since:

____a. You can't remain overnight because you and your passenger have to be at the office in the morning.

____b. The required fuel reserves are overly conservative.

____c. The winds will probably diminish for the return trip.

____d. You don't want to admit to your lack of planning in front of anyone else.

____e. It's not your fault the airport services are not available; you will just have to try to make it home.

Situation 5. You have been cleared for the approach on an IFR practice flight with a friend acting as safety pilot. At the outer marker, ATC informs you of a low-level windshear reported for your intended runway. Why do you continue the approach?

____a. You have to demonstrate to your friend that you can make this approach in spite of the wind.

____b. It has been a perfect approach so far; nothing is likely to go wrong.

____c. These alerts are for less experienced pilots.

____d. You need two more approaches to be current and want to get this one completed.

____e. The tower cleared you for the approach, so it must be safe.

Situation 6. You are about to fly some business associates in a multi-engine aircraft. You notice a vibration during run-up of the left engine. Leaning the mixture does not reduce the vibration. You take off without further diagnosis of the problem. Why?

____a. You need to be at your destination by five o'clock and are behind schedule. The aircraft can be checked there.

____b. You have encountered the vibration before without any problem.

____c. You don't want your business associates to think you can't handle the aircraft.

___d. The requirement for two perfectly smooth running engines is overly conservative.

___e. The shop just checked this plane yesterday. The mechanics would not have released it if there were a problem.

Situation 7. You are in IMC and receiving conflicting information from the two VOR receivers. You determine that the radios are out-of-tolerance and cannot determine your position. You believe ATC will soon suggest that you are off course and request a correction. You are thinking:

___a. Try to determine your position so ATC won't find out that you are lost.

___b. You will continue to navigate on the newer VOR receiver. It should work just fine.

___c. You will get out of this jam somehow; you always do.

___d. If ATC calls, you can be noncommittal. If they knew all, they would only make things worse.

___e. Inform ATC immediately that you are lost and wait impatiently for a response.

Situation 8. During an instrument approach, ATC calls and asks how much fuel you have remaining. You have only two minutes before reaching the missed approach point, and wonder why they have inquired as to your fuel status. You are concerned about severe thunderstorm activity nearby and assume that you might be required to hold. You believe that:

___a. Your fuel status is fine, but you want to land as soon as possible before the thunderstorm arrives.

___b. You are in line with the runway and believe that you can land, even in any crosswind that might come up.

___c. You will have to complete this approach; the weather won't improve.

___d. You won't allow ATC to make you hold in potentially severe weather; it's not their neck.

___e. The pilot who landed ahead of you completed the approach without any problems.

Situation 9. You are a new instrument pilot conducting an instrument flight of only 20 miles. The turn coordinator in your airplane is malfunctioning. The visibility is deteriorating, nearing approach minimums at your destination. You continue this trip thinking:

___a. You've never had a need to use the turn coordinator.

___b. You recently passed the instrument flight test and believe you can handle this weather.

___c. Why worry about it; ATC will get you out of a bad situation.

___d. You had better get going now before you get stuck here.

___e. Back up systems are not needed for such a short trip.

Situation 10. You encounter clear-air turbulence. You are not wearing a shoulder harness and do not put it on. Why not?

____a. Putting on a shoulder harness might give the appearance that you are afraid—you don't want to alarm your passengers.

____b. Shoulder harness regulations are unnecessary for en route operations.

____c. You haven't been hurt thus far by not wearing your shoulder harness.

____d. What's the use in putting on a shoulder harness, if it's your time, it's your time.

____e. You need to maintain aircraft control, there's no time for shoulder harnesses.

Interpreting your attitude inventory

Transfer the numbers from your questionnaire onto the Attitude Inventory Scoring Key (Table 2-1). The higher scores indicate the thought patterns and attitudes that you are susceptible to expressing. Remember, they do not indicate how your attitudes compare with anyone else, and it in no way represents a personality test.

Table 2-1. The attitude inventory scoring key.

Situation	Scale I	Scale II	Scale III	Scale IV	Scale V	Total
1.	a.	d.	e.	b.	c.	15
2.	d.	c.	b.	a.	e.	15
3.	c.	b.	e.	d.	a.	15
4.	b.	a.	c.	d.	e.	15
5.	c.	d.	b.	a.	e.	15
6.	d.	a.	b.	c.	e.	15
7.	d.	e.	c.	a.	b.	15
8.	d.	e.	c.	a.	b.	15
9.	e.	d.	a.	b.	c.	15
10.	b.	e.	c.	a.	d.	15
Total						150

The sum of your scores across should be 15 for each situation. If it is not, go back and make sure that you transferred the scores correctly and check your addition. The grand total should be 150.

Adapted from the Jensen FAA/DOT Study: Aeronautical Decision-Making—Cockpit Resource Management, January 1989.

The five hazardous thought patterns

Scale I: Anti-authority

This attitude is found in pilots who resent any external control over their actions. They have a tendency to disregard rules and procedures. "The regulations and SOPs are not for me."

Scale II: Impulsivity

This attitude is found in pilots who act too quickly. They tend to do the first thing that pops in their head. "I must act now; there's no time to waste."

Scale III: Invulnerability

This attitude is found in pilots who act as though nothing bad can happen to them. Many pilots feel that accidents happen to others but never to them. Those who think this way are more likely to take chances and run unwise risks. "It won't happen to me." Famous last words.

Scale IV: Macho

This attitude is found in pilots who continually try to prove themselves better than others. They tend to act with overconfidence and attempt difficult tasks for the admiration it gains them. "I'll show you. I can do it."

Scale V: Resignation

This attitude is found in pilots who believe that they have little or no control over their circumstances. They might feel, "What's the use?" These pilots might also deny that a problem is as it appears and believe, "It's not as bad as they say." It's unlikely that they would take charge of a situation, and they might even go along with unreasonable requests just to be a nice guy. Another common feeling is, "They're counting on me, I can't let them down."

Countering hazardous thought patterns and attitudes. Granted, we all have bits and pieces of these nasty thought patterns and attitudes; that's what, in part, makes up our individual personalities. But the degree to which we display these patterns, especially in the cockpit, is where the real problem lies. This is how the attitude inventory fits in. One of the best ways to begin eliminating, or at least alleviating, these patterns is by simply recognizing what thought pattern and attitude you are most vulnerable to.

The hazardous thought pattern exercise. Now that you have a better understanding of the five hazardous thought patterns, try this exercise to test your ability to spot each specific pattern. As we discussed earlier, being able to recognize these attitudes in yourself and others is half the battle. This exercise is also an excellent guide in helping you develop and enhance good judgment abilities. Remember, sound judgment leads to correct decisions.

The anti-authority hazardous thought pattern. From the five choices following each situation, pick the ONE choice that is the best example of an anti-authority hazardous thought pattern. Check your answers before you continue to the next situation. If you don't choose the correct answer, select another until you choose the correct one.

Situation 1. You do not conduct a thorough preflight. On takeoff you notice that the airspeed indicator is not working, nevertheless, you continue the takeoff roll. Your passenger feels strongly that you should discontinue the flight and return to the airport. You then become upset with your friend. Which of the following options best illustrates the ANTI-AUTHORITY reaction?

a. You tell your passenger to "cool it" for butting in.

b. You start banging on the airspeed indicator to get it working.

c. You think that the preflight check is something thought up by bureaucrats just to waste a pilot's time.

d. You tell the passenger that nothing dangerous will happen on the flight.

e. Your passenger continues to become more upset, but you do nothing, because you feel there is no use trying to calm the guy down.

Response options:

a. Macho. By acting in a superior way, you are being macho. "I can do it." Go back to Situation 1 and select another option.

b. Impulsive. By becoming upset and banging on the airspeed indicator, and by not thinking about the situation, you are being impulsive. "Quick! Do something!" Go back to Situation 1 and select another option.

c. Anti-authority. Yes. You selected the correct response. Looking at rules and procedures as just a waste of time, instead of taking them seriously, is an indication of an anti-authority attitude. Go on to Situation 2.

d. Invulnerable. Thinking that nothing will happen to you shows an invulnerable tendency. Go back to Situation 1 and select another option.

e. Resignation. By assuming that what you do has no effect on the passenger, the pilot is illustrating a tendency towards resignation. Go back to Situation 1 and select another option.

Situation 2. You have been cleared for an approach to a poorly lighted airport. You are not sure if this is the airfield where you want to land. The surrounding buildings do not look familiar, but it has been more than a year since your last visit. A much larger, more familiar airport is 15 miles away. Which of the following options best illustrates the ANTI-AUTHORITY reaction?

a. You decide to land anyway, thinking, "Of course I can handle this situation."

b. Rather than confuse yourself by thinking about options, you decide to land and get the flight over with.

c. You feel nothing will happen since you have gotten out of similar jams before.

d. You decide to land since the controller cleared you.

e. You decide to land because the regulations do not really apply in this situation.

Response options:

a. Macho. Thinking that you can handle the situation even when there is reason to be concerned, is an example of a macho attitude. Go back to Situation 2 and select another option.

b. Impulsive. "Quick! Do something! Anything!" Go back to Situation 2 and select another option.

c. Invulnerable. Thinking that nothing will happen to you, even in a problem situation, is illustrating a tendency towards invulnerability. Go back to Situation 2 and select another option.

d. Resignation. A pilot with the belief that "the controller is watching over me" has just relieved himself of duty as pilot-in-command. That pilot has given in to the resigning thought of, "What's the use?" Go back to Situation 2 and select another option.

e. Anti-authority. Well done. You chose the correct response. Go on to the next hazardous thought exercise.

The impulsivity hazardous thought pattern. From the five choices following each situation, select the ONE option that is the best example of an impulsivity hazardous thought pattern. Check your answers from the response list, and keep selecting until you have made the correct choice.

Situation 1. As you enter the pattern, you normally lower the flaps. The tower suddenly changes the active runway. Distracted, you forget to use the before-landing checklist. On short final you find yourself dangerously low with a high sink rate. Glancing back, you realize that you forgot to extend the flaps. Which of the following options best illustrates the IMPULSIVITY reaction?

a. You feel that nothing is going to happen because you've made intentional no-flap landings before.

b. You laugh and think, "Boy, this low approach will impress people on the ground."

c. You think that using a checklist is a stupid requirement.

d. You immediately grab the flap handle and add full flaps.

e. You think, "It's all up to whether I get an updraft or downdraft now."

Response options:

a. Invulnerable. "Nothing bad can happen to me." Go to Situation 1 and select another option.

b. Macho. When you're thinking more about impressing people on the ground than flying the airplane, look out. Go back to Situation 1 and select another option.

c. Anti-authority. Thinking that checklists are stupid is an invitation for disaster. Go back to Situation 1 and select another option.

d. Impulsivity. You're right. Immediately adding full flaps without thinking about the consequences is a clear example of an impulsive thought pattern. Go on to Situation 2.

e. Resignation. The answer's not blowing in the wind, go back to Situation 1 and select another option.

Situation 2. Landing at an unfamiliar airport for fuel, you tell the lineman to "fill it up," and run inside the terminal to use the rest room. Returning, you pay the bill and take off without checking the aircraft, the fuel caps, or the fuel. Which of the following options indicates an IMPULSIVITY reaction?

a. You feel that it's a silly requirement to preflight an aircraft which you've just flown.

b. You just want to get underway, quickly.

c. You know that you have skipped preflights before and nothing bad ever happened.

d. You have every confidence that a pilot with your skill level could handle, in flight, anything that might have been overlooked on the ground.

e. You feel that since you paid top dollar for the fuel, it's the responsibility of the lineman to ensure the airplane was refueled properly.

Response options:

a. Anti-authority. Thinking that regulations requiring a preflight inspection are nonsense, suggests a definite anti-authority attitude. Go back to Situation 2 and select another option.

b. Impulsivity. Bingo. Having that itch to get a move on shows great impulsivity. Go on to the next hazardous thought exercise.

c. Invulnerability. Just because you got away with it before doesn't mean that it's safe. "It won't happen to me." Go back to Situation 2 and select another option.

d. Macho. Even though you might think, "I can do it," you'll find yourself turning gray by your 25th birthday. Go back to Situation 2 and select another option.

e. Resignation. Feeling that everything is up to someone else, is a resigning attitude. Go back to Situation 2 and select another option.

   The invulnerability hazardous thought pattern. From the five choices following each situation, select the ONE option that is the best example of an invulnerability hazardous thought pattern. Check your answers with the appropriate response list, and keep selecting until you have made the correct choice.

   Situation 1. You are making a pleasure flight with four friends, all of whom are drinking. You refuse to drink, but your friends remind you that you have flown this route many times and that the weather conditions are excellent. They begin to tease you for not drinking with them. Which of the following options best illustrates the INVULNERABILITY reaction?

a. You decide to drink, thinking that a little liquor will not have any bad effect on you.

b. You believe that the government is far too rigid in its regulations about drinking and flying.

c. You resent your friends' insults and start drinking, saying to yourself, "I'll show them."

d. You bend to their will saying to yourself, "If my time is up, it's up whether I drink or not."

e. You suddenly decide to take a drink.

Response options:

a. Invulnerability. You are correct. Liquor affects everybody, and pilots who believe that it will not bother them, consider themselves invulnerable. Go on to Situation 2.

b. Anti-authority. Considering the authority of the government as too rigid is another way of thinking, "Those rules are much more strict than they need to be, so I can disregard them." Go back to Situation 1 and select another option.

c. Macho. The need to prove yourself, or show off to total strangers, is definitely a macho thought pattern. Go back to Situation 1 and select another option.

d. Resignation. Thinking that you have nothing to do with the outcome of a flight is a resigning attitude. Go back to Situation 1 and select another option.

e. Impulsivity. Making the foolishly sudden decision to drink is an impulsive thought pattern. Go back to Situation 1 and select another option.

Situation 2. The control tower advises you to land on a runway other than the one you prefer. You see larger planes using the runway of your choice and wonder why you have been denied permission. Since the tower-recommended runway is on the far side of the airport, you radio the tower and ask for reconsideration. Which of the following options best illustrates the INVULNERABILITY reaction?

a. Before you receive a reply, you start making your approach to the unauthorized runway.

b. You feel that if other pilots can land their airplanes on the other runway, so can you.

c. You think that nothing dangerous will occur because you believe wake turbulence is very unlikely.

d. Regardless what the tower tells you, you are going to do what you want to.

e. You figure there is no sense in waiting for instructions because the tower is going to do whatever it pleases, regardless of your wishes.

Response options:

a. Impulsivity. Rushing into an action without thinking about the consequences is an impulsive attitude. Go back to Situation 2 and select another option.

b. Macho. Thinking that you can do anything, anytime, anywhere, with any configuration is a macho attitude. Go back to Situation 2 and select another option.

c. Invulnerability. This is the correct response. Disregarding a potentially hazardous situation, like wake turbulence, and thinking there's nothing to worry about is an invulnerable thought pattern. Go on to the next hazardous thought exercise.

d. Anti-authority. "I'll do what I want to do" regardless of the consequences is an anti-authority thought pattern. Go back to Situation 2 and select another option.

e. Resignation. Believing that nothing you do will make any difference is a resigning attitude. Go back to Situation 2 and select another option.

The macho hazardous thought pattern. From the five choices following each situation, select the ONE option that is the best example of a macho hazardous thought pattern. Check your answers from the appropriate response list, and keep selecting until you have made the correct choice.

Situation 1. Visibility is barely more than three miles in blowing snow with a 1000 foot ceiling. Earlier you cleared the airplane of snow, but takeoff has been delayed for 15 minutes. Snow and ice are forming again, and you wonder if you will be able to take off. Which of the following options best illustrates the MACHO reaction?

a. You feel that there is no use getting out and removing the snow since it's only going to form again.

b. You believe that you can take off in these conditions and think of how impressed your friends will be when they hear of it.

c. You take off immediately, thinking that any further delay will worsen the problem.

d. You reason that you can do it because other pilots have done it and nothing happened to them.

e. You resent being delayed 15 minutes and decide you are not going to clear the snow and ice again for anybody.

Response options:

a. Resignation. When you don't think what you do affects what happens, you are displaying a resigning thought pattern. Go back to Situation 1 and select another option.

b. Macho. Correct. You want to show off to others and want to prove yourself. Definitely a macho attitude. Go on to Situation 2.

c. Impulsivity. You take off immediately. No thinking and no planning show a great impulsive thought pattern. Go back to Situation 1 and select another option.

d. Invulnerability. "Nothing happened to them, so nothing will happen to me." Go back to Situation 1 and select another option.

e. Anti-authority. Pilots who resent using appropriate safety precautions show an anti-authority attitude. Go back to Situation 1 and select another option.

Situation 2. The weather forecast calls for freezing rain. En route you notice ice accumulating on the wings. You are not sure what to do because you have never encountered this problem before. Because the airplane is still flying well, you are tempted to do nothing. A passenger suggests you might radio for information. Which of the following options best illustrates the MACHO reaction?

a. You feel that there probably will not be any problem since you have always come out of difficult situations rather well.

b. You feel that there is nothing you can really do because radio information won't change the weather conditions.

c. You quickly tell the passenger to stop butting in.

d. You tell the passenger that you are the boss and will handle the problem your way.

e. You radio for information but decide to ignore the advice since the airplane continues to fly well enough.

Response options:

a. Invulnerability. When you think that since nothing has ever happened before, nothing will happen in the future, you're displaying an invulnerable thought pattern. Go back to Situation 2 and select another option.

b. Resignation. "What's the use?" Go back to Situation 2 and select another option.

c. Impulsivity. Acting without thinking is impulsive. Go back to Situation 2 and select another option.

d. Macho. This is the correct answer. "We'll do it my way" is a good indication of a macho attitude. Go on to the next hazardous thought exercise.

e. Anti-authority. Those who ignore information or advice show an anti-authority attitude. Go back to Situation 2 and select another option.

The resignation hazardous thought pattern. From the five choices following each situation, select the ONE option that is the best example of the resignation hazardous thought pattern. Check the answers from the appropriate response list, and keep selecting until you have made the correct choice.

Situation 1. You would like to arrive early for an important business meeting. If you stick to your flight plan, you will just about make it, assuming there are no problems. Or, you can take a route over the mountains, which will get you there much earlier. If you choose the route through the mountain passes, it means you might encounter low hanging clouds while good weather prevails over the planned route. Which of the following options best illustrate the RESIGNATION reaction?

a. You take the mountain route even though the weather briefer has advised against it.

b. You take the mountain route, thinking that a few clouds in the passes will not cause any trouble for this flight.

c. You feel it will be a real victory for you if you can take the mountain route and arrive early.

d. You tell yourself that there is no sense sticking to the planned route because, "There's nothing else to do to be sure to make it early."

e. You quickly choose the mountain route, deciding that you just must get there early.

Response options:

a. Anti-authority. Not accepting the advice of a weather briefer, is an example of an anti-authority attitude. Go back to Situation 1 and select another option.

b. Invulnerability. "It won't happen to me." Go back to Situation 1 and select another option.

c. Macho. Making potentially dangerous situations into personal challenges is a macho thought pattern. Go back to Situation 1 and select another option.

d. Resignation. Good choice. Thinking that there is nothing you can do is an example of a resigning thought pattern. Go on to Situation 2.

e. Impulsivity. A quick decision isn't always the right decision. Go back to Situation 1 and select another option.

Situation 2. The weather briefer advises you of possible hazardous weather conditions at your destination, but you elect to go anyway. En route you encounter a brief snowstorm and increasingly poor visibility. Although you have plenty of fuel to return to your departure point, you have a hunch that the weather will improve before you reach your destination. Which of the following options best illustrates the RESIGNATION reaction?

a. You feel there is no need to worry about the weather since there is nothing one can do about mother nature.

b. You immediately decide to continue, and block the weather conditions out of your mind.

c. You feel nothing will happen to you since you have plenty of fuel.

d. You think that the weather people are always complicating your flights, and sometimes, such as now, it's best to ignore them.

e. You fly on, determined to prove that your own weather judgment is better than the forecaster's.

Response options:

a. Resignation. You picked the correct answer. A "what will be, will be," attitude in the cockpit can get you killed.

b. Impulsivity. Quickly blocking important matters out of your mind is impulsive. Go back to Situation 2 and select another option.

c. Invulnerability. Having plenty of fuel does not mean that all is well in the world. Go back to Situation 2 and select another option.

d. Anti-authority. Disregarding sound advice is the same as ignoring regulations. They're both an anti-authority thought pattern. Go back to Situation 2 and select another option.

e. Macho. Showing off again. Go back to Situation 2 and select another option.

THE DECISION-MAKING PROCESS

A DECIDE model has been used in many disciplines for a lot of years, but the FAA has been able to adapt this technique to fit flying scenarios. This should not be confused with a logic tree, which forces the participant to think in a robotic, unrealistic flow. Rather, it breaks down the natural thinking pattern to illustrate the steps involved in a decision-making process.

You'll find this to be an interesting and thought-provoking exercise for analyzing an accident. It can also be a valuable tool for evaluating how errors are made during the course of any flight.

The steps to take

- **D—Detect:** The pilot detects the fact that a change has occurred that requires attention.
- **E—Estimate:** The pilot estimates the significance of the change to the flight.
- **C—Choose:** The pilot chooses a safe outcome for the flight.
- **I—Identify:** The pilot identifies plausible actions to the change.
- **D—Do:** The pilot acts on the best options.
- **E—Evaluate:** The pilot evaluates the effect of the action on the change and on the progress of the flight.

A DECIDE model example

The following case study is an example of a DECIDE model exercise. As you review the information, refer to Table 2-2.

Air Illinois Flight 710

Reported weather at time of accident was instrument meteorological conditions. Cloud base at 2000 feet msl, with cloud tops at 10,000 msl. Visibility below 2000 feet was 1 mile in rain. Scattered thunderstorms reported in the area.

On 11 October 1983, the Air Illinois Hawker Siddley 748-2A, departed Springfield, Illinois, around 2020 CDT. The IFR-filed flight was already 45 minutes late. About 1½ minutes later, the crew of Flight 710 radioed Springfield departure control and reported they had just experienced a slight electrical problem. However, they informed ATC that they were continuing to their destination of Carbondale, Illinois, about 40 minutes away.

Although their intended cruise altitude was supposed to be 5000 feet, they told departure control, "We'd like to stay as low as we can." Flight 710 was then cleared to maintain 3000 feet, and the controller asked if he could be of any assistance. The crew responded, ". . . we're doing okay, thanks."

At 2023, the first officer told the captain that "the left [generator] is totally dead, the right [generator] is putting out voltage, but I can't get a load on it." Seconds later, the first officer reported, ". . . zero voltage and amps on the left side, the right [generator] is putting out . . . [volts] . . . but I can't get it to come on the line." Shortly thereafter he again told the captain that the battery voltage was going down "pretty fast."

By 2026, Flight 710 had left Springfield departures control's jurisdiction and had called into Kansas City Center with an "unusual request." The captain asked clearance to descend to 2000 feet, ". . . even if we have to go VFR." He also asked the controller, ". . . to keep your eye on us if you can." However, because 2000 feet was below Kansas City Center's lowest usable altitude and because the controller couldn't guarantee that he could maintain radar contact, the captain decided to remain at 3000 feet.

Table 2-2. DECIDE model example exercise.

Change	D	E	C	I	D	Action	E
Left gen fails after TO	Y	Y	N	N	Y	CP misidentifies failed gen and disconnects good gen	Y
CP tells Dep Con "slight" electrical problem	Y	Y	Y	Y	Y	Dep Con offers return to Springfield airport	Y
Crew gets Dep Con offer to return to Springfield	Y	Y	N	N	Y	Capt rejects offer and continues to Carbondale	Y
Right gen doesn't take electrical load	Y	Y	Y	Y	Y	CP tells Capt of loss right gen	Y
CP tells Capt of right gen failure	Y	Y	N	N	Y	Capt requests lower altitude for VFR conditions	Y
CP tells Capt bat voltage is dropping fast	Y	N	N	N	Y	Capt tells CP to put load shedding switch off	Y
CP reminds Capt of IFR weather at Carbondale	Y	N	N	N	N	No reaction	N
CP turns on radar to get position	Y	Y	N	N	Y	CP tells Capt about dropping voltage	Y
CP tells Capt bat volt is dropping	Y	Y	Y	Y	Y	Capt turns off the radar	Y
CP warns Capt about low battery	Y	Y	Y	Y	Y	Capt starts descent to 2400	Y
Cockpit instruments start failing	Y	Y	Y	N	Y	Capt asks CP if he's got any instruments	Y

According to the chart, the "Ys" indicate that the task was completed and the "Ns" indicate that the task was not completed. The quick interpretation of the findings is as follows: A number of "Ns" are found in the "I" column suggesting that there was a failure of the crew to identify the correct action to counter the change. However, the crucial "Ns" occurred when the first officer reminded the captain of IFR weather at Carbondale and got no response until it was too late. The first officer appeared to have the answer that may have avoided the accident, but did not offer it to the captain nor did he voice his concerns about the action the captain was assertively pursuing.

Derived from DOT/FAA/PM-86/46, "Aeronautical Decision-Making—Cockpit Resource Management." R. Jensen, 1989

Two minutes later, the captain said, "Beacon's off . . . ," followed by, "Nav lights are off." The first officer then reminded the captain that Carbondale had a 2000 foot ceiling, and the visibility was two miles with light rain and fog.

The flight attendant eventually came forward to report that only a few lights were operating. The captain instructed her to inform the passengers that he had turned off the excess lights because the airplane had experienced, ". . . a bit of an electrical problem . . ." and that they were proceeding to Carbondale. Their estimated time of arrival was in 27 minutes.

Only a few minutes after the flight attendant went back to the cabin, the first officer began to explain to the captain what he had found while troubleshooting the system. ". . . when we . . . started losing the left one [generator], I reached up and hit the right [isolate button] trying to isolate the right side . . . "cause I assumed the problem was the right side but they [the generators] both still went off."

At 2045, Flight 710 had 20 volts left in their battery. But six minutes later, the first officer told the captain, "I don't know if we have enough juice to get out of this." The captain then asked the first officer to: "Watch my altitude, I'm going to go down to twenty-four hundred [feet]." He then asked the first officer if he had a flashlight, and if so, have it ready. At 2053, the first officer reported, "We're losing everything . . . down to thirteen volts . . . ." A minute later the aircraft was at 2400 feet, and the captain asked the first officer if he had any instruments. The captain again repeated, "Do you have any instruments? Do you have a horizon [attitude director indicator]?"

About 2051, Kansas City Center had lost radar contact with Flight 710—presumably because they had already dipped below 2000 feet. And sometime after 2054:16, Flight 710 crashed 40 nm north of their destination airport. Three crewmembers and seven passengers were killed in the accident.

The NTSB determined that the probable cause of the accident was the captain's decision to continue the flight after the loss of dc electrical power from both airplane generators. The captain's decision was adversely affected by self-imposed psychological factors which led him to inadequately assess the airplane's battery endurance after the loss of generator power. He also did not weigh the magnitude of risks involved in continuing to the destination airport.

Also contributing to the accident, according to the NTSB, was the failure of the airline management to provide a satisfactory company recurrent flightcrew training program. This subsequently led to the inability of the captain and first officer to cope promptly and correctly with the airplane's electrical malfunction. The failure of the FAA to assure such a program existed was also cited as a contributing factor.

Lessons learned. This crew displayed poor judgment when they willingly flew into instrument conditions with a known electrical problem that quickly turned into a full-blown emergency. Although you might be thinking, "I wouldn't have accepted the airplane in the first place, let alone fly it into night IFR," there *are* lessons that can be learned from this accident that translates to other situations.

1. Aviate, communicate, navigate. This is a basic piloting concept when faced with an emergency. From the outcome of Flight 710, it's easy to see what can happen if those steps are not followed. ALWAYS: Fly the airplane. Talk to ATC. Stay on course.

2. Avoid "head-lock." When pilots become engrossed with an emergency, they tend to keep their heads down and focused on the problem. The actual flying of the airplane soon goes unchecked.

3. Keep ATC in the loop. In this case, the crew never declared an emergency or asked the controller for assistance. If the situation warrants, then by all means declare an emergency. Otherwise, notify ATC of your problem and ask for help. If the crew of Flight 710 had indicated the seriousness of the problem, the controller might have been able to suggest a closer airport.

DECIDE model exercise

Closely study the events that led to the following accident. Then enter the appropriate responses on the provided DECIDE Model Exercise chart (Table 2-3).

GP Express Airlines Flight 861

Pilot experience: The captain had been hired, for that position, nine days prior to the accident. He had been a U.S. Army helicopter pilot with 1611 military flight hours, and a civilian flight instructor with 857 civilian flight hours. He had a total of 87 hours in

Table 2-3. DECIDE model exercise.

Change	D	E	C	I	D	Action	E

Read the accident analysis of GP Express Flight 861. Use the example exercise as a guideline.

Derived from DOT/FAA/PM-86/46, "Aeronautical Decision-Making—Cockpit Resource Management," R. Jensen, 1989.

the Beech 99, 76 of which were simulated. The day of the accident was his first day on the job. The first officer had 1234 total flight hours, and had been flying the Beech 99 for about five weeks. He had flown approximately 90 hours as a GP Express first officer, all on Midwest routes. The day of the accident was his first day flying the airline's southern route structure.

Reported weather at time of accident: 700 feet scattered, estimated 1500 feet broken, 9000 feet overcast. Visibility was three miles in fog and haze.

On 8 June 1992, the crew of the GP Express Beech 99 had been on duty since 0400 CDT. By the time Flight 861 departed 27 minutes late from Atlanta, Georgia, for Anniston, Alabama, the captain and first officer had already flown two legs for a total of one hour and 20 minutes. The flight time between Atlanta and Anniston was estimated to be 50 minutes.

After the departure out of Atlanta, Flight 861 was cleared to maintain 6000 feet. Throughout the cruise phase, the captain and first officer had quite a bit of difficulty understanding each other because of noise on the intercom system. Additional conversations indicated that the first officer had noted several problems with the airplane's autofeather system and the battery. He even seemed to have trouble being able to set the radio frequencies.

At about 0841, Atlanta Center cleared Flight 861 to, ". . . descend at pilot's discretion, maintain five thousand [feet]." The captain asked the first officer, to which he did not reply, "Does he want us to resume own navigation?" The captain then said, "I heard him say that. As far as I'm concerned, I'm still on vectors two eight zero [heading]." The first officer responded, "Yeah, two eight zero's fine. Because we're on course anyway, so let's just hold it." The captain remarked that he thought they were, ". . . slowly drifting off [course]."

After a short conversation over whether the airplane was on course, the captain again asked, "What's the course?" The first officer replied, ". . . zero eight five inbound." The captain finally concluded that, ". . . we're way off course." In fact, 085 degrees was the outbound course from the VOR that they had been tracking. The inbound course, to which the aircraft was headed, was the reciprocal of 085, or 265 degrees.

Moments after this discussion, Atlanta Center informed the crew that radar contact was terminated, and they were to go over to Birmingham approach control. The first officer complied, but during the investigation, he testified that he believed the flight was continuing to receive radar vectors from ATC, even though Atlanta Center had terminated radar services.

At 0843, Birmingham approach instructed the flight to descend and maintain 4000 feet, and to proceed direct to the [Talladega] VOR. They were also told to expect a visual approach into Anniston if they were able to see the airport. If that wasn't the case, then the crew was to set up for an ILS approach to runway 5, from over the BOGGA approach fix. The first officer acknowledged the call.

Several minutes later, after receiving the latest Anniston weather, the crew decided to take the ILS. Before they could make that request, the controller notified them to, "Proceed direct BOGGA, maintain four thousand 'till BOGGA, cleared . . . ILS runway 5 approach." Rather than ask the controller the distance to BOGGA, the first officer mentally computed the distance as being five miles.

As the first officer tuned in the localizer, he made the comment to the captain, "Didn't realize that you're going to get this much on your first day, did ya?" By the time he had tuned in the correct frequency, the airplane had gone ". . . right through it." At that, the crew thought they were, ". . . right over BOGGA . . . four and a half [miles] out . . . ." The first officer told the captain to, ". . . go ahead and drop your gear . . . speed checks."

Seconds later, the captain stated that the glideslope wasn't even alive. He then asked, "What's the minimum altitude I can descend to 'till I'm established?" The response was, ". . . twenty-two hundred [feet]." Because the company had provided only one set of approach charts in each airplane, only one pilot could review them at a time. Therefore, the Board believed that was why the captain asked the first officer the question concerning the minimums.

By 0850, the weather was moving in their direction and was reported to be only two miles from the localizer outer marker. The first officer notified the controller that the, ". . . procedure turn inbound [was] complete." But less than a minute later, the captain told the first officer, ". . . we gotta go missed [approach] on this." The first officer though, replied, ". . . there you go . . . there, you're gonna shoot right through it again . . . keep 'er goin' . . . you're okay." Within seconds, the captain called out, ". . . there's the glideslope." Followed by the first officer saying, "We can continue our descent on down. We're way high."

Almost immediately after they started their descent, they lost the glideslope. They were at 1100 feet, when the captain asked the first officer to confirm that the correct ILS frequency had been tuned in. The answer was yes. But by this time, the captain had decided to go around. He asked the first officer, "What's our missed approach point now?" His reply was, "Twelve hundred [feet] . . . coming up . . . ." One second later was the sound of impact.

The aircraft struck terrain a little over seven miles from the airport, at about the 1800 level. Although there were no witnesses to the crash, the area near the accident site was reported to be shrouded in fog and low-lying clouds. The captain and two passengers were killed, and the first officer and the other two passengers received serious injuries.

The NTSB determined that the probable causes of this accident were the failure of senior management of GP Express to provide satisfactory training and operational support for the startup of the southern routes, which resulted in the assignment of an inadequately prepared captain with a relatively inexperienced first officer. Additionally, the failure of the flightcrew to use approved instrument flight procedures, which resulted in a loss of situational awareness and terrain clearance. Contributing to the accident was GP Express' failure to provide approach charts to each pilot and to establish stabilized approach criteria. Other contributing factors were the inadequate crew coordination, and a role reversal on the part of the captain and first officer.

Lessons learned.

1. Monitor and cross-check. Both crewmembers thought they were on course, yet moments later the captain realized they were flying the reciprocal heading for the inbound course. Always verify headings and altitude.

2. Avoid complacency. After the captain asked, "what's the course," the first officer replied, "zero eight five inbound." In fact, that was the *outbound* course.

3. Get out of a bad situation. From the start, the flight was not going well. The crew was flying the wrong way, chasing ILS needles, and too high, all under the threat of poor weather. *Before* a rapidly deteriorating situation exceeds your capabilities, go around. Ask for ATC assistance and try again.

4. Use FAA-approved instrument procedures. When operating in IMC, this will help with situational awareness and ensure terrain clearance.

5. Avoid role reversals. As pilot-in-command, you must be mentally and technically prepared for the job. A feeling of "I guess I'm ready" has no place in a cockpit, especially in the left seat. First officers who face the dilemma of a captain who is in over his head must stay alert and vigilant with each task. Monitor and cross-check everything.

A FINAL LESSON IN GOOD JUDGMENT

So far we've discussed the potentially disastrous results that can happen when a pilot's negative thought patterns and attitudes interfere with the flight. However, it's important to realize that other crewmembers or passengers can also negatively influence your judgment if you're not prepared. The following scenario is an example of how one pilot remained in control and avoided those pitfalls.

A recently hired commercial pilot was flying a light plane with her new boss to attend a business meeting. The weather conditions at their destination, which was at a high elevation, was socked in with sky obscured, and visibility at less than a ¼ mile. Their alternate was clear, but 50 miles away. The boss desperately wanted to get to the meeting and had already told the pilot several times to, "Go ahead, give it a try," "Let's go lower," and "It doesn't look that bad to me." The controller had informed them earlier that the only other airplane that had tried a landing that day had declared a missed approach.

The pilot was already well aware of her boss's macho attitude, but after thinking through the situation of bad weather, mountainous terrain, busy traffic area in marginal VFR, and clear alternate, she decided en route to proceed to her alternate. She was also concerned that her very vocal passenger would interfere with her decisions, or possibly pull rank and insist she land. So instead of risking a power struggle while foolishly attempting an unnecessary approach, she made the correct and timely decision to make a safe landing at her alternate.

The pilot proved three valuable points. One, she "perceived" the situation: bad weather and the boss being a pain in the neck. Two, she "distinguished" between her available options: "I could try to land in unsafe conditions" or, "I have plenty of fuel so I could safely land at my alternate." And three, she processed the information, did not succumb to external pressure, remained in command, and reached a successful conclusion.

CASE STUDY REFERENCES

I-2, I-3, I-4, I-6, I-7, I-8, II-2, II-3, III-2, III-5, IV-1, IV-3, IV-4

CHAPTER REVIEW

Good judgment

- Awareness.
- Observation.
- Recognition.
- Understand differences between correct and incorrect alternatives to a solution.

Hazardous thought patterns

- Macho. Must prove themselves.
- Invulnerability. "Nothing bad can happen to me."
- Impulsivity. Acts too quickly.
- Anti-Authority. Resents external control.
- Resignation. "What's the use?"

DECIDE model

- DECIDE model helps analyze the causes of errors and unsafe situations.

CHAPTER REFERENCES

Judgment Training Manual for Student Pilots. 1983. Federal Aviation Administration. Transport Canada. General Aviation Manufacturer's Association.

Jensen, Richard S. January 1989. *Aeronautical Decision Making, Cockpit Resource Management.* The Ohio State University Research Foundation, Columbus, Ohio.

Jensen, Richard, Janeen Adrion. 1988. *Aeronautical Decision Making for Commercial Pilots.* Federal Aviation Administration.

Manning, C.K. "Creeping Complacency." *Human Factor.* May 1988: 22–24.

Nagel, David C. *"Human Error in Aviation Operations." Human Factors in Aviation. Ed.* Wiener, Earl L. and David C. Nagel. San Diego: Academic Press, 1988: 263–301.

National Transportation Safety Board. Aircraft Accident Report: Air Illinois Hawker Siddley, HS 748-2A, N748LL, near Pinckneyville, Illinois, October 11, 1983. 5 March 1983. Washington, D.C.

National Transportation Safety Board. Aircraft Accident Report: Controlled Collision with Terrain. GP Express Airlines, Inc., Flight 861. A Beechcraft C99, N118GP, Anniston, Alabama, June 8, 1992. 2 March 1993. Washington, D.C.

Orasanu, Judith M. October 1990. *Shared Mental Models and Crew Decision Making*. Princeton University.

Wickens, Christopher D. and John M. Flach. "Information Processing." *Human Factors in Aviation*. Ed. Wiener, Earl L. and David C. Nagel. San Diego: Academic Press, 1988: 111–155.

3
Distractions in the cockpit

AFTER COMPLETING THIS CHAPTER, YOU SHOULD BE ABLE TO:

1. Explain the types of distractions that can occur during nonflight operations.
2. Explain the types of distractions that can occur during flight operations.
3. Learn ways to prevent unnecessary cockpit distractions.

Distractions will always be part of a cockpit environment, but how the pilot handles them, and to what extent they are allowed to continue, can often mean the difference between a safe flight or an accident.

PILOTS TELL IT LIKE IT IS

A few years ago, the NASA-Ames Research Center initiated a study on distraction because air carrier pilots identified that as being the most frequent cause for operational errors. The data were taken from 169 pilot reports and broken down into the following two groups: nonflight operations and flight operations. Nonflight operations consist of paperwork such as crew logs, engine logs, block/air times, and public address announcements, passenger problems, and inappropriate cockpit conversations. Flight operations includes crew coordination, completing checklists, and of course, actual flying tasks.

Distractions that can occur during an actual hands-on flight scenario show great similarity with those associated with the nonflight operations. In both types, you'll see how routine tasks and activities can be quickly mishandled or forgotten altogether.

The bottom line is that simple distractions, such as checklists, air-ground communications, PA announcements, and minor system failures, can take a perfectly good pair of eyes and hands away from flying the airplane.

NONFLIGHT OPERATIONS

The following excerpts are actual pilot reports submitted to the ASRS.

1. We were cleared to 11,000 feet by departure control. Once the workload diminished, I started to complete the logbook and time sheets. The F/O was flying and the aircraft leveled off and picked up speed. As I finished the paperwork, Center called and asked our altitude. I then noticed it was 10,000 feet.

    NASA's take on the situation was that the captain was obviously at ease with the skill of the first officer, and was therefore comfortable enough with the situation to become involved with paperwork. The first officer might have misunderstood the clearance, or had gotten distracted with other tasks, but for those brief moments, that aircraft was not where it was supposed to be.

2. We were climbing out of XYZ airport. The first officer was flying. I acknowledged a 7000-foot restriction, then went back to my paperwork. I didn't see the F/O set 17,000 in the altitude select window. As we passed 12,000, Center called, wanted to know where we were going.

    The captain correctly heard the clearance and assumed the first officer did also. But the first officer obviously misunderstood the clearance. The captain went back to his paperwork and didn't monitor the other cockpit activities.

    A simple task that is completed several times each day can easily become a distraction, one that is overlooked as inconsequential to the safety of the flight. Yet from this example, it was apparent that a lack of communication and crew coordination resulted in a 5000-foot altitude bust.

3. The copilot was on the public address telling the passengers about the thunderstorm deviation. While climbing through FL 270 [assigned FL 280] the number 4 generator tripped the line. I asked the F/E to monitor the fault panel. The problem turned out to be a Generator Control Unit . . . When I looked back at the instrument panel, our altitude was 28,000. [We leveled off at 28,700.]

    Watch out for simultaneous distractions. In this case, the attention of all three crewmembers was directed away from flying the airplane. As the crew flew through their assigned altitude, unbeknownst to them an aircraft on a head-on course was converging towards them at FL 290.

4. Flight attendant discussing a cabin situation with captain. Clearance was received by first officer for flight to cross 15 DME at or below FL 230. Captain crossed 15 DME at FL 240 . . . [the] F/O failed to mention correct altitude.

5. We were at FL 230 and told to descend and cross ABC at 18,000. I hurried to fill out the engine readings. Just then a flight attendant came up front with a request for a wheelchair. Center asked us for our altitude. We were just west of ABC and still at FL 230.

   About 20 percent of the distractions experienced during nonflight operations were caused by pilot-flight attendant conversation. Most of these cases occurred at the descent phase and pertained to travel connections, cabin conditions, and general passenger problems. There were numerous errors made in overshooting or undershooting restricted altitude crossings. One such interruption resulted in a pilot misreading the altimeter by 10,000 feet.

6. We were coming into XYZ. We checked in with approach and were told to expect ILS. While F/O was calling in range to the company, I thought I understood Center having cleared us down to 4000. During the descent, I was informed that I did not have clearance.

   Altitude deviations caused by company communications were reported nearly 30 percent of the time. The mistakes are similar in nature as those occurring during PA announcements. It takes one pilot completely out of the ATC loop. Therefore, misunderstood clearances are quite common. Pilots who mistakenly departed their assigned altitude went unchecked for several minutes because the nonflying pilots were on the radio talking to OPS. The numbers are significant. Nonflying pilots caught an altitude deviation only 40 percent of the time.

A NASA Air Carrier Flight Operations policy recommends not to use the PA below 10,000 feet in the terminal area. Nonessential paperwork should also be delayed until the cruise segment of the flight.

FLIGHT OPERATIONS

1. We were cleared for an ILS approach and advised to contact the tower at the outer marker. At this time the crew became involved with checklists and inadvertently forgot to contact the tower prior to our landing.

2. We were cleared to descend to 5000. I was doing the approach checklist. Suddenly, I saw the altimeter going through 4200. Before I could do anything, a light airplane came over the top of us. We missed him by 200 feet.

   Nearly 30 percent of the distractions reported during flight operations were attributed to completing checklists and studying approach plates. There was also a close connection between checklist activity and other cockpit duties. During climb-out and descent, the busier times of a flight, pilots were rushed to get through the checklists, which made them feel overloaded and stressed with tasks.

3. We had a light airplane and obtained a high rate of climb. Due to other distractions . . . rechecking SID, looking outside for traffic, resetting climb power, after takeoff checklist, changing frequencies, and selecting radials . . . we inadvertently passed through our assigned altitude.

Altitude deviations at critical phases of flight accounted for a little more than 40 percent of the reported cases.

4. During climbout from XYZ we were assigned 6000 feet. At 5000 the bell and light altitude reminder worked as planned. The 1000-to-level call was made. Climb checklists were being completed, navaids tuned and identified. Center . . . and radar continuously monitored . . . The 6000-foot altitude was missed. Reminder: A system is only as good as its vigilant operator.

5. We broke out . . . at FL 190 but had been cleared to only 16,000, and immediately lowered the nose . . . to 370 knots . . . rate of sink increased to 3000 to 4000 fpm. Crew began to troubleshoot anti-ice [malfunction] . . . noise level was high . . . didn't hear altitude warning bell. [Flying] pilot was troubleshooting switches . . . [nonflying] pilot didn't make the 1000-foot-to-level-off callout. An altitude overshoot of 2000 feet occurred before the captain noted the altimeter.

Of the 19 system malfunctions reported during the study, all were considered minor. But these minor distractions were enough to cause 12 flights to deviate from their altitude clearances, 3 flights to alter their courses, 2 flights to land without clearance, and 1 flight to penetrate restricted airspace. Only 1 flight was successful and had no deviations. All of the 18 crews admitted being distracted to the point that they no longer monitored the assigned flight path.

6. We were cleared to 8000 feet, passing 6000 . . . advised . . . VFR traffic at twelve o'clock, 4 miles. My copilot and I strained to see traffic but were unsuccessful . . . asked for vector away from traffic . . . [never saw traffic]. I [caught us] passing through 8700 feet.

7. An earlier traffic advisory had drawn the attention of the first and second officers toward one o'clock, when a westbound light aircraft passed over us from our nine o'clock position.

It was often reported that traffic advisories cause pilots to stop flying the airplane. Altitude and heading busts are quite common.

A FEW FINAL THOUGHTS

Granted, there will be times when you can't prevent an interruption. Communication problems, deviations for weather, passenger and crew interaction, and systems malfunctions are just a few of the numerous routine distractions pilots face every day. There are many valuable lessons that can be learned from our fellow aviators' own words. As a result, I found it appropriate to add the following section as this chapter's conclusion.

LESSONS LEARNED AND PRACTICAL APPLICATIONS

1. Fly the airplane. No matter what, always remain in control of the airplane.
2. Be decisive with the delegation of responsibilities.
3. Don't complete nonessential tasks during critical phases of flight. This not only takes a good pair of eyes and ears away from traffic detection, it also removes a crewmember from monitoring the instruments. Remember those altitude busts.

4. Keep everyone in the loop. Immediately following ATC communications, make sure the flying pilot knows the exact clearance or message. Then cross-check.

5. Remain in the loop. Immediately following company communications or a PA announcement, ask what clearances or messages had been received while you were busy. Then cross-check the instruments for accuracy.

6. Maintain situational awareness. Be aware of the activities going on around you. This is particularly important when a momentary distraction takes you away from the tasks at hand. When the interruption has ended, promptly get back in the loop.

7. Prioritize your interruptions. If you know the conversation is nonflight essential, and you're in the middle of something important, like finishing a checklist, ask the person to wait until you have time. Don't get suckered into thinking there's no harm in a 30-second conversation/distraction.

8. Isolate a distraction. Make sure one distraction doesn't spread into other parts of the operation.

9. Don't allow a minor distraction to turn catastrophic.

10. Maintain optimum time management. You would be surprised how much time you actually have for completing checklists when you don't allow unnecessary distractions to take over. Hack the clock.

11. Maintain vigilance at all times. Not just when you hear, "Traffic at twelve o'clock, one mile."

12. Keep your head moving. Whether your head is in the full up or full down and locked position, remember, your situational awareness has just been cut in half.

CASE STUDY REFERENCES

I-6, I-7, III-2, III-5, IV-3, IV-4

CHAPTER REVIEW

• Distractions can negatively affect situational awareness, decision making, and judgment.
• Nonflight operations include paperwork, PA announcements, passenger problems, and inappropriate cockpit conversations.
• Flight operations include crew coordination, and actual flying tasks.

Distractions: nonflight operations

Paperwork keeps pilots' heads in the cockpit, preventing effective instrument and traffic scans. Common flight errors associated with paperwork include:

• Altitude deviations
• Misunderstood clearances

- Forgotten clearances
- Breakdown in crew communication

NASA Air Carrier Flight Operations Policy recommends that nonessential paperwork be delayed until the cruise segment of a flight.

PA announcements remove pilots from the ATC communication loop. Common errors associated with conducting public address messages include:

- Altitude deviations

- Misunderstood clearances

- Missed clearances

- Nonflying pilot is unavailable to cross-check altitude or heading assignments. NASA Air Carrier Flight Operations Policy recommends not to use the PA below 10,000 feet in the terminal area.

- Common errors associated with nonflight-essential cockpit conversation include:

- Altitude deviations

- Misunderstood clearances

- Missed clearances

- No one flying the airplane

Professional cockpit discipline should be maintained at all times.

Most pilot-flight attendant interaction occurs during the descent phase. Common errors associated with flight attendant interruptions include:

- Misread instruments
- Altitude deviations
- Overshooting/undershooting altitude crossing restrictions
- Breakdown in crew communication

Common errors associated with company communication include:

- Altitude deviations
- Misunderstood clearances
- Missed clearances
- Nonflying pilot is unavailable to cross-check altitude and heading assignments

Distractions: flight operations

Common errors associated with checklist activity include:

- Altitude deviations
- Missed clearances
- Landing without clearances
- Excessive workload

Common errors associated with systems malfunctions include:

- Altitude deviations
- Route deviations
- Landing without clearances
- Penetration of restricted airspace
- No one flying the airplane

Troubleshooting can also lead to a reduced state of vigilance.

Traffic advisories. ATC-prompted traffic advisories can cause target fixation. Pilots tend to focus on the perceived threat aircraft and disregard the entire scan pattern. Pilots tend to suddenly become alert when a, "Traffic at twelve o'clock" advisory is issued while disregarding all other cockpit duties. Common errors associated with traffic advisories include:

- Altitude deviations
- Nonstabilized approaches
- Landing without clearances
- Near misses with unknown aircraft
- Not requesting a vector away from threat aircraft when visual detection is not possible

Heads in the cockpit. Inadequate and nonexistent visual search is the primary cause for near misses. Common errors associated with head-lock in the cockpit include:

- Near-misses
- Altitude deviations

Conclusion

- Watch for sources of distractions that can suddenly and unexpectedly appear
- Flight hours and experience do not guarantee a distraction-free flight
- Distractions can lead to confusion and chaos in the cockpit

CHAPTER REFERENCES

Foushee, H. Clayton, and Robert L. Helmreich. "Group Interaction and Flight Crew Performance." *Human Factors in Aviation.* Ed. Wiener, Earl L. and David C. Nagel. San Diego: Academic Press, 1988: 189–227.

Kantowitz, Barry H. and Patricia A. Casper. "Human Workload in Aviation." *Human Factors in Aviation.* Ed. Wiener, Earl L. and David C. Nagel. San Diego: Academic Press, 1988: 157–187.

Nagel, David C. "Human Error in Aviation Operations." *Human Factors in Aviation.* Ed. Wiener, Earl L. and David C. Nagel. San Diego: Academic Press, 1988: 263–303.

National Aeronautics and Space Administration. *Aviation Safety Reporting System: Ninth Quarterly Report.* 1978. Ames (NASA) Research Center, Moffett Field, California.

Smith, Ruffell. 1979. *A Simulator Study of the Interaction of Pilot Workload With Errors, Vigilance, and Decisions.* National Aeronautics and Space Administration.

4
Cockpit discipline

AFTER COMPLETING THIS CHAPTER, YOU SHOULD BE ABLE TO:

1. Discuss the benefits of a positive corporate philosophy to the overall safety of flight operations.
2. Explain the nonflight-related issues that can generate company policy.
3. Discuss the areas affected by procedural deviation in the cockpit.

A recent series of NASA studies has shown a dramatic increase in airline accidents caused by procedural deviation in the cockpit. In one of the earlier reports, researchers analyzed 93 jet-transport crashes that occurred between 1977 and 1984. Thirty-three percent of those were caused by a pilot deviation from basic operational procedures, making this the leading crew-induced error. The remaining top three factors included: inadequate cross-check by second crewmember, crews not conditioned for proper response during abnormal condition, and pilot did not recognize need for go-around.

A similar study of Part 121 air carriers was conducted in 1991 that illustrated a significant rise from the previous statistics. The lack of cockpit procedural behavior accounted for 69 percent of crew errors, more than three times that of the second ranking category of poor decision-making.

The research continued in 1994 with a plan to study the connection between flight safety and procedural conduct in the cockpit. The authors of the report presented four areas of concentration: philosophy, policies, procedures, and practices. Each point is considered a link to sound cockpit discipline.

PHILOSOPHY

A company's philosophy, or corporate culture, is influenced tremendously by the senior level management. Although most airline managers are unable to clearly state their philosophy, especially with regards to flight operations, it is, however, generally understood by crews through the presence, or absence, of certain conduct. A feeling of personalized employee appreciation, few contract disputes, a solid training department, an excellent quality assurance program in the maintenance facility, a no-nonsense flight safety office, and a reasonable latitude of crew duties and responsibilities are all examples of how an unspoken philosophy can affect safety in a positive way. Conversely, an underlying tone of management-pilot strife or inappropriate decisions made at key levels can, over time, become a serious detriment to flight operations.

On the other hand, as of 1991, Delta Airlines was believed to have been the first carrier to develop a one-page formal statement of philosophy on the use of flight-deck automation. Distributed to the company's pilots, it detailed the importance of system knowledge and proficiency, and the significance of a new generation aircraft fleet in doing business.

POLICY MAKING

Company policies are usually generated by the philosophy of operations, in combination with economic factors, pubic relations campaigns, newer aircraft, and major organizational changes. It is those policies that spell out the manner in which management expects the day-to-day operations to be accomplished.

Oftentimes, cockpit procedures are driven by unique policies that might have nothing to do with flying. For instance, one carrier's new public relations policy called for more interaction between the cockpit crew and the passengers. It was recommended by the marketing department that at each destination, the captain stand at the cockpit door and greet the departing passengers, especially those in first-class. To ensure the captain was in place by the time the cabin door was opened, dictated a procedural change in the "Secure-Aircraft" checklist. Normally, the checklist would have been completed by both crewmembers, but because of the new policy, the first officer was left behind to finish the final cockpit duties.

PROCEDURES

Standard operating procedures (SOP) are a set of guidelines that serve to provide a common ground for a crew who are usually unfamiliar with each other's experience and technical capabilities. As airline mergers and acquisitions continue to be the mainstay of the industry, standardization becomes increasingly more important for flight safety. As a result of those changes, pilots from diverse employment backgrounds and exposure to various good or bad corporate philosophies are routinely crewed together. Without a collection of SOPs to follow, crew coordination can be seriously compromised.

PRACTICES

At this point, our discussion has come full circle. A strong company philosophy, excellent policies, and logical procedures can all go out the window if a pilot chooses to not effectively put them into practice. Provided the airline has a solid operational foundation, the pilot is the last person who can either break the final link or strengthen it.

As part of this study, crews were observed during actual flights. In one case, the first officer mentioned a mandatory taxi procedure and the captain replied: "I just don't do that procedure." That way of thinking is not rare in today's cockpits.

According to the report, pilots sometimes deviate from an SOP due to one of several reasons: individualism, complacency, laziness, or frustration.

INDIVIDUALISM

Pilots will often put a personal touch to a particular procedure, but as the study suggested that's not necessarily bad. If the change maximizes cockpit efficiency or adds an extra safeguard to the process, then a slight deviation from an SOP probably will not compromise an acceptable margin of safety. However, a pilot who has a practice of making small changes must be incredibly disciplined to ensure that these deviations do not cross the line and interfere with the overall safety of the flight.

COMPLACENCY

A pilot who has encountered few emergencies is often tempted to relax cockpit vigilance. So, too, is the pilot who has been flying for decades and no longer feels challenged, and true even for the young aviator who flies the same airplane on the same route, week after week. These are just a few examples, but complacency can creep up on any pilot who has become overly self-confident or believes that accidents happen only to other pilots. A serious deviation of cockpit procedures can many times be traced back to muddled thinking and a negative mindset.

LAZINESS

With few exceptions, problems associated with laziness closely resemble complacency. The difference is usually seen in nonstandard phraseology. A perfect example of this was noted by researchers who witnessed a potentially dangerous situation between crewmembers on one particular air carrier flight. The captain was flying the airplane during takeoff. According to SOP, the first officer was supposed to make standard airspeed calls of V1, Vr, and V2. Instead, he combined the first two into a nonstandard call of "V-one-r," and he referred to V2 as, "two of 'em." Apparently, the captain knew what the first officer meant and never discussed it afterwards.

Based on similar instances on other flights, the report suggested that when either a captain or first officer allows nonstandard verbiage to be used, or important parts of SOPs to be disregarded, they unwittingly create an atmosphere of tolerance. This might lay the foundation for more serious procedural deviations later in the flight.

FRUSTRATION

Many pilots feel they've been driven to procedural nonconformity due to inconvenient policies. For instance, the researchers found that most pilots don't use their oxygen mask, as required above FL 250, when one pilot leaves the cockpit. The complaints ranged from the mask being too uncomfortable to wear to difficulty in deflating and returning it to its holding container.

One crew was observed skirting the issue entirely while cleverly obeying the regulation. In a two-pilot aircraft, the captain left the cockpit briefly as the airplane was climbing unrestricted to FL 330. At about FL 200, the first officer called ATC and requested a level off at FL 250. He maintained that altitude until the captain returned. Since the aircraft was never *above* FL 250, the first officer never had to don his mask. Once the crew was back in place, they asked ATC to continue their climb. Although this trick benefited the crew, it actually cost the airline more money in fuel costs, due to the interrupted climb profile. It might have also caused some problems with traffic separation in the ATC system.

SAFETY ISSUES AND PROCEDURAL CHANGES

Over the years, there have been numerous factors that have caused airlines to develop new cockpit procedures. Some have been related to technological advancements, labor relations, and various safety hazards. The following are a few examples that you might find interesting.

1. The introduction of the Traffic Alert and Collision Avoidance System (TCAS) into the cockpit required a new set of procedures to be created "from scratch" since there was no established precedent for airborne collision warning devices.

2. In the 1980s, airlines began the practice of powering back their aircraft from the gate. The procedure was developed in order to side-step labor contracts that required mechanics be used to push back the airplanes.

3. In 1990, USAir was experiencing a high rate of altitude busts, an average of four per month. A thorough study resulted in a change in procedure whereby the nonflying pilot would enter the command altitude in the altitude-alerter window. The pilot would then repeat the altitude out loud, while keeping a finger on the knob. Immediately thereafter, the flying pilot would duplicate the procedure. In subsequent months, FAA-reported cases dropped to less than one per month.

CASE STUDY REFERENCES

I-1, I-2, I-3, I-4, I-5, I-6, I-7, I-8, III-1, III-2, III-5, IV-1, IV-3, IV-4

CHAPTER REVIEW

A breakdown in cockpit discipline can lead to:

- Pilot deviation from basic operational procedures
- Inadequate cross-check by second crewmember
- Crews not conditioned for proper response during abnormal condition
- Pilot not recognizing need for go-around
- Poor decision-making

Philosophy

- A productive and positive corporate culture can enhance overall safety
- Poor management-labor relations can be detrimental to safety

Policy making

- Company philosophy
- Economic factors
- Public relations
- New generation of aircraft
- Major organizational changes

Procedures

- Standard operating procedures are important guidelines for unfamiliar crews

Practices

- Provided the company has sound philosophy, policies, and procedures, it's up to the pilot to put into practice

SOP deviation caused by:

- Individualism
- Complacency
- Laziness
- Frustration

CHAPTER REFERENCES

Degani, Asaf and Earl L. Wiener. June 1994. *On the Design of Flight Deck Procedures*. NASA CR 177642.

Edwards, Elwyn. 1988. "Introductory Overview." *Human Factors in Aviation*. Ed. Wiener, Earl L. and David C. Nagel. San Diego: Academic Press: 20.

Stone, Richard and Gary Babcock. 1988. "Airline Pilot's Perspective." *Human Factors in Aviation*. Ed. Wiener, Earl L. and David C. Nagel. San Diego: Academic Press: 550.

CASE STUDY I-1: USAir Flight 1493 and Skywest Flight 5569

Safety issues: Situational awareness, distraction, role of ATC, cockpit discipline

On 1 February 1991, a USAir 737-300 while landing on runway 24L at Los Angeles, California, International Airport collided with a Skywest Metroliner.

Probable cause

The NTSB determined that the probable cause of this accident was the failure of the Los Angeles Air Traffic Facility Management to implement procedures that provided redundancy comparable to the requirements contained in the National Operational Position Standards. It also found a failure of the FAA Air Traffic Service to provide adequate policy direction and oversight to its ATC facility managers. These failures created an environment in the Los Angeles control tower that ultimately led to the failure of the local controller 2 (LC2) to maintain an awareness of the traffic situation, culminating in the inappropriate clearances and subsequent collision of the USAir and Skywest aircraft. Contributing to the accident was the failure of the FAA to provide effective quality assurance of the ATC system.

History of flights

USAir Flight 1493 was a regularly scheduled passenger flight from Columbus, Ohio, to Los Angeles (LAX). The 737 departed Columbus at 1317 Eastern Standard Time with 89 passengers and 6 crewmembers onboard.

Skywest Flight 5569 was a regularly scheduled passenger commuter flight from Los Angeles to Palmdale, California. The Metroliner departed the gate area around 1758 PST with ten passengers and two crewmembers onboard.

Pilot experience

The USAir captain had 16,300 total flight hours, 4300 in the 737. He was upgraded to 737 captain in 1985, and had completed his last proficiency check the month before the accident.

The USAir first officer had 4316 total flight hours, 982 in the 737. He had been with the airline since 1988, and had completed his last proficiency check two months prior to the accident.

The Skywest captain had 8808 total flight hours, 2107 in the Metroliner as pilot-in-command. He had been with the company since 1985, and had completed recurrent training and proficiency checks two months before the accident.

The Skywest first officer had 8000 total flight hours, 1363 in the Metroliner as second-in-command. He had flown for Skywest since 1989, and completed his last proficiency check about seven months before the accident.

Weather

A special local weather observation was taken at 1816. The reported conditions were 30,000 feet thin scattered and visibility at 15 miles. Official sunset at LAX occurred at 1723. The official end of twilight was at 1748.

The accident

The nearly five-hour USAir flight was uneventful, and the crew set up for the approach into LAX. At 1759, the captain told the controller that he had the airport in sight. The airplane was about 25 miles from the field, and the first officer was at the controls. Seconds later, the crew was "cleared visual approach runway two-four left . . . ." The captain acknowledged the clearance. Close to a minute later, the captain reconfirmed the correct runway with ATC, and at 1803 the crew was told to contact Los Angeles tower.

The first officer said he remembered the horizon was dark during the approach and landing. He lined up visually for runway 24L, and used the ILS glideslope for 24R for initial vertical flight path guidance since there was no operating ILS or VASI for 24L. He recalled configuring the airplane for landing approximately 12 miles from the threshold, and told the captain that he had the runway in sight.

About five minutes later, at 1758, the Skywest crew began their taxi to runway 24L. Shortly thereafter, the ground controller instructed them to, ". . . turn right on Tango [taxiway] and then at Forty-Five [taxiway intersection] transition to Uniform [taxiway], taxi to runway two-four left . . . ." At 1803:44, the crew changed to the tower frequency of 133.9, and advised the LC 2 that, "Skywest . . . at forty five, we'd like to go from here if we can." The LC 2 responded, ". . . taxi up to and hold short of two-four left." The clearance was acknowledged.

Nearly one minute later, the USAir captain called the LC 2 (Local controller, number 2 position) on 133.9 and told her that aircraft's location. Although the transmission was received, the LC 2 did not respond. She did, however, tell the Skywest crew to, ". . . taxi into position and hold runway two-four left, traffic will cross downfield." At 1804:49, the Skywest crew made a last transmission and replied, "Okay, two-four left position and hold, Skywest . . . ."

Another Metroliner was waiting to cross runway 24L, but the crew had inadvertently tuned to a different frequency, preventing the LC 2 from issuing the clearance to them. The crew eventually returned to the tower frequency, and was given permission to cross the runway at 1805:16. Meanwhile, the Skywest flight remained at the intersection of taxiway 45 and the center of runway 24L.

At 1805:29, the USAir captain made a second call to the LC 2 and said, "USAir . . . for . . . two-four left." The controller responded with, "USAir . . . cleared to land runway two-four left." Followed by the captain's reply, "Cleared to land two-four left . . . ." That was Flight 1493's last transmission.

The LC 2 then began working with other departing traffic. At 1806:08, a Wings West Metroliner was ready for takeoff, but the controller could not locate its flight progress strip. The strip had been misfiled and was finally found about 22 seconds later.

The USAir first officer recalled hearing a conversation between the tower and another airplane, concerning its location on the field. He did not remember hearing a hold or takeoff clearance for any aircraft for runways 24L or 24R. He told investigators that he looked down to the runway and saw the lights and overall landing environment. He noted that the cockpit interior lighting was normal, and that he had not been distracted during the approach.

The first officer further described the approach as stable, and heard the captain call out "500 feet." He confirmed that the landing light switches were "ON," and that the aircraft crossed the threshold at approximately 130 knots. The main landing gear touched down about 1500 feet from the approach end of the runway and on the centerline. He deployed the thrust reversers, but was not sure if they had fully deployed before the collision. As he lowered the nose of the airplane, he saw the Metroliner directly in front and below him. He noticed the red light on its tail as the jet's landing lights reflected off the propellers.

Although the first officer tried to apply the brakes before the accident there was not enough time for evasive action. He believed that the initial point of impact was directly on the nose of the 737 and the Skywest's tail. There was an explosion and fire at the moment of contact.

Impact and wreckage path

The left underneath side of the 737 crushed a major portion of the Metroliner. Refer to Fig. I-A. Both aircraft skidded 600 feet down runway 24L before veering to the left an-

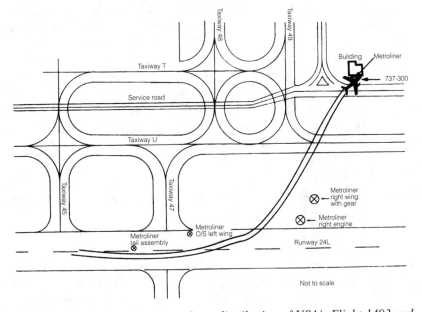

Fig. I-A. *Ground track and wreckage distribution of USAir Flight 1493 and Skywest Flight 5569.* Adapted from NTSB

59

other 600 feet into a vacant fire station. A total of 19 gouge marks was made from the propellers, along the collision route. Parts of the Metroliner were scattered along the wreckage path, but the only sections of the jet that separated were the nose cone, nose gear doors, and left pitot tube.

The 737, however, was destroyed by fire when the aft fuselage collapsed. The impact of the jet with the building caused heavy damage to the left side of the cockpit, the left engine, and the leading edge of the left wing. Several propeller slashes were on the lower right side of the jet's fuselage, near the forward galley door.

Accident survivability. The collision was not survivable for the occupants of the Metroliner. The USAir captain, one flight attendant, and 21 passengers aboard the 737 sustained fatal injuries.

Many survivors said that the cabin filled with thick black smoke within seconds of hitting the building. There was a delay in opening the right overwing exit because a nearby passenger "froze" which prompted an altercation between two other passengers. Eleven victims, including the flight attendant, were found lined up in the aisle from 4½ to 8 feet from the overwing exits. According to the Board, they most likely collapsed from smoke and particulate inhalation while waiting to evacuate.

Witnesses agreed that both airplanes were ablaze shortly after initial contact on the runway. Investigators found the 737's crew oxygen cylinder, which was installed in the forward cargo compartment, depleted, and the low-pressure oxygen supply line broken. Refer to Fig. I-B. The discovery suggested that the oxygen contributed to the fire because there were several holes found in the fuselage near the cylinders. Boeing confirmed that a full cylinder would bleed down in about 90 seconds.

Survivors told investigators that thick, black smoke filled the cabin within 45 seconds. Based on that information, as well as data obtained from other accidents, the Board requested the FAA conduct "burn tests" to determine the effects of compressed gaseous oxygen on cabin fires. Baseline tests were initially conducted that did not include the introduction of compressed gaseous oxygen. In those cases, the fire and smoke spread into the cabin in about five minutes. However, the release of such elements was proven to "exacerbate the rate at which the fire and smoke spread into the cabin." In two sets of tests, the forward cabin area became totally engulfed by flames and smoke in less than two minutes.

The investigation

The Safety Board analyzed ATC-related factors and flightcrew performance with regards to this accident.

ATC procedures. In an effort to reduce workload at the ground control position, LAX ATC procedures did not specify the use and handling of flight progress strips at that duty level. As a result, aircraft could request intersection departures directly from the local controller. The ground controller was thereby relieved from coordinating with the local controller and marking flight strips accordingly. Although intended to lessen the ground controller's workload, the procedures eliminated redundancies that were built into the system and consequently increased the local controller's workload.

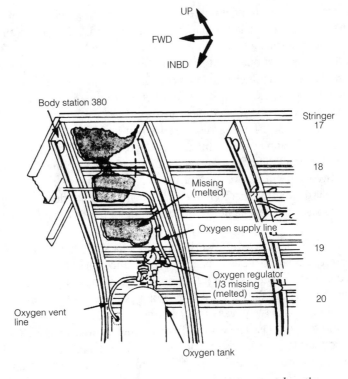

UP
FWD
INBD

Body station 380

Stringer 17

18

Missing (melted)

Oxygen supply line

19

Oxygen regulator 1/3 missing (melted)

20

Oxygen vent line

Oxygen tank

Fig. I-B. *Fuselage damage at crew oxygen system location.*
Adapted from NTSB

Without the information from the flight progress strip, the only way the local controller could keep track of taxiing aircraft was by contacting the flightcrew and relying on memory or observation as to their particular movement on the field. If the controller was unable to remember the details of an aircraft's intended taxi route, or could not distinguish which airplane was assigned what flight number (a real problem when working similar aircraft from the same airline), then the probability for serious errors rose significantly.

A review of the communications transcript of the LC 2 provided valuable insight to the Safety Board as to the controller's activities just prior to the accident. At 1803:38, the Skywest crew told her that they were, " at forty five, we'd like to go from here." In later testimony, she stated that she did not hear the "at forty five" portion of the radio call. The Board was unable to conclusively determine whether or not she heard the transmission in its entirety. However, investigators noted from her subsequent taxi instructions to other aircraft, that she was aware of the Skywest's position on runway 24L as late as 1805:16.

Unnecessary distractions. Between 1804:11 and 1804:52, the LC 2 attempted to contact the Wings West Metroliner four times. The crew had mistakenly turned off the active ground control frequency as they waited clearance to cross runway 24L. Not until 1805:09 was the controller able to resume communication with the crew. The Board believed that this generated additional workload for the LC 2, and the subsequent unnecessary and extraneous conversation with them created a distraction. This was evident from the fact that at one point she identified the Wings West flight as an aircraft that she had cleared to another runway more than four minutes earlier. The Board further believed that as the LC 2 worked to correct the problem, she became preoccupied and forgot that the Skywest was on the runway.

She appeared to still be a bit confused when at 1806:08, the Wings West crew called for takeoff clearance. She immediately asked them, "you at forty seven [intersection] or full length?" The Board noted that instead of clarifying everyone's position, she became involved with searching for the flight's progress strip. This situation created yet another distraction that took the LC 2 away from her duty to scan the runway. If the progress strip had been at her station to begin with, this diversion would not have occurred.

As a result of the demanding workload and a lack of "memory aids" such as the progress strip, the Board believed that she "forgot" that Flight 5569 was on the runway. To further complicate matters, she misidentified the Wings West aircraft for the Skywest. When she saw the Wings West Metroliner pass in front of her on taxiway Uniform, she thought the runway was empty, so she cleared the USAir flight to land.

ATC noncompliance. The Safety Board believed that the LC 2's performance was related to facility procedures that did not allow for human error. The LC 2 was required to assume full responsibility for flight progress strip marking and position determination, in addition to departure and arrival sequencing. As a result, the situation created an "abnormal burden" on the controller.

Furthermore, the Board discovered a discrepancy between the FAA's Operational Position Standards order (referred to as National OPS) and the LAX Facility OPS guide. In Chapter 23 of the National OPS, it states, "[the GC position] shall handle the flight progress strips . . . and mark the runway the aircraft is assigned." The LAX OPS manual states that "strips are not required," for the GC position. The assistant division manager of the Air Traffic Terminal Procedures Branch in Washington, D.C., testified that LAX was in compliance with the National OPS because the handbook states that the progress strips will be forwarded to the "appropriate position." Since the LAX ATC management decided that the appropriate position was at the LC level, they believed the facility was "in compliance with the intent of the National Order." However, the Board noted that the authors of the National OPS recognized the potential for unique circumstances to arise (such as gate holds) that would preclude the established progress strip procedures. It was under those conditions that a facility was allowed to modify a procedure. It was not considered a cause to permanently change it for routine operations.

The Board believed that the decision made by LAX management to remove the GC from the flight progress strip marking and forwarding loop, caused a breakdown in redundancy for aircraft tracking. Based upon that information, the Board concluded that the LAX tower was not in compliance with the National OPS Order.

Airplane conspicuity. As part of the investigation, an airplane conspicuity test was conducted for runway 24L in night VMC. An identical Metroliner was positioned at the same location as Flight 5569. The runway edge and centerline lighting were set on low intensity. The tests revealed that during visual approaches, cockpit observers found it difficult to distinguish between the Metroliner and the runway environment.

The design of the Metroliner's anticollision beacon added to the problem. Although it is positioned on top of the vertical stabilizer, the rudder cap obstructs the light when viewed from behind. A representative for the aircraft's manufacturer testified that as the USAir descended below 100 feet over the runway surface, "it is very possible he couldn't see the beacon." When investigators asked the first officer why he was unable to see the Metroliner earlier, he replied, "It wasn't there. It was invisible."

Test results. The Safety Board noted that an aircraft should hold about 3 feet off the runway centerline lighting for best rear-view detection. The use of high-energy strobe lights was also considered beneficial in these cases.

Flightcrew vigilance. Because runway incursions are relatively uncommon, the Board expressed concern that flightcrews might relax their scanning vigilance for ground traffic. A state of shared responsibility between pilot and controller might also be a factor. In any event, the Board reminded pilots that they need to pay attention to all ATC communication, not just those calls that directly pertain to them.

The Skywest aircraft had been holding on runway 24L for nearly two minutes before the USAir flight had been cleared to land. In that time, there were about 20 ATC and pilot transmissions on the tower frequency, many of which concerned the Wings West flight. Therefore, the Board noted that the Skywest crew should have contacted the LC 2 when seemingly less important matters were being discussed—after all, they were the ones sitting on an active runway. They should also have been more cognizant of the other radio calls, and been immediately alerted when they heard, ". . . cleared to land . . . 24L."

Communication phraseology. In review of the LC 2's communication transcripts, the Board found numerous instances where pilots used vague and ambiguous terms. For example: "We'll take forty-seven [intersection 47], "We'd like to go from here," and "For the left side, two-four left." Since the LC 2 testified that she did not hear the Skywest crew tell her they were at the taxiway 45 intersection, the Board believed that more standard phraseology might have prevented that from occurring. They recommended the following: "Cessna 12345 request intersection takeoff from runway 24 left at taxiway 45." The controller's response would then be, "Cessna 12345, taxi into position and hold runway 24 left at intersection 45."

Lessons learned and practical applications

1. Be vigilant about situational awareness. This is especially critical in a high traffic environment, and without question while sitting on an active runway. In this case, the phrase, "cleared to land two-four left," was spoken twice over the radio. Yet, there was no response from the Skywest crew. Listen to the radio.

2. Don't be programmed to hear only your call sign. Many pilots tune out the background chatter, responding exclusively to their own flight number. A method that was taught to pilots for years was to listen to every fourth word. Not a good practice around busy airport traffic areas, or while taxiing on an expansive airfield.

3. Use standard phraseology. When everyone is speaking the same language, operational errors and oversights are dramatically reduced. Avoid made-up phrases.

4. When in doubt, ask. If you're in a critical phase of flight, or sitting on an active runway, and you *think* you heard a suspicious radio call, immediately check with ATC.

5. Communicate clearly and specifically. Review the Board's suggestions mentioned under the "Communication phraseology" section in this case study.

Case study reference

National Transportation Safety Board. 22 October 1991. Aircraft Accident Report: Runway Collision of USAir Flight 1493, Boeing 737 and Skywest Flight 5569, Fairchild Metroliner. Los Angeles, California. February 1, 1991. Washington, D.C.

CASE STUDY I-2: Air Florida Flight 90

Safety issues: CRM, cockpit discipline, procedural deviation, intimidation, judgment, ADM, winter flight operations, engine icing, role of ATC

On 13 January 1982, Flight 90 took off from Washington, D.C., National Airport during a snowstorm, struck the 14th Street Bridge and plunged into the icy Potomac River.

Probable cause

The NTSB determined that the probable cause of this accident was the flightcrew's failure to use engine anti-ice during ground operation and takeoff. The crew's decision to take off with snow and ice on the airfoil surfaces of the aircraft, and the captain's failure to reject the takeoff during the early stages when his attention was called to anomalous engine instrument readings.

History of the flight

Flight 90 was a regularly scheduled passenger flight from Washington National to Ft. Lauderdale, Florida, with an intermediate stop in Tampa, Florida. The 737-222 departed National at 1415 Eastern Standard Time with 74 passengers and 5 crewmembers onboard.

Pilot experience and performance history

The captain had 8300 total flight hours, 1100 as 737 captain. In May 1980, he failed a line check for unsatisfactory marks in the following areas: adherence to regulations, checklist usage, and flight procedures. As a result, his captain qualifications were suspended. Three months later, he successfully completed the line check and was reinstated as a captain. In April 1981, he failed his recurrent proficiency check when he showed deficiencies in memory items, knowledge of aircraft systems, and aircraft limitations. He passed the recheck three days later. About three months before the accident, the captain completed a 737 simulator course in lieu of a proficiency check.

The first officer had 3353 total flight hours, 992 in the 737. He had been with Air Florida since 1980. His previous experience was as a U.S. Air Force F-15 flight examiner and instructor pilot and had logged 669 hours of fighter time.

Weather

Earlier in the day, a SIGMET had been issued from 1340 to 1740 that included the District of Columbia and Virginia. Moderate occasional severe rime or mixed icing in the clouds was reported by aircraft throughout the mid-Atlantic, and the freezing level was from the surface to 6000 feet.

At 1558, the surface observations were updated to: Ceiling indefinite, 200 feet obscured. Visibility ½ mile. Snow. Temperature/dewpoint, 24 degrees F/24 degrees F. Wind 010 degrees/11 knots. Runway visual range 2800 feet, variable to 3500 feet.

The accident

Snow was falling when the aircraft arrived at National around 1329. A moderate to heavy snowfall continued, forcing the airport to close from 1338 to 1453. The scheduled departure of Flight 90 had already been delayed 1 hour and 45 minutes.

At 1450 (70 minutes before takeoff) the captain requested the jet be deiced. Twenty minutes later, the passengers were boarded and the crew was ready for pushback. However, because of the combination of ice, snow, and glycol on the ramp, the tug became stuck and was unable to move the jet. Snow was still falling heavily when the crew decided to push themselves back by using their thrust reversers. While operating their reversers for an estimated 30 to 90 seconds, much of the snow and slush on the ground blew back onto the aircraft surfaces. The attempt was unsuccessful and they called in a second tug. At 1535, Flight 90 was finally pushed back without further difficulty. Fifteen aircraft were already lined-up for departure.

While taxiing, the crew began the after-start checklist. When the first officer called, "anti-ice" the captain replied "off." As they approached the active runway, they entered a second deicing area and intentionally parked close behind another jet. The purpose was to use that airplane's exhaust to help in their own deicing. Although both pilots had previously commented on the severity of the weather, they placed little importance to this last deicing. The captain was concerned about having "his windshield deiced" but remarked, ". . . don't know about my wings." The first officer then replied, ". . . all we need is the inside of the wings [presumably, the leading edge devices] . . . the wingtips are gonna speed up on eighty [knots during takeoff roll] . . . they'll shuck all that other stuff [as in, slush, snow and ice?]." A minute later they discussed the condition of the wings as they were being deiced. The captain noted "I got a little on mine." While the first officer added, ". . . this one's got about a quarter to half an inch on it all the way." The Board believed that they were talking about accumulation of ice, snow, or both.

Seconds later, the first officer asked the captain, "See this difference in that left engine and right one?" The captain responded with a "yeah" and no further comment. Meanwhile, the first officer continued to question the problem, "I don't know why that's different . . . 'less it's hot air going into that right one [engine], that must be it . . . from his exhaust . . . it was doing that at the chocks awhile ago . . . ." The captain completely ignored the first officer's comments, and even changed the subject. One minute later, the first officer again expressed concern over the same discrepancy by saying, "This thing's settled down a little bit, might'a been his hot air going over it."

The instrument to which the first officer was referring was the engine pressure ratio (EPR), which is the primary gauge to set and monitor thrust on JT8D engines. The EPR is equal to the ratio of the pressure measured at the engine discharge (Pt7) to the pressure measured at the compressor inlet (Pt2). The EPR target value for takeoff thrust is a predetermined setting dependent on existing conditions. Therefore, if the airflow is disrupted going through the Pt2 probe, the EPR readout will be erroneous.

At 1553, the first officer again discussed the deicing operation and added, "Boy . . . this is a losing battle here . . . trying to deice those things, it gives you a false sense of security, that's all it does." The captain replied, ". . . it satisfies the Feds."

As the crew taxied towards the active runway, they continued their pretakeoff checklist, including verification of the EPR setting of 2.04. The first officer then asked the captain if he had any suggestions for taking off on a slushy runway, but he offered none.

At 1559, Flight 90 was hurriedly, ". . . cleared for takeoff . . . no delay on departure . . . traffic's [an Eastern 727] two-and-a-half out for the runway." Almost immediately after they started their takeoff roll the first officer, who was flying, exclaimed, "God, look at that thing, that don't seem right, does it?" Followed by, ". . . that's not right . . . ," to which the captain responded, "Yes it is, there's eighty." The first officer pressed, "Naw, I don't think that's right . . . maybe it is . . . I don't know." Two seconds after the captain called V2, the stickshaker activated, followed by the captain yelling, "Forward, forward, just barely climb." At 1601, the aircraft struck the heavily congested northbound span of the 14th Street Bridge, and plunged into the frozen Potomac River.

Impact and wreckage path

As the aircraft lost altitude, it struck seven occupied vehicles on the bridge before tearing away a 41-foot section of the bridge wall, and 97 feet of its railings. Refer to Fig. I-C and Fig. I-D. Except for fragments of the right wing, the remainder of the wreckage sank in 24 to 30 feet of water. The airplane came to rest about three-quarters-of-a-mile from the departure end of runway 36, and broke into several major pieces. Refer to Fig. I-E. The nose section and cockpit, the center section of the fuselage, the fuselage-to-wing-intersection, and the empennage were mostly confined to the south side of the river between the 14th Street and Memorial Bridges. Refer to Fig. I-F. There was no post-impact fire.

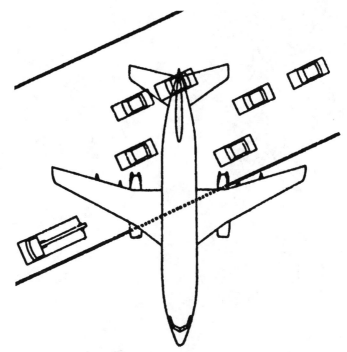

Fig. I-C. *Top view of impact point with 14th Street bridge.*
Adapted from NTSB

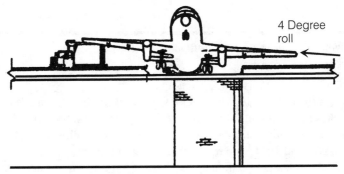

Fig. I-D. *Front view of impact point with 14th Street bridge.*
Adapted from NTSB

Accident survivability. The captain, first officer, one flight attendant, and 70 passengers sustained fatal injuries, along with four motorists on the bridge. Trauma to the head and neck was the leading cause of death among those on Flight 90. The remaining five passengers and one flight attendant received incapacitating injuries due to secondary impact forces, and were unable to reach the shoreline without assistance.

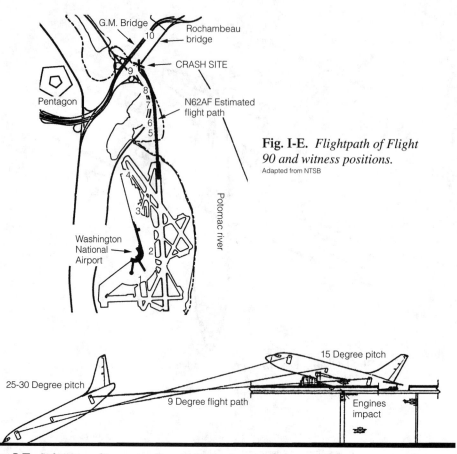

Fig. I-E. *Flightpath of Flight 90 and witness positions.* Adapted from NTSB

Fig. I-F. *Side view of impact points and attitudes.* Adapted from NTSB

One of the survivors later drowned as they all tried to cling to pieces of floating debris. The water temperature at 4 feet was 34 degrees F, and it took between 22 and 35 minutes for the remaining survivors to be rescued. Although firefighters first arrived at the scene around 1611, due to the frozen conditions they had great difficulty retrieving the survivors. About 10 minutes later, a U.S. Park Police helicopter pilot assisted with the rescue and was able to pluck all of the survivors, one by one, out of the Potomac.

The investigation

Investigators found that ice accumulation had blocked the inlet of the Pt2 probe on both engines, which was the direct result of the crew's failure to turn on the anti-ice. According to the 737 manual, ". . . engine inlet anti-ice system shall be on when icing conditions exist . . . ." Another factor overlooked from the manual was the known abrupt pitchup characteristics the 737 has with wing leading edge contamination. The advisory to pilots states: "If leading edge roughness is observed or suspected for any reason, care should be exercised to avoid fast/over rotation." The Board has since recommended, ". . . prohibition of takeoff if leading edge contamination is observed or suspected."

Flight 90 went into a stall shortly after rotation. Because the crew believed that thrust was already at the takeoff limit (from the EPR setting), they chose to concentrate their efforts on pitch attitude. It should have been apparent from the continuation of the stickshaker and the steady decrease in airspeed that the jet was not recovering. Therefore, the Board believes the crew should have responded immediately with a thrust increase, regardless of their assumption that EPR limits would be exceeded.

The Safety Board was confused as to why a crew would not turn on the engine anti-ice in the middle of a snowstorm, so they sent the CVR tape to the FBI voice lab for detailed analysis. In its final report, the FBI confirmed that the crew's checklist response of "off" seemed to be without hesitation, and there was no discussion regarding the existing weather conditions. Therefore, the Board concluded that there was no evidence that suggested the crew ever considered the use of engine anti-ice during their taxi to the runway.

As Flight 90 left the second deicing area, the Board believed that the Pt2 probe had already iced-over. The situation was further compromised when, at the gate, slush was blown back onto the surface of the jet, including the intakes. Furthermore, the exhaust blast from the other airplane in the deicing area caused some of the ice that had already formed in the Pt2 probe to melt. When the jet then taxied away from Flight 90, the probe refroze into a solid block of ice.

The Board noted the interesting comment that the first officer made when he was trying to determine the reason for the EPR fluctuation. Although he eventually convinced himself the problem was due to the "hot air going into the right one," he revealed that, "it was doing that at the chocks." Obviously, Flight 90 did not have exhaust blowing in the intakes while at the gate and in the chocks. Neither crewmember picked up on that observation, nor did they discuss the discrepancy any further.

During the takeoff roll, the first officer questioned the readouts of the engine instruments four times. Performance calculations revealed that the crew would have been able to safely reject the takeoff as late as 18 seconds before V2. The EPR was set at 2.04, but the N1 (low pressure) was 80 percent, Exhaust Gas Temperature (EGT) read 4000 degrees C, N2 (high pressure) was 85 percent and the fuel flow was at 5500 lbs/hour. However, the instruments should have read: N1: 92 percent; EGT: 4500 degrees C; N2: 90 percent, and fuel flow: 8000 lbs/hour. It was later determined that the EPR was at 1.70—not 2.04.

ATC performance. According to the ATC Handbook, a controller is required to "separate a departing aircraft from an arriving aircraft on final approach by a minimum of two miles, if separation will increase to a minimum of three miles within one minute after takeoff." With regard to Flight 90, because of the limited visibility, the local controller could not see the airplane when he cleared the crew for takeoff. Investigators found that the distance from Flight 90 and a landing Eastern 727 was between 1500 feet and 4000 feet. Discrepancies in the 727's FDR precluded more precise calculations. In any event, the controller cleared the Eastern jet to land without determining the location of Flight 90, whereby violating separation minimums.

Lessons learned and practical applications

1. Know your airplane's operations manual. In this case, the 737 manual clearly states:

   ". . . the Pt2 probe will ice up in icing conditions if anti-ice is not used; and an erratic EPR readout may be an indication of engine icing."

   The manual further explains that when the Pt2 probe becomes solidly blocked, EPR fluctuations will cease. The first officer noticed both anomalies. Stay current with reading your OPS manual and aircraft bulletins. As you review the material, think of possible emergencies you might encounter—even unlikely ones. Then "chair-fly" your decisions at home. This way you will have the time to think carefully through your reactions, and mentally prepare yourself before the emergency.

2. Fly the airplane. Do what it takes to clear obstacles and remain airborne. Because the crew thought they were already at the engine limits, they didn't add power to climb. Instead, they pulled back on the yoke, increasing angle of attack (AOA) and decreasing airspeed. If what you're doing isn't working, try something else. In this case, when the jet failed to recover, the Board believed that the crew should have increased thrust, regardless of their assumption that it would cause engine damage. If you can, climb with your throttles—not with the stick. For specific aircraft procedures, read your OPS manual.

3. Know what you're looking for during your instrument cross-check. Multi-engine pilots might get into a habit of just glancing at "matched needles" instead of actually reading the indications. Flight 90 had matched needles—they were just all the wrong readout. Matched needles don't automatically mean that everything's O.K. First *understand* what you're looking for, then *read* the instruments.

4. Watch out for your fellow pilots. Flightcrews who saw Flight 90 before and during takeoff observed unusually heavy accumulation of snow and ice on the jet. Even more disturbing was the airline crew who, when taxiing past Flight 90, saw, ". . . almost the entire length of the fuselage with snow and ice . . . including the top and upper side of the fuselage above the passenger cabin windows." They never called Air Florida with this information. One can only wonder what would have happened if that crew had alerted Flight 90 by simply saying, "Air Florida, you're covered with snow and ice . . . ."

5. Use clear communication. Concise and direct communication is vital for a safe flight. None of us can argue with that—but why don't we do it as a standard practice? The cockpit transcript of Flight 90 is filled with choppy and incom-

plete sentences, half-developed ideas, and discussions nonessential to flight. The Board suggested that the captain never truly listened to the first officer, nor did he sense the urgency in his voice. The first officer never effectively communicated his distress. There were many "maybes" and "I don't knows." Not only *listen* to what is said, but *how* it's said. This also applies during your interaction with controllers, weather folks, maintenance personnel, and anyone else you come in contact with during your flight.

6. Be assertive. Why wasn't the first officer more assertive when telling the captain his concerns? Was it fear of "stepping on the captain's toes?" Perhaps he was just taking his cue from the captain—after all, the captain was the more experienced airline pilot. The first officer's previous flying experience included being a single-seat F-15 check and instructor pilot. This individual obviously was used to calling his own shots in a very aggressive environment, yet he did not speak up on that fateful day. If this can happen to a qualified flier like that, it can happen to anyone. The next time things aren't right, or the hair stands up on the back of your neck, or you have a funny feeling—TAKE ACTION!

7. Be decisive. Both pilots fell into this trap at a critical time—during takeoff. According to the Air Florida Training Manual: "Any crewmember will call out any indication of engine problems affecting flight safety." However, "The captain ALONE makes the decision to REJECT a takeoff." It appears that the first officer was waiting for that decision from the captain, but it never came. Even if you only suspect a problem—MAKE A DECISION! Who's going to argue with you when you tell them your reason was for safety?

8. Avoid complacency. Take time to read your checklist and listen to the answer. What is also very helpful is to touch the switch or light after the callout, and before you proceed to the next checklist item.

9. Resist intimidation. Intimidation comes in many expressions. A captain looking disinterested, not answering questions, and sitting in silence are common forms. But most pilots don't think of being intimidated by ATC, but it happens. Flight 90 had not even taxied onto the runway when the tower told them, ". . . no delay . . . in departure." It was probably obvious to the crew that traffic was tight, and nobody wants to be "the jerk that screwed up departures and arrivals" on a busy day. But when you're feeling rushed or know the inbound traffic is too close, hold short. Remember who's pilot-in-command. Takeoff clearance is a clearance not an order.

Case study reference

National Transportation Safety Board. 10 August 1982. Aircraft Accident Report: Air Florida, Inc., Boeing 737–222, N62AF. Collision with 14th Street Bridge, Near Washington National Airport, Washington, D.C., January 13, 1982. Washington, D.C.

CASE STUDY I-3: Avianca Flight 052

Safety issues: CRM, fuel management, role of ATC, traffic flow management, non-standard phraseology, foreign carrier operations

On 25 January 1990, an Avianca 707 crashed while making a second attempt to land at John F. Kennedy (JFK) International Airport, New York.

Probable cause

The NTSB determined that the probable cause of this accident was the failure of the flightcrew to adequately manage the airplane's fuel load. Furthermore, they did not communicate an emergency fuel situation to ATC before fuel exhaustion occurred. Contributing to the accident was the flightcrew's failure to use an airline operational control dispatch system to assist them during the international flight into a high-density airport in poor weather. Additional factors included the inadequate traffic flow management by the FAA and the lack of standardized, understandable terminology for pilots and controllers for minimum and emergency fuel states.

History of flight

Avianca Flight 052 operated as a regularly scheduled passenger flight from Bogota, Colombia, South America, to JFK International Airport, New York, with an intermediate stop at Medellin, Colombia, South America. The 707 took off from Medellin at 1508 EST with 149 passengers and 9 crewmembers aboard.

Pilot experience

The captain had 16,787 total flight hours, 1534 in the 707. He had been a 707 captain since June 1987. The first officer had 1837 total flight hours, 64 in the 707. His initial line check for the airplane was one month prior to the accident. The second officer had 10,134 total flight hours, 3077 in the 707. He requalified in the jet three months prior to the accident.

Weather

According to the National Weather Service (NWS), at 0700 a deep low-pressure area was centered over northeastern Illinois. A stationary front extended eastward through Indiana and Ohio, curved northeastward through parts of Maryland, over Long Island, New York, and into eastern Massachusetts. Another stationary front was positioned from central Georgia to as far north as coastal Virginia. As a result, the skies were reported as overcast with rain over all of the mid-Atlantic states from southern Virginia to extreme southeastern New York.

An International Airdrome Forecast (IAF) for JFK was activated at 1300, and remained valid for 24 hours. The weather conditions were expected to be: Visibility one mile; Light rain; Ceiling 400 feet. Winds southeast at 15 knots with gusts to 25 knots.

8/8 stratus. An IAF for Boston-Logan International Airport (Flight 052's filed alternate), was also activated for the same period as the JFK report. The forecast included: Visibility one mile; Light rain; Ceiling 800 feet; Winds southeast at 15 knots. 8/8 nimbostratus.

Actual weather conditions. Less than 45 minutes prior to the accident, surface observations at JFK and Boston-Logan were as follows:

JFK 2100: Special; Ceiling indefinite; 200 feet obscured. Visibility ¼ mile; Light drizzle and fog; Wind 190 degrees at 21 knots.

Boston 2050: Ceiling indefinite; zero feet obscured; Visibility ⅛ mile; Light drizzle and fog; Wind 100 degrees at 9 knots.

The accident

Flight 052 arrived in Medellin at 1404 following an uneventful 54-minute flight from Bogota. The aircraft was refueled and departed at 1508 for the four-hour and 40-minute flight to JFK.

Due to the poor weather in the New York area, there were numerous and lengthy flight delays into JFK. The northeast corridor was so congested with traffic that Flight 052 was instructed to enter a 19-minute hold over Norfolk, Virginia. At 1943, Flight 052 was again cleared to hold, this time for 29 minutes at the BOTON intersection, near Atlantic City, New Jersey. The third and final holding pattern that the crew had been instructed to enter was for 29 minutes at the CAMRN intersection, 39 nm south of JFK.

At 2044, while Flight 052 was holding at the CAMRN intersection, New York center advised the crew to, "expect further clearance [EFC] at 2105." This had been their third EFC since they had begun holding at CAMRN. The first officer notified center that, ". . . I think we need priority . . ." The controller promptly asked, ". . . roger, how long can you hold and what is your alternate [airport]?" The first officer responded, ". . . we'll be able to hold about five minutes. That's all we can do." He then added, ". . . we said Boston, but . . . it is . . . full of traffic, I think." When the controller asked him to repeat his alternate, the first officer replied, ". . . it was Boston, but we can't do it now, we . . . don't . . . we run out of fuel now."

A hand-off controller, who was assisting the center controller by monitoring the radio, overheard the fuel situation that Flight 052 had described, and called the New York TRACON, the next control facility that the crew would contact. At 2046, a TRACON controller was notified that, "Avianca 052 just coming on CAMRN, can only do five more minutes in the hold. Think you'll be able to take him, or I'll set him up for his alternate?" After a brief discussion concerning the aircraft's current speed, the TRACON controller told the hand-off controller to, ". . . slow him to one eight zero knots and I'll take him. He's . . . radar three [nm] southwest of CAMRN." The entire discussion took only 20 seconds.

After the center controller was informed of the coordination with the TRACON, he advised Flight 052, ". . . cleared to the Kennedy Airport via heading . . . maintain one one thousand, speed one eight zero." After the first officer acknowledged the clearance, the flight was handed off to Kennedy approach. Once Flight 052 made initial contact with ATC on the new frequency, the crew was given routine radar services, including altitude and heading changes. At 2054, the controller cleared Flight 052 to a new heading and told the crew that, ". . . I'm going to have to spin you [360 degree turn] . . . ." A couple of minutes later, Flight 052 was advised of a ". . . windshear . . . increase of ten knots at fifteen hundred feet . . . increase of ten knots at five hundred feet reported by seven twenty seven." The advisory was acknowledged.

At 2103, Flight 052 was handed off to the Kennedy final controller. The second officer, apparently concerned about the fuel state, reviewed the "go-around procedure with a maximum of 1000 pounds of fuel in any tank." About five minutes later, as Flight 052 descended to 3000 feet, the crew discussed their landing priority status. There seemed to be some confusion on the part of the captain, and he pressed the issue with the first and second officers. At 2109, the first officer told the captain that ". . . they [ATC] accommodate us." The second officer added, "They already know that we are in bad condition . . . they are giving us priority."

When the final controller advised Flight 052 that they were, ". . . one five miles from the outer marker . . . cleared ILS two-two left," the captain told the first officer, "select the ILS on my side." For the next few minutes, the crew continued the landing checklist.

At 2115, they contacted JFK tower and were informed that they were "number three to land." Shortly thereafter, the crew was instructed to increase their airspeed by ten knots. The captain then made a request to the first officer: "Tell me things louder because . . . I'm not . . . hearing it." By 2119, they were nearing the outer marker, had intercepted the glideslope, and had lowered the gear. Less than a minute later JFK tower cleared Flight 052 to land.

As the crew completed the checklist, the captain requested, "give me fifty [degrees of flaps]," and asked, "are we cleared to land?" The first officer replied, ". . . we are cleared to land." He added that they were "below glideslope." Seconds later, the controller asked, ". . . can you increase your airspeed one zero knots at all?" The first officer answered, "yes, we're doing it."

Less than 3 miles from the approach end of runway 22L, the aircraft encountered windshear. The first and second officers made several "glideslope" and "sink rate" callouts as the ground proximity warning system (GPWS) sounded 11 "whoop pull up" voice alerts. Four "glideslope deviation" voice alerts immediately followed. By this time, they were 1.3 miles from the approach end of 22L and at an altitude of 200 feet. The captain urgently asked, "the runway, where is it?" The first officer replied, "I don't see it." The captain commanded, ". . . landing gear up" and began to execute a missed approach.

The final controller instructed Flight 052 to, "climb and maintain two thousand [feet], turn left heading one eight zero." The first officer acknowledged the clearance, as the crew discussed their surprise over not seeing the runway. About 30 seconds later, the controller asked Flight 052 if they were turning to the new heading, but by that time the captain realized the seriousness of the fuel situation, and told the first officer to, "tell them we are in emergency." At 2124, the first officer erroneously repeated the heading clearance ("right turn," instead of a left turn) back to ATC and added, ". . . we'll try once again, we're running out of fuel." The controller replied, "okay." Seconds later, the captain pressed the first officer to, "advise him we are emergency," and, "did you tell him?" The first officer replied, "yes sir, I already advised him."

At 2124:55, the first officer contacted the TRACON and informed the controller that, ". . . we just missed a missed approach . . . maintaining two thousand . . . ." The controller cleared them to "climb and maintain three thousand." The captain again told the first officer to, "advise him we don't have fuel." At 2125:10, the first officer acknowledged the clearance to, "climb and maintain three thousand," and added, "we're running out of fuel." When the captain asked the first officer if he had, ". . . advise[d] that we don't have fuel," he answered, "Yes sir. I already advise him hundred and eighty on the heading. We are going to maintain three thousand feet, and he's going to get us back." The captain replied, "Okay."

About one minute later, the final controller told Flight 052, ". . . I'm going to bring you about 15 miles northeast and then turn you back on for the approach. Is that fine with you and your fuel?" The first officer responded, "I guess so, thank you very much." However, by 2129:11 the aircraft's fuel supply was so low that the first officer asked the controller, ". . . can you give us a final now?" The controller immediately replied "affirmative," and gave him a new heading. Several seconds later the controller told Flight 052 to, "climb and maintain three thousand," but the first officer answered, ". . . negative . . . we just running out of fuel, we okay three thousand now, okay?" As the crew received yet another heading change, they proceeded to set the flaps and monitor the flight instruments.

At 2131:01, the controller advised Flight 052 that they were, . . . number two for the approach. I just have to give you enough room so you make it without . . . having to come out again." Just more than a minute later the second officer exclaimed, "flame out, flame out on engine number four. Flame out on engine number three, essential on number two or number one." At 2132:49, the captain said, "show me the runway."

The first officer immediately told ATC that, ". . . we just . . . lost two engines and . . . we need priority, please." The final controller advised him that they were 15 miles from the outer marker, and to "maintain two thousand until established on the localizer, cleared for ILS two-two left." At 2133:23, the first officer replied, "it is ready on two." That was the last radio transmission from Flight 052.

Impact and wreckage path

The airplane crashed on an up-sloping hill in a wooded residential area. Refer to Fig. I-G. The main section of the fuselage came to rest 21 to 25 feet after impact. The cockpit and forward cabin separated from the remainder of the fuselage and stopped 90 feet in front of the main wreckage.

The fuselage was found partially separated into three sections. Debris from the cabin, including passenger seats, parts of the galley, and overhead bins, was scattered

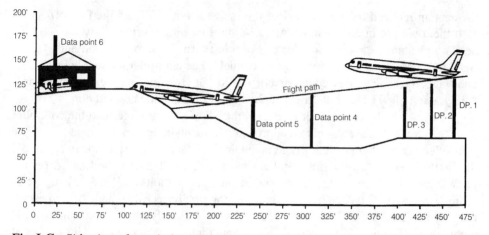

Fig. I-G. *Side view of terrain impact.* Adapted from NTSB

along the wreckage path. Some of the debris was found as far as 100 feet beyond the cockpit section.

The cockpit was substantially damaged when the right side struck a huge oak tree (with a 42-inch diameter), which penetrated the flightcrew compartment. There was also considerable damage near the mid-section of the cabin, caused by a fracture in the longitudinal floor track beam. Consequently, the floor dropped three inches on the right side, shearing the outboard legs of numerous seat assemblies. Those passenger seats were found outside of the cabin, just forward of the wing. The damage inside the aft cabin section was extensive. The cabin had rolled slightly to the left and had cracked open. Most of the seat assemblies had separated from the floor tracks and were thrown from the fuselage.

Accident survivability. The flightcrew sustained fatal injuries, along with 5 flight attendants, 65 passengers, and 1 infant. The Board could not determine where all the passengers were seated at the time of impact because the airline assigned seats to only a small percentage of passengers. It was also noted that since the aircraft was not full, passengers freely changed their seats throughout the flight. For those reasons, the Board was unable to accurately develop an individual injury diagram for each passenger.

Of the 74 survivors, most of those who received serious injuries sustained multiple fractures of the lower extremities, hips, and spine. They also suffered head trauma, bone dislocations, lacerations, and contusions. According to investigators, there were three nearly simultaneous events that occurred at the time of impact that would account for these types of injuries. Most likely, as the passengers' legs were jolted upward against the bottom of the seat units in front of them, the seats collapsed and twisted downward. The seat assemblies then separated from their floor attachments, which caused each seat to be pushed forward. While this domino effect was still in motion, passengers were thrown into other passengers, seat units, and various wreckage debris.

Of the ten surviving infants onboard, eight sustained serious injuries. The Board determined that all of the infants were either being held by an adult, or belted into the same seat with the passenger. Those adults who were holding an infant at the time of impact stated that the child was ejected from the adult's grasp and were unable to be located in the dark.

According to the lead flight attendant, there was no warning from the cockpit crew regarding the low fuel status, loss of engines, or impending emergency landing. Therefore, the passengers had not been recently briefed on brace positions, nor were they given pillows and blankets to help cushion the force of the impact. The Board believed that if either of these procedures had been carried out, some passengers might not have suffered as severe injuries.

FAA vs. ICAO standards. Since 1980, the FAA has required cockpit seats to be equipped with combined seat belts and shoulder harnesses; however, the International Civil Aviation Organization (ICAO) failed to address those types of restraint systems. Seats for the captain and first officer on Flight 052 did not have shoulder harnesses. All three crewmembers died from blunt force head and upper torso trauma.

In 1988, the FAA *required* the installation of emergency path lights on the cabin floor, whereas, ICAO had only *recommended* installation of such lights. According to the Board, those lights might have been useful during the rescue operation of Flight 052, since the piles of debris and the dark cabin hampered emergency workers' efforts.

The investigation

The evidence gathered during the early days of the investigation confirmed that Flight 052 crashed due to fuel exhaustion. Beyond the obvious clues left by the crew on the CVR (cockpit voice recorder) tapes, the Board noted that there was no fire at the accident site, and there was only residual fuel found in the airplane. There was no rotational damage to any of the four engines from impact forces, which indicated that they had stopped operating before hitting the ground. Furthermore, the investigation team observed no engine or fuel system component malfunction that would have caused either premature fuel exhaustion or an interruption of the fuel supply to the engine.

Weather planning. Although weather data provided to the flightcrew before departure from Medellin was nine to ten hours old, it still showed that JFK and Boston-Logan were forecast to have low ceilings and restricted visibility for that evening. The Syracuse and Buffalo, New York, airports were expecting weather at the required minimums; however, neither was listed as possible alternates on the documentation that the crew received. During the investigation, it was learned that Boston was chosen as the alternate as part of a computer-generated flight plan created for all flights bound for JFK without regard to weather forecasts.

Because Avianca was a foreign carrier and its Flight 052 operated in U.S. airspace, the crew was required to comply with all applicable ICAO standards and FARs. According to FAR Part 121.625: "No person may list an airport as an alternate . . . unless the appropriate weather reports or forecasts . . . indicate that the weather conditions will be at or above the alternate weather minimums specified in the [airline's] operations specifications for that airport when the flight arrives." Similarly, ICAO Annex 6, 4.3.6.1 states, in part: "A flight shall not be commenced unless taking into account both the meteorological conditions and any delays that are expected in flight . . . ."

The operations' specifications issued to Avianca by the FAA provided detailed criteria for standard weather minimums at alternate airports. According to the document, ". . . weather minima applicable to [flights] designated for dispatch . . . are 600 [ceiling] and 2 [visibility in sm] at airports served by precision approach procedures." However, the airline's policy manual stated, ". . . when an afternoon or evening takeoff, with a night landing is scheduled, the requirements for the destination, alternate . . . airports are a 1000 foot ceiling and 10 km (6.2 sm) visibility." JFK and Boston-Logan did not meet any of the minimum criteria throughout the entire day. The Board believed that although the flightcrew should have taken a more active role in determining their planned alternate, the inadequacies of Avianca's dispatch services might have affected the crew's performance.

Fuel planning. United States federal and ICAO regulations are similar in content and are very specific about fuel supply requirements. FAR Part 121.645 indicates:

> . . . no person may release for flight or takeoff a turbine-engine powered airplane, unless, considering wind and other weather conditions expected, it has enough fuel:
>
> (1) To fly to and land at the airport to which it is released;
>
> (2) After that, to fly for a period of 10 percent of the total time required to fly from the airport of departure to, and land at, the airport to which it was released;
>
> (3) After that, to fly to and land at the most distant alternate airport specified in the flight release;
>
> (4) After that, to fly for 30 minutes at holding speed at 1500 feet above the alternate airport.

Furthermore, an excerpt of FAR Part 121.621(b) states, ". . . the weather conditions at the alternate airport must meet the requirements of the air carrier's operations specifications."

The ICAO Annex 6, 4.3.6.4 closely parallels the regulations in FAR Part 121 when it states: "In computing the fuel and oil required . . . the following shall be considered: meteorological conditions forecast . . . expected air traffic control routings and traffic delays . . . one instrument approach at the destination aerodrome, including a missed approach and . . . any other conditions that may delay the landing of the airplane or increase fuel and/or oil consumption."

Airline personnel stated that pursuant to standard operating procedure, a dripstick, in addition to the fuel bay and cockpit fuel panel gauges, was used to ensure that the requested fuel was properly loaded into the tanks.

According to the Board, the airplane had sufficient fuel to complete its flight safely, provided there were no extensive delays. But as we already know, Flight 052 encountered weather and traffic delays totaling one hour and seven minutes.

Flightcrew communications. It is important to note that all of the intracockpit conversation was spoken in the flightcrew's native Spanish language. Although their radio communication with ATC was in English, it was mostly regarded as broken. This was especially true as the crew's stress level rose, and critical information had to be discussed in a hurried manner.

The first indication that the flightcrew had some concerns about weather, and possibly the fuel state, occurred about 2009. After being in a holding pattern for 26 minutes, the crew asked the Washington center controller about delays into Boston. He told them that Boston was open and accepting traffic, but they could expect an additional 30 minutes of holding in the New York center airspace.

Unfortunately, the CVR tape saved only the final 40 minutes of intracockpit conversations prior to the accident, therefore, the Board was unable to learn whether the crew discussed their fuel situation as they held at CAMRN. However, it was apparent from the air-to-ground communication, that by 2045, the crew knew they could no longer hold and asked for "priority." The first officer responded to the New York center controller inquiries by informing him that they were, "able to hold about five minutes, that's all we can do." He added that Boston was their alternate but, "we can't do it . . . we run out of fuel now."

Although the center and TRACON controllers coordinated a quick hand-off for Flight 052, they were unaware of the criticality of the aircraft's fuel because the crew never directly told them how serious it was. At 2054, the crew was given a routine 360 degree turn for sequencing and spacing with other arrival traffic, which, in the Board's opinion, should have alerted the crew that they were not getting priority handling. At that time if they had declared an emergency, or at least requested direct routing to the final approach, the Board believed that Flight 052 would have been able to arrive with an acceptable minimum fuel level.

At 2103, without prior discussion, the second officer led the review of the missed approach procedure with less than 1000 pounds of fuel in any tank. About six minutes later the first officer made the comment, "they [ATC] accommodate us," followed by the second officer replying, "they already know we are in bad condition." When the captain questioned the descent clearance, the first officer told him that they were cleared to "one thousand feet," to which the captain replied, "ah, yes." The Board believed that from his response and the tone of his voice, he had understood the controller to be giving them special handling. The second officer reinforced this notion when he answered, "they are giving us priority." This conversation suggested to the Board that the flightcrew thought ATC was aware of their critical situation, and that they were receiving "priority." However, from the time that had elapsed and the seemingly routine nature of the vectoring, it should have been apparent to the flightcrew that they were not receiving an expeditious approach clearance. No direct inquiries, though, were ever made to verify their priority status.

Shortly after the captain initiated the missed approach, he told the first officer, "tell them we are in emergency." But instead, the first officer explained to ATC that, ". . . we'll try once again, we're running out of fuel." Seconds later, at 2124, the captain again told the first officer to, "advise him we are emergency." When the captain pressed him for an answer, "did you tell him?" he replied, "yes sir, I already advised him." The first officer never used the word "emergency" as instructed by the captain, so therefore, he failed to communicate the urgency of the situation.

Less than a minute later, the captain told the first officer, "advise him we don't have fuel." After a second inquiry, "did you already advise that we don't have fuel?" the first officer said, "yes sir, I already advise[d] him . . . and he's going to get us back." At 2126, the approach controller asked the crew if they could accept a base leg of 15 miles northeast of JFK, to which the first officer replied, "I guess so." However, three minutes later, and without prompting from the captain, the first officer said to the controller, ". . . can you give us a final now . . ." From the lack of vital information associated with the sudden request, the controller continued to provide Flight 052 with routine vectors. The crew never challenged those vectors or told ATC of their emergency fuel status.

According to the Safety Board, the intracockpit conversations indicated a "total breakdown in communications by the flightcrew in its attempts to relay the situation to ATC." The Board noted that even though the crew had obvious limitations in their ability to use the English language, they were also unable to communicate effectively among themselves in Spanish.

Although the captain repeatedly told the first officer to tell ATC they had an emergency situation, the first officer never conveyed that message. The evidence strongly suggested that the captain was unaware, at times, of the content of the first officer's transmissions, and that he did not hear or understand the ATC communications. The captain might have been preoccupied with flying the airplane and paid little attention to the first officer's radio calls. Regardless, the Board believed it was more likely that his limited command of the English language prevented him from effectively monitoring the content of the transmissions. Furthermore, the Board believed that this deficiency might have been a factor in the accident, particularly if the captain thought the first officer had adequately expressed the criticality of the fuel state when they left the holding pattern at CAMRN.

Another point of significance was the crew's possible confusion between the term "priority" and "emergency." Avianca flightcrews were trained primarily with the use of Boeing manuals. In one such bulletin it stated, ". . . during any operation with very low fuel quantity, priority handling from ATC should be requested." Similar procedures were published in the airline's own manuals. During the investigation, several Avianca pilots testified that the Boeing-trained flight and ground instructors conveyed the impression that the words "priority" and "emergency" were interchangeable with standard ATC phraseology. That assumption might have explained why the first officer said to the approach controller, ". . . we just . . . lost two engines . . . we need priority, please." It might also have accounted for the first officer's positive responses to the captain when asked if he had advised ATC of their emergency situation.

Flightcrew performance. According to the Board, three key factors contributed to the inability of the captain to remain on the glideslope when he attempted the first ILS approach. First, the prevailing weather conditions and subsequent low-level wind shear activity near 22L caused abnormally high headwind components at Flight 052's final approach fix altitudes. The wind speed was at least 60 knots at 1000 feet, about 50 knots at 500 feet, and almost 20 knots at the surface. As the captain intercepted the glideslope he called for 40 degrees of flaps followed less than a minute later with 50 degrees of flaps. He also made thrust and pitch adjustments to establish a rate of descent that would have been normal for a light headwind component rather than the actual and more severe conditions. As a result of these inappropriate configurations, the airplane immediately descended below the glideslope. The FDR (Flight Data Recorder) revealed that the captain delayed in his reaction and "chased" the glideslope with progressively greater control inputs. From this data it was determined that the captain never established a stabilized descent.

Second, the Board believed the captain showed signs of fatigue due to the rigors of such a tedious flight. It was also possible that maintenance problems with the autopilot and flight director might have caused the crew to manually fly the aircraft from Medellin, and forced the captain to attempt the first ILS approach without the aid of a flight director. It was observed that from the time Flight 052 was on the final vector to the localizer until the missed approach, there were nine distinct incidents of the captain asking for instructions to be repeated, or for confirmation of the airplane's configuration. There was also one request from the captain for the first officer to speak louder.

And third, the captain's performance might have been compromised from the steady increase in aggravated stress levels. The critical fuel state obviously caused stress to the entire flightcrew, however, the captain had made repeated attempts to ensure that the first officer was advising ATC of their situation. The captain became progressively insistent that ATC be told, "we have emergency," and because of his limited English-speaking skills, he might have felt restricted in having to rely on the first officer for all of the communications.

A bilingual breakdown in CRM. Regardless of the fact that the flightcrew had a language barrier with ATC, the final 40 minutes of intracockpit conversations revealed a breakdown in effective communication in their own native tongue. According to the transcripts, the captain never led a discussion on any decisive measures that would be necessary to ensure a safe landing. He depended on the first officer for all of the communications and clearance information. Even though the flight did not seem to be receiving expeditious handling, the captain did not perform an adequate cross-check of the first officer's duties. And, although the captain repeatedly told the first officer to use the word "emergency," he never once recited that command to ATC. Therefore, the first officer appeared to have assumed a slightly more influential role in the cockpit than the captain. There was also no reference in the transcripts of the second officer providing the captain with important fuel burn calculations—nor was he ever asked.

Granted, the crew's limited command of the English language was a detriment to the safety of the flight, however, their own inability to effectively communicate as a cohesive team, with a strong leader, most likely created a weak CRM environment. It was possible if the crew had actively worked together to make a clear plan of action, that message could have been conveyed to ATC, even if it could only be delivered in broken English. Vague and ambiguous statements ("Can only do five more minutes." "[Can't do] Boston, full of traffic, I think." "Running out of fuel.") and inappropriate requests ("We need priority, please.") would most likely have been replaced with something to the effect of, "Emergency! Low fuel! Must land now!"

ATC communications. Considering the information ATC personnel had received from Flight 052, the Board concluded that air traffic control had handled the flight in an appropriate manner. During the investigation, the Board had some concern over the seemingly lack of significance many controllers placed on the term "priority." Although the controllers stated that the word "priority" does not require them to provide emergency responses, the Board noted that it is defined in the ATC Handbook as a request that provides, "precedence, established by order of urgency or importance."

The New York center controller testified that he felt he had adequately assisted the flight since the crew said they could only hold for five more minutes, and he coordinated the hand-off to the TRACON in less than a minute. He had interpreted Flight 052's statement to mean that they needed to leave the holding pattern within five minutes. He further testified that he did not hear the crew add, ". . . we can't do it now [fly to Boston], we run out of fuel now." Consequently, when the hand-off controller contacted the TRACON about accepting the aircraft for JFK, he advised that Flight 052 was near the CAMRN intersection and that they could only "do five more minutes in the hold." Therefore, when the TRACON controller set up Flight 052 for the approach he was unaware of any fuel problems or requests for special handling.

The Board believed that, although the TRACON and JFK tower controllers should have clarified the crew's comments about "running out of fuel," put in context, the statements were not reinforced by any indication of an emergency situation. In fact, at 2126, when the TRACON controller was providing Flight 052 with vectors for the second ILS approach, he asked the crew, "I'm going to bring you about 15 miles northeast and then turn you back for the approach. Is that fine with you and your fuel?" The first officer replied, "I guess so, thank you very much." Moments later, the first officer refused a vector to climb and said, ". . . we . . . running out of fuel . . . we okay three thousand . . . ." Even at that point, he did not convey the situation clearly to ATC. In less than three minutes the engines began to flame out.

EFC confusion. The Board also believed that the flightcrew might have misunderstood the purpose of the three EFCs issued by ATC. An EFC is strictly a time reference so a pilot knows when to expect the next clearance. However, in the case of Flight 052 it was possible that they assumed they would receive approach vectors instead of additional holding clearances at the specific times of their issued EFCs.

Flow control and the weather. The Board discovered that on the day of the accident, the FAA's Central Flow Control Facility had a traffic management program in effect for JFK that was designed to accept 33 arrivals per hour on 22L/R. Although engineered performance standards for the airport revealed that number was quite high in IFR conditions, it was noted that it was based upon the assumption that the volume of expected flight cancellations would offset the acceptance rate of 33 aircraft.

Meanwhile, the NWS failed to advise the traffic management personnel of the severe wind conditions at JFK, which prevented ATC from providing appropriate separation in the approach control airspace. Therefore, the combination of ill-informed controllers and the deteriorating weather caused extensive airborne holds of more than one hour.

The Board determined that even though numerous aircraft had executed missed approaches on 22L because of poor weather, flow control personnel still allowed an acceptance rate of 33 airplanes per hour. In the Board's opinion, flow control did not react appropriately or timely enough to prevent the large numbers of aircraft from being stacked in holding patterns. When flow control did implement a ground stop for traffic destined for JFK, the action was too late to alleviate the airborne holding problem that had already begun.

The pilot and controller connection. Although the pilot-in-command is the one who is ultimately responsible for the safety of a flight, air traffic controllers also share in that responsibility. In this case it was apparent to ATC that they were working with a foreign carrier, and talking with a pilot who spoke in broken English. Between 2044 and shortly before impact, the first officer asked for "priority" twice. On four separate occasions, during the same period, he advised ATC that they were low on fuel, yet no controller directly asked him to clarify his statements. Regardless of the fact that ATC was very busy that evening, it would seem reasonable and logical for a controller to question a foreign pilot who makes the statement, "we're running out of fuel."

There were numerous accumulative factors that proved significant in the events that led to this accident, one of which might have been a breakdown in the pilot-controller team effort. Quite simply, the philosophy of this concept is to look out for each other, and when one drops the ball the other is there to pick it up. In the case of Flight 052, it was especially critical because half of the "team" had a limited English vocabulary, and was using nonstandard ATC phraseology.

Lessons learned and practical applications

1. Communicate clearly and directly. Don't mince words. If you have an emergency or precautionary situation, then say so. Let ATC know the seriousness of your problem and make specific requests.

2. Use standard phraseology. This prevents misinterpretations and second-guessing.

3. Speak up. If you've already followed lesson #1, and you still think ATC is not getting the message, tell them again until you're sure they understand.

Case study reference

National Transportation Safety Board. 30 April 1991. Aircraft Accident Report: Avianca, the Airline of Columbia, Boeing 707–321B, HK 2016. Fuel Exhaustion. Cove Neck, New York, January 25, 1990. Washington, D.C.

CASE STUDY I-4: FAA Beech Super King Air 300

Safety issues: controlled flight into terrain, CRM, pilot judgment, ADM, cockpit discipline, airmanship, intimidation, FAA surveillance

On 26 October 1993, a Super King Air (BE-300) turboprop that was owned and operated by the FAA, crashed into mountainous terrain near Front Royal, Virginia.

Probable cause

The NTSB determined that the probable cause of this accident was the failure of the pilot-in-command to ensure that the airplane remained in VMC over mountainous terrain. The executives and managers of the FAA flying program also failed to: (1) establish effective and accountable leadership and oversight of flying operations; (2) establish minimum mission and operational performance standards; (3) recognize and address performance-related problems among the organization's pilots; and (4) remove from flight operations duty pilots who were not performing to standards.

History of flight

The purpose of the flight was to inspect FAA airways facilities at several airports during a scheduled 5-day duty assignment. Shortly before 1540 EDT, the BE-300 departed the Winchester Regional Airport, Virginia, for a routine point-to-point flight to Newport News/Williamsburg International Airport, Virginia. Two pilots and an electronic technician were onboard.

Pilot experience and performance history

The pilot-in-command (PIC) had 6000 total flight hours, 2000 of which were in the BE-300. His last proficiency check in the airplane was the month before the accident. He completed a mission check in February 1993 and the BE-300 simulator pilot refresher course two months later.

While serving in the U.S. Air Force, he obtained his commercial pilot certificate and flight instructor rating. The PIC was hired by the FAA in 1983 as an air traffic assistant, and two years later he earned his instrument rating. The following year he attained an ATP certificate on his second attempt. In October 1987, he was selected as an airspace system inspection pilot in the Atlantic City, New Jersey, Flight Inspection Area Office (FIAO). He was originally assigned to the procedures section where, in addition to developing instrument procedures, he also served as a second-in-command (SIC) for flight-inspection duty.

In the two-and-a-half years that the PIC spent in that division, his supervisor told investigators that he was slow in developing instrument procedures and appeared uninterested in the work. The supervisor added that there were vocal objections when the PIC was upgraded to that position, and that several of the SICs did not want to fly with him. His flight records also revealed that he had failed his first two check rides for his

BE-300 type rating, including one for unsatisfactory instrument procedures. He later received approval to attend the upgrade course again, and in April 1989 his third attempt at the type rating was successful. He finally became a BE-300 PIC in November 1990.

During the investigation, fellow crewmembers told the Safety Board that the PIC had often demonstrated poor judgment on previous flights. Among the numerous problems they had witnessed were:

1. Continued on a VFR positioning flight into IMC.
2. Conducted VFR flight below clouds at less than 1000 feet agl in marginal weather conditions.
3. Informed ATC that the flight was in VMC when it was in IMC.
4. Conducted departures without the crew's knowledge of essential flight planning information.
5. Disregarded checklist discipline on many occasions.

In January 1993, the PIC received a letter of reprimand from the flight operations section supervisor which stated:

> . . . while readying the aircraft for flight, you failed to follow required standard operating procedures which resulted in substantial damage to the left engine . . . Specifically, you failed to follow the appropriate checklist. Additionally, there appears to have been a lack of communication and coordination between you and your second-in-command . . . .

According to the supervisor, the PIC was upset with the reprimand and believed that he should not be held responsible. In his [PIC] opinion, it was the SIC's duty to start the engines, and that he [PIC] was looking elsewhere when the damage occurred.

The PIC was involved in yet another incident a few months later while on temporary assignment at the Oklahoma City, Oklahoma, FIAO. Although no disciplinary action was taken against the flightcrew, it was believed that the PIC had pulled a circuit breaker which caused the airplane's brakes to overheat during a long taxi.

The Board reviewed the PIC's medical records and found that he had received two DUI convictions for alcohol in 1987 and 1991. The latter DUI resulted in the suspension of his driver's license. Although he noted the conviction on his renewal application for his medical certificate, he did not report it to the FAA's Civil Aviation Security Division as required under FAR 61.15. In doing so, the PIC might have lost his airman certificate, and personnel action could have been taken against him. Investigators also found that his driver's license was suspended in January of 1993 for nonpayment of the automobile insurance surcharge, and again two months later because he failed to comply with the state alcohol program. His license was still suspended at the time of the accident.

The SIC of the accident flight had 13,800 total flight hours, 1000 in the BE-300. He had held that position since 1989 and had completed his last proficiency check ten months before the crash. He passed the BE-300 simulator pilot refresher course in May 1993. He earned an ATP rating in 1973 and spent several years as a corporate pilot and flying with the National Guard. According to his flight records, all of his check rides and evaluations were satisfactory. He received a DUI conviction for alcohol in 1992.

Weather

The prevailing weather at the time of the accident was a moist, easterly flow of air over northern Virginia and Maryland with widespread low ceilings, fog, and scattered light rain. About seven minutes before the crash, the reported weather at the Winchester Regional Airport was as follows: Ceiling 1900 feet scattered, 2600 feet broken, 4000 feet overcast; Visibility 10 miles; Temperature/dewpoint; 61 degrees F/55 degrees F.

The accident

The crew was scheduled to begin a five-day flight assignment on 25 October, but due to a maintenance problem with the airplane their duty was postponed to the following day. The PIC had planned a morning departure to inspect the ILS localizer at Winchester, however, another maintenance delay had threatened the cancellation of the flight. The PIC told the FAIO manager that if they were given permission to depart Atlantic City by 1400, they could still finish the mission and get back on schedule. The PIC was also concerned that because the ILS ground technician for the Winchester Airport had a six-hour round trip drive, and that he had already been inconvenienced the day before due to the canceled mission, he felt it appropriate to complete the flight that day. The FAIO manager gave the PIC verbal approval for one hour of overtime in order to finish the mission.

The flight finally departed Atlantic City around 1332 after the PIC had received a weather briefing from the Direct User Access Terminal System (DUATS), and had filed an IFR flight plan to Winchester. The PIC made initial contact with the Washington-Dulles south approach control at 1408. From listening to the ATC tapes, the FAIO manager identified the voice on all of the transmissions as that of the PIC. He added that he believed that the PIC would have been seated in the right seat, and that the SIC would therefore, have been the pilot flying from the left seat.

For the next several minutes there were numerous conversations between the PIC and the controller concerning the type of approach and weather conditions. During one of the calls, the controller advised the PIC that the minimum vectoring altitude in the area of the Winchester facility was 4000 feet. After a brief discussion, the PIC told the controller that they would execute the full ILS approach to Winchester. The flight was then cleared for the approach, and ten minutes later the PIC canceled their IFR clearance and said, "we're going to maintain two thousand and . . . provide us VFR advisories at two thousand feet going back and forth across the localizer." The controller complied.

The Dulles approach control area of responsibility divides in the vicinity of the Winchester Airport. Therefore, when the flight reached the edge of the south controller's sector, it was switched to the Dulles west arrival controller. When the PIC checked in on the new frequency at 1444, the flight was still operating under VFR. Around 1450, the PIC asked the controller, "what's the lowest altitude IFR you can give us?" The controller responded, "the lowest . . . is three thousand and . . . that's only from where you are for a little while . . . south of you is four thousand, is my minimum vectoring altitude." The PIC was then given an IFR clearance to 4000 feet. Shortly thereafter, the flight landed and completed the inspection.

Following a brief ground-time, the crew departed Winchester and according to witnesses remained clear of the clouds and appeared to be in VMC. Around 1551, the PIC contacted Dulles approach control to obtain an IFR clearance for their next destination. The west arrival controller advised the flight to, "maintain VFR for right now, it's going to be about five minutes before I can get to you. I'm extremely busy at the moment." At 1549, the PIC reported, "we're over Linden VOR [17 nm SSW of the Winchester Airport] at 2000. Can you get us a little higher, VFR on top and we'll be on our way." The controller answered, "standby, I have traffic just over the VOR right now descending to five [thousand feet], he's out of seven point five [7500 feet]. . . ." There was an unintelligible transmission, presumably from the PIC, because the controller replied, "O.K. thanks . . . I'll have an IFR clearance for you in just a moment." The controller added, "maintain VFR . . . contact Dulles . . . you're just about to enter his airspace . . . ." The west and south controllers had coordinated an IFR altitude assignment and clearance to the flight's destination, but the PIC never made contact. Several witnesses in the area of Front Royal observed the airplane orbiting in and out of the clouds. One noted that the tops of the hills were covered with fog. A witness driving very close to the accident site remembered hearing a "smooth" noise getting louder, followed by a "swoosh" for three or four seconds, then a sudden explosion.

Impact and wreckage path

The initial point of impact was on top of a ridge line at 1770 feet msl. The surrounding trees had a number of slash marks that were consistent with high-power propeller rotation. The wreckage was scattered in down-sloping terrain for a distance of about 1300 feet. Both of the wings and the engines had separated, and the majority of the fuselage was consumed by fire.

Accident survivability. According to the Safety Board, the force of impact was not survivable.

The investigation

Based upon the examination of the airplane's flight controls, powerplants, propellers and structural integrity, the Board determined that the crew did not experience a mechanical malfunction. The weather information that the flightcrew received, including the AIRMET warning of IFR conditions and mountain obscuration, was found to have been accurate. The Board also believed that the ATC handling of the flight was appropriate and not causal to the accident. Therefore, flightcrew performance became a factor of particular interest to investigators.

Flightcrew performance. The Board centered its investigation on the fact that the PIC took off during marginal weather conditions without obtaining an IFR clearance, flew into an area of mountainous terrain, and at an altitude too low for en route flight. Investigators also noted that facilities at the Winchester Airport provided pilots with easy access to Dulles approach control, yet the PIC made the "deliberate decision not to use the ground communication [facilities] to obtain an IFR clearance before take-off." The Board further believed that some of the PIC's actions reflected that "flight safety considerations did not appear to be a high priority to him."

In the Board's probe of the crew's performance histories, it became clear that the SIC, not the PIC, displayed a more balanced approach between the needs of the mission and those of flight safety. The PIC, however, dominated the cockpit environment. The SIC was an experienced pilot who could have refused to follow the PIC's directions, but the Board believed that the negative atmosphere that had been created by the PIC, combined with the poor FAIO supervisory staff, "probably impeded such action by the SIC until the accident was unavoidable." As a result, investigators believed that the SIC had little or no role in the decision-making that led directly to the accident.

The Board concluded that the accident was caused, in part, by three critical decisions made by the PIC:

1. Not to obtain an IFR clearance for the flight to Newport News/Williamsburg while on the runway at Winchester, even though the communications facilities were available.

2. To take off and attempt visual flight into an area of mountainous terrain while encountering marginal VFR conditions.

3. To continue to remain aloft, at a low altitude, with insufficient distance from the clouds to maintain visual flight; and to proceed towards their destination (and the nearby mountains) under VFR, while waiting for an IFR clearance.

FAIO supervision. The Board learned that three turbine-powered airplanes had been destroyed during FAIO missions between 1983 and 1993. Two of the accidents resulted in fatal injuries to the flightcrews onboard, and were caused by PIC judgment and decision-making related to weather. Interestingly, the Atlantic City FAIO operated the airplanes involved in both of the fatal accidents, and the crewmembers were supervised by the same Flight Operations manager (FO/SS).

Investigators learned that the Atlantic City FAIO management had witnessed a number of safety-related concerns regarding the PIC for several years before the accident. Some of these concerns included failed check rides, seconds in command refusing to fly with him, and deviation of SOPs and checklists. Just two weeks prior to the accident, the same SIC had requested a formal investigation into what he charged was the PIC's deliberate violation of FAIO procedures by performing a below-glideslope maneuver close to the ground in IMC. The Board discovered that this report was just one of many against the PIC that was never acted upon by the FO/SS.

The Board added that mounting evidence indicated that the PIC had a record of noncompliance with checklists, "displayed an impatient and arrogant attitude," and exhibited "poor judgment and decision-making in the air and on the ground." His "deliberate disregard for authority" was apparent by the DUI convictions and the several license suspensions that he never reported through the proper channels.

Given the PIC's attitudes and behavior, the Board examined the nature of the FAIO's direct supervision of him. They found that when the FO/SS received complaints about the PIC's flying performance, no action was taken. In the six years that the PIC was assigned to Atlantic City, he had only one recorded reprimand. According to the Board, the primary reason the January 1993 letter was issued was because the aircraft had sustained a costly engine repair and the incident could not be suppressed at the FO/SS level.

Despite the known problems with the PIC, the FO/SS not only failed to take the necessary corrective action but, in fact, did the opposite. In the most recent performance appraisal before the accident, he gave the PIC a positive rating. The FO/SS marked "proficient" with interpersonal skills, and added the comment, "get[s] along well with his fellow workers."

According to the Safety Board, the FO/SS had ample evidence to remove the PIC from his flight status, but by not taking appropriate action, he failed to carry out his responsibilities to ensure the safety of his department and, in part, "also caused the accident." Furthermore, when the FO/SS disregarded the complaints of the SIC and others against the PIC, the Board believed he sent a poor message to the FAIO's SICs that he would not investigate legitimate concerns that might jeopardize the safety of a flight.

The Board also noted that the weak supervision extended from the FO/SS to the FAIO manager. The manager was in that position for seven months prior to the accident, and told investigators that because he didn't want to "micromanage," he stayed out of the flight operations' business. Therefore, his failure to properly oversee the flying unit was, in part, causal to the accident.

And lastly, the Board believed that the supervisor of the FAIO manager also failed to address the known problems at the Atlantic City office. The manager of the Airspace Systems Assurance Division in Oklahoma City was directly responsible for all of the FAIOs. Ironically, he had been the FAIO manager in Atlantic City immediately prior to his new position in Oklahoma. He told investigators that he was "well aware of the PIC's arrogant attitude," and that he was fully knowledgeable with the problems at the FAIO. However, the Board found no evidence to suggest that he ever insisted on stronger supervision or reprimands of the PIC. As a result, it was concluded that his inaction was also considered a cause to this accident.

Lessons learned and practical applications

1. Take your responsibilities seriously. Don't look the other way or allow "office politics" to interfere with the safe operation of any flight. Stand up for what's right.

2. Don't fly. Drastic measures must be taken for extreme circumstances. In this case, the SIC had gone up his chain of command, but without satisfactory re-

sults. If he had refused to fly with this PIC, as did the majority of his colleagues, the mission would have been scrubbed. This might have forced management to take appropriate action. It's a personal and professional risk, but the alternative might be crashing into the side of a mountain. Your choice.

3. Be a vigilant flier. If you *do* find yourself in a potentially dangerous situation, stay on top of *every* cockpit activity. Cross-check and monitor.

Case study reference

National Transportation Safety Board. 12 April 1994. Aircraft Accident Report: Controlled Flight into Terrain. Federal Aviation Administration Beech Super King Air 300/F, N82. Front Royal, Virginia. October 26, 1993. Washington, D.C.

CASE STUDY I-5: Northwest Airlines Flight 255

Safety issues: CRM, cockpit discipline, procedural deviation

On 16 August 1987, a Northwest DC-9-82 crashed shortly after takeoff from the Detroit Metropolitan Airport, Michigan.

Probable cause

The NTSB determined that the probable cause of this accident was the flightcrew's failure to use the taxi checklist to ensure that the flaps and slats were extended for takeoff. Contributing to the accident was—for unknown reasons—the lack of electrical power to the airplane's takeoff warning system.

History of flight

Flight 255 was a regularly scheduled passenger flight between Saginaw, Michigan, and Santa Ana, California, with intermediate stops in Detroit, Michigan, and Phoenix, Arizona. The flight departed Detroit at 2044 Eastern Daylight Time with 149 passengers and 6 crewmembers onboard.

Pilot experience

The captain had 20,859 total flight hours, 1359 in the DC-9-82. He had initially upgraded to captain in 1972, and became a DC-9-82 FAA designated check airman from 1979 to 1984. Beginning in 1986 he flew as a 757 captain, but due to fleet restructuring the airline stopped operating them. The captain then completed his DC-9-82 requalification training, including a line check, two months before the accident.

The first officer had 8044 total flight hours, 1604 in the DC-9-82. He had held that position for more than three-and-a-half years, although he was originally hired by North Central Airlines in 1979. During the North Central-Republic-Northwest mergers, he was never furloughed.

Weather

At the time of the accident, the Detroit weather was: Ceiling 2500 feet scattered, 4500 feet broken, 10,000 feet overcast; Visibility six miles, haze; Wind 180 degrees at 7; Cumulonimbus west through northwest moving east.

The accident

Although the flightcrew did not follow company standard procedures, they completed the "Before Engine Start" checklist as the aircraft was being pushed-back from the gate at 2032. Three minutes later the west ground controller cleared the aircraft to taxi and told the crew to expect runway 3C for takeoff. As the crew proceeded with the taxi, the controller gave them additional information, including a new radio frequency. The first officer repeated the clearance, except for the frequency change, which was later found not to have been tuned in.

Shortly thereafter the first officer told the captain that he was switching radios, "to get the new ATIS" (Automatic Terminal Information Service). According to the tower supervisor, there had been a change in runways and weather conditions, prompting an ATIS update at 2038. The active runways were no longer 21L or 21R, but the shorter 3C. At 2037, the captain asked the first officer if they were within the takeoff weight limits for 3C. After consulting the company's weight chart manual, they both agreed that it was safe to use that runway.

A couple of minutes later, the first officer changed frequencies and contacted the east ground controller. The controller informed the crew that the latest ATIS information, including the windshear advisories, was still current. At 2042:11, the local controller cleared Flight 255 to taxi into position and hold on runway 3C. He also explained that there would be a three-minute delay in order to get the required "in-trail separation behind traffic just departing." At 2044:04, the flight was cleared for takeoff.

Shortly after applying engine power, the crew discussed not being able to engage the autothrottle system. They quickly corrected the problem, and at 2044:45, the first officer called 100 knots. Seconds thereafter, the first officer called "rotate," which was followed by the stickshaker activating. Warnings generated by the supplemental stall recognition system (SSRS) continued until the end of the CVR tape. Investigators noted that between 2044 and 2045, the takeoff warning system did not sound. After Flight 255 became airborne, it began rolling to the left and right. The airplane struck several objects and crashed in a rental-car lot.

Impact and wreckage path

According to an on-duty Northwest first officer, who was sitting in an aircraft parked on a nearby taxiway, Flight 255 was intact until the left wing struck a light pole in a rental-car lot. He saw a "four-to-five foot chunk of the wing section" fall from the airplane. He said there was no sign of fire until after it hit the pole, and then he saw "an orange flame" coming from around the left wing tip.

The airplane continued across the first car lot, clipped a second light pole and smashed into yet another rental-car lot. As it slid across the ground it hit a railroad embankment and two interstate highway overpasses. The aircraft disintegrated as fire erupted along the 3000-foot-long wreckage path. Except for two fairly large fuselage sections, nothing else remained intact. All of the passenger seats had separated from the cabin area, and were scattered in the debris.

Accident survivability. One passenger, a four-year-old child, was the only survivor. She was found in the wreckage beneath one of the highway overpasses. According to the passenger manifest, she had been assigned seat 8F.

Two persons on the ground were killed, and four others received various degrees of injuries.

The investigation

Since a windshear alert was in effect at the airport, and the stickshaker on Flight 255 activated shortly after rotation, the Board initially investigated the possibility of a windshear encounter. However, after examination of the CVR and DFDR readouts, that was ruled out. The aircraft did not experience any meteorological factors that would have decreased its airspeed. On the contrary, at liftoff its airspeed was about 169 KIAS and quickly accelerated to 184 KIAS.

The data revealed, however, that Flight 255 had not been correctly configured for takeoff. During the 14-second flight, the aircraft's stickshaker remained activated, and its SSRS sounded four times. With the flaps set at 11 degrees and the slats in the mid-sealed position, the DC-9's stall speed was about 121 KIAS. If the flaps were retracted and the slats remained in the mid-sealed position, the stall speed would increase to 128 KIAS. However, the DFDR indicated that the airplane was flying at speeds between 169 KIAS and 184 KIAS, but the stall warnings continued. Therefore, the Board determined that the only wing configuration that could have produced such a condition was if the slats and flaps were retracted.

Airplane performance study. The DFDR data showed that both engines were operating at or above takeoff power. If the airplane had been configured for takeoff, it should have lifted off the runway between a six degree and eight-degree noseup attitude. In the case of Flight 255, however, the aircraft rotated to an 11-degree noseup attitude, stabilized at that pitch angle, then accelerated to a higher airspeed before actual liftoff. That information provided further evidence to investigators that confirmed the DC-9 was not properly configured. If it had been, the V2 speed should have been 163 KIAS by the time it climbed through 35 feet agl. However, it took Flight 255 an additional six knots just to liftoff.

The Board also examined the climb profiles for the jet to determine its ability to clear obstacles at the end of runway 3C. If the flaps had been retracted, the aircraft would have come dangerously close to the first light pole. Other profiles revealed that with the slats in the mid-sealed position and flaps at 11 degrees, or the flaps retracted and the slats in the mid-sealed position, the aircraft would have cleared the light pole by 400 to 600 feet. The Board believed that the results of this study corroborated the DFDR readouts.

Physical evidence. Investigators also found reliable physical evidence from the wreckage that supported their conclusion. First, investigators found the flap handle was still in the up/retract (UP/RET) detent position before impact. And second, they noted the damaged inboard flap roller tracks and the breaks in the drive cables to the number 5 slat transition drum. Those findings proved that the flaps and slats were fully retracted when the damage occurred.

Central aural warning system. In the DC-9, when the aircraft is not properly configured for takeoff, a warning tone sounds to alert the crew. According to the CVR tape, that did not occur. Therefore, the Safety Board analyzed the central aural warning system (CAWS) and found the failure was caused by the lack of 28 volt dc input power from the airplane's power supply-2 unit. The CAWS receives this power from the left dc bus and through the P-40 circuit breaker. There was no evidence that suggested a failure in the bus itself, so through the process of elimination, investigators believed the power interruption came from the P-40 circuit breaker. Unfortunately, the circuit breaker was so badly damaged during the accident that it was impossible for the Board to positively determine its preimpact condition.

Cockpit discipline. According to the Board, the CVR tape proved that the crew neither called for nor completed the "taxi" checklist. The first item on that checklist required both pilots to orally verify that the flaps and slats were positioned correctly for takeoff. The autothrottles were also required to be set, but since the crew was initially unable to engage them during their takeoff roll, this further indicated to investigators that the crew had not completed the checklist.

Northwest procedures clearly defined the crew's duties and responsibilities as to how checklists were to be started and accomplished. After each checklist, the first officer was required to identify the checklist by name and state that it was "complete." Those requirements were met only once during the pretakeoff procedures. The captain did not call for the after start, taxi, or before-takeoff checklists, nor did the first officer ask the captain if he was ready to begin those checklists.

Substandard performance. Based on the examination of the crew's performance patterns during the previous flight into Detroit and its subsequent continuation, the Board believed that the crew's "performance was below the standards of an air carrier flightcrew." They cited four examples to substantiate their conclusion:

1. After landing at Detroit, the flightcrew taxied by the entrance to their assigned gate and had to turn 180 degrees to return to the gate.
2. The airplane's weather radar is normally turned off during the after-landing checklist, which is usually completed shortly after the crew clears the runway. However, the aircraft's radar was still on after a prolonged taxi and in the proximity of the gate area. The Board believed that most likely the crew had not yet performed the checklist or had missed turning it off when they came to that item on the checklist.

3. During the taxi-out at Detroit, the ground controller directed the crew to taxi to runway 3C and to contact ground control on a new radio frequency. The first officer did not change frequencies, and ground control was unable to raise the crew after the aircraft passed its assigned taxiway.

4. The first officer told the ATC taxi clearance and runway information to the captain at least twice. Although the captain was familiar with the airport, he failed to turn off at the designated taxiway and expressed doubt as to where it was located.

Lessons learned and practical applications

1. Complete all checklists. Complacency, distractions, discipline, and fatigue are just a few reasons why pilots fail to accomplish checklists. Remember, when you allow a hazardous thought pattern or an unsafe practice to enter the cockpit, it trickles down to every other area of the operation. On the other hand, if you're normally conscientious about checklists, and you were unavoidably distracted, take a moment to make sure everything is in order before you move on to other tasks. Incomplete checklists should be a problem reserved for CFIs and their primary students—not for licensed pilots and professional flightcrews.

2. Maintain vigilant cockpit discipline. Set the tone for a strict work environment (this also applies for single-pilot flights). Discipline yourself to not allow unnecessary conversation to distract you (or other crewmembers) from duties and responsibilities. Lead by example.

3. Avoid procedural deviation. This is a critical element in maintaining an overall safe flight operation. Following SOPs will also eliminate the chance that you might forget important tasks.

Case study reference

National Transportation Safety Board. 10 May 1988. Aircraft Accident Report: Northwest Airlines, Inc., McDonnell Douglas DC-9–82, N312RC. Detroit Metropolitan Wayne County Airport, Romulus, Michigan, August 16, 1987. Washington, D.C.

CASE STUDY I-6: Continental Express Flight 2286

Safety issues: Substance abuse, CRM, airmanship, FAA surveillance, preemployment verification

On 19 January 1988, a Continental Express Metro III crashed while the crew attempted a VOR/DME approach to runway 20 at the Durango-La Plata County Airport, Colorado.

Probable cause

The NTSB determined that the probable causes of this accident were the first officer's flying and the captain's ineffective monitoring of an unstabilized approach, which resulted in flying below the published descent profile. Contributing to the accident was the degradation of the captain's performance due to his use of cocaine before the accident.

History of flight

Flight 2286 was a scheduled passenger flight from Denver, Colorado, to Cortez, Colorado, with an intermediate stop in Durango, Colorado. The aircraft and crew were assigned to Trans-Colorado Airlines, and operated as a Continental Express flight under a marketing agreement with Continental Airlines. The Metro III departed Denver-Stapleton at 1820 Mountain Standard Time with 15 passengers and 2 crewmembers onboard.

Pilot experience

The captain had 4184 total flight hours, 3028 in the Metro III. He served as captain for 1707 of those hours. He had previously been employed by Pioneer Airways as a Metro III captain, and when that company went out of business he was hired by Trans-Colorado Airlines in May 1986. Due to his background as a captain, he was upgraded to that position one month later.

The first officer had 8500 total flight hours, 305 in the Metro III. In 1981, after flying for Pioneer Airways for about one year, he was fired because he demonstrated a lack of proficiency in his attempt to upgrade to captain. He joined Trans-Colorado in 1987.

Weather

The 1700 surface weather map, prepared by the NWS, showed a large, low-pressure area over Missouri that affected virtually all of the weather east of the Rocky Mountains. A trough extended south through Colorado, which produced variable light winds, overcast skies, and snow showers. The 700 millibar (10,000 feet msl) map indicated foggy conditions over southwestern Colorado.

Actual weather conditions. Approximately 15 minutes before the accident, the surface observations at the Durango-La Plata County Airport were as follows:

1905: Ceiling partial obscuration, estimated 800 feet overcast; Visibility 5 miles. Winds calm. Snow showers, intensity unknown all quadrants.

The accident

NOTE: At the time of the accident, the FAA did not require the 19-seat Metro III to be equipped with a CVR or an FDR, therefore, no intracockpit conversations or data points were recorded. The Safety Board, however, determined that the first officer was at the controls of the aircraft, and the captain conducted all of the ATC communications.

Due to the poor weather conditions, Flight 2286 took off from Stapleton at 1820 MST, a delay of 40 minutes from their scheduled departure time. The first officer made a normal climbout to the assigned cruise altitude of 23,000 feet, and according to the surviving passengers, the en route portion of the flight was uneventful.

At 1900:40, a Denver center controller asked Flight 2286 if they would, "rather shoot the ILS or . . . will the . . . DME approach to runway two-zero be . . . sufficient?" The captain replied that they would plan on the VOR/DME approach. The controller then gave the flight clearance to, "proceed direct to the Durango zero-two-three radial, eleven mile fix." The captain acknowledged the call. Less than three minutes later, Flight 2286 was cleared to descend at pilot's discretion to 16,000 feet.

The station agent at Durango told investigators that around 1905, the captain called her on the company radio to inform her that they were 25 minutes out, and would not be needing to refuel. She then provided Durango weather information to the flightcrew.

About nine minutes later, the center controller told the flight to cross the Durango 023 degree radial, 11-mile fix, at or above 14,000 feet. He then cleared them for the VOR/DME runway 20 approach. The captain did not reply immediately and the clearance was repeated. At 1914:28, the captain responded that they were, "down to 14 [14,000 feet] and we're cleared for the approach." Shortly thereafter, the controller informed the flight that radar coverage had been terminated, and the captain acknowledged the call with a "Wilco." That was the last communication from Flight 2286.

Impact and wreckage path

According to the surviving passengers, the airplane briefly leveled off during its descent, then hit something hard. They felt an abrupt pitch up and an increase in engine power, followed by the airplane rolling laterally several times before it hit the ground and slid to a stop. The aircraft came to rest about five miles from the airport. There was no post-impact fire.

The investigation revealed that the wreckage path extended about 1000 feet on an approximate heading of 198 degrees. The airplane initially clipped several trees about six to eight feet below the top of a hill. It then flew over the hill, and hit the ground near the bottom of the other side.

The fuselage was crushed and fragmented from the radome to the first cabin window, and again just aft of the trailing edge of the wing. The wing itself, (the Metro III has a single wing that is mated to the fuselage) had separated from the fuselage at the attachment fittings, and was lying inverted above the cabin. The right engine was found hanging downward from the wing, and the left engine had been torn away from its mounting and was buried in the snow.

Accident survivability. The flightcrew and seven passengers sustained fatal injuries. Injuries to the survivors ranged in severity from a fractured vertebrae to muscle strains, and one case of first degree frost bite. Six of the ambulatory survivors, including a two-year-old child who was being carried, left the crash site to get help. One lone passenger came upon a house, and told the owner to call the police because he had just been in an airplane accident. Meanwhile, the other five survivors walked a mile-and-a-half—for more than 90 minutes—to a highway where they flagged a motorist. The group had been driven just a short distance, when they passed responding rescue workers who stopped and took them to a nearby hospital.

At 2004, airport personnel notified the Durango central dispatch that Flight 2286 was 25 minutes overdue. Although the Civil Air Patrol (CAP) was also advised of the situation, they did not initiate a search because the location of the airplane had not yet been determined. However, at 2032, a CAP official from Denver informed the central dispatch that the last center radar contact with the flight showed it at a point six miles east of Durango. Two minutes later, the local resident who had assisted one of the survivors called the central dispatch to notify them of the accident.

The crash site was located at 2226, but because of the inclement and remote conditions emergency units did not reach the airplane until 48 minutes later. Rescue efforts at the site were often hampered by the snow, darkness, and extreme cold, and therefore, extrication of the remaining survivors took more than an hour. The last survivor was transported from the scene at 0030.

The investigation

Once the Safety Board had concluded that there were no signs of preexisting defects in the airplane systems, powerplants, or airframe, the focus of the investigation centered on the crew's performance during the approach. Subsequent findings indicated poor airmanship, a deficiency in piloting skills, a lack of situational awareness, illegal drug use, and inadequate company procedures were directly linked to the events which led to this accident.

The mishandled approach. At 1915, when Flight 2286 was at the 11 DME fix on the 023 degree radial, the aircraft was at an altitude of 14,000 feet with a ground speed of 195 knots (143 KIAS). At that point, according to the Board, the first officer should have been at 10,400 feet in order to descend 3200 feet to the MDA at a more appropriate speed of 135 KIAS. Without considering wind velocity or direction, the descent rate would have been about 900 fpm. But the evidence indicated that because the first officer began his approach from 14,000 feet, the ground speed ranged from 180 to 142 KIAS as he maintained a descent rate of nearly 3000 fpm, all the way to just before impact. Adding to the difficulties in flying such a steep approach was the presence of a constant 10 to 15 knot tailwind.

The Board determined that from the outset, the flightcrew flew the approach at an altitude that was too high to fly it safely within the established parameters. In any event, the standard VOR/DME approach to runway 20 was considered to be a challenge, especially in poor weather conditions and with a tailwind. Therefore, the Board believed that the captain should have closely monitored the first officer's conduct and ability to ensure that the proper altitude was maintained.

A consensus of perspectives. As part of the investigation, the Board interviewed 11 Trans-Colorado pilots and asked them to describe their procedures for flying the VOR/DME approach to runway 20. There were surprisingly few consistencies among their answers, except for those concerning the demanding nature of the approach, and the additional time required to fly the *published* approach when arriving from a northerly direction. One pilot had flown the VOR/DME approach approximately 30 times; most of the respondents reported flying it about seven times. The subsequent comments provided to the Board a revealing perspective of the level of skill necessary for this particular approach. In part, the statements included, ". . . because of the high descent rate required, a pilot must plan for the approach way ahead," ". . . the biggest difficulty in flying the approach is getting the airplane slowed up and properly configured by the 11 DME fix," "[I was] usually too high when [I] reached the runway and had to circle to land," ". . . approach was safe, as long as you're set up in advance and there's a minimal tail wind component."

The second series of comments pertained to the problem of trying to stay on schedule, while flying the published approach. One pilot noted that the, "[VOR/DME] approach saved ten minutes of flying time when arriving from the north." He also believed that pilots flew that particular approach to stay on schedule since only 70 minutes was allotted for the flight from Denver to Durango. From those remarks, the Board suggested that Trans-Colorado pilots might have been discouraged from flying the published approach when they were either arriving from the north or already late. Since Flight 2286 was inbound from the northeast and behind schedule, flying the 11-mile DME arc would have forced them to backtrack, causing the noted delay of up to ten minutes.

Crew performance history. The Board investigated each pilot's professional and personal histories to determine if their flying records or behavioral traits might have been a precursor to this accident. Although some of the findings might appear to fall into the "dirty laundry" category, the Board considered this information to be significant and relevant to this accident.

First officer. Even though the first officer had accumulated nearly 8500 flight hours, the investigation revealed that his performance records disclosed a history of failed proficiency checks, mediocre comments from various flight instructors, and job terminations. In 1981, after a year of flying the right seat for Pioneer Airways, he was fired because he showed a lack of proficiency in his attempt to upgrade to captain. According to one of the airline's flight instructors, the first officer, "demonstrated periods of inaction as the flight regime required changes in the aircraft's configuration, attitude, or change of phase . . . ."

The first officer continued to fly for five years as an instructor at an FBO in Denver before a short stint as an Alaskan charter pilot. It was during the latter job that he failed a proficiency check due to difficulties involving ILS and NDB approaches. He finally passed a VFR check ride, which restricted him to fly only during visual operations. He eventually quit the Alaskan job, and nine months later was hired by Trans-Colorado.

Although his ground school at the airline was without incident, comments from his training records included, "average performance," and "more time spent on cockpit procedures would be beneficial." His flight training reports noted, however, that his performance was, "okay/weak," and "weak but improving." Just prior to his proficiency check in July 1987, an instructor remarked that he "overcorrected and chased needles during ILS." Nevertheless, a few days later the first officer passed his check ride, which included demonstration of ILS and VOR approaches.

The Board also probed his personal record and uncovered a drinking problem from 1972 to 1983, which resulted in two alcohol-related driving convictions. In 1984 an FAA medical examiner interviewed him with regard to those offenses, and reported that he no longer had an alcohol problem. His subsequent records indicated that to be true. However, the Board noted that the first officer had regularly attended Alcoholics Anonymous meetings.

Captain. Interviews conducted during the investigation revealed that many Trans-Colorado pilots considered the captain to be a highly skilled aviator, but with a reputation for rushing and taxiing at excessive speeds. An airline dispatcher stated that the captain was known for taking an airplane that was behind schedule and getting it back on schedule by the end of the day. There was at least one instance when the captain turned an aircraft around in seven minutes. On two consecutive days in 1987, he violated company procedures. First, he personally refueled the Metro III he was in command of because he was behind schedule. And second, he boarded a late passenger with one of the airplane's engines running.

It was also noted that in 1983 the captain was involved in a Cessna 182 accident. At the time, the Board had determined the probable causes were the pilot-in-command's selection of the wrong runway, improper compensation for wind conditions, misjudging distance, and delaying a go-around. Following this mishap, he satisfactorily passed an FAA-required check ride.

According to the Board, the captain had a poor driving record, including a license suspension in 1980. Because the state in which the violation occurred no longer maintained those records after a certain number of years, the Board could not determine the reason for the suspension. They did, however, discover that while he had a suspended license he moved to Colorado, and by withholding information from his past record, was able to obtain another driver's license. Furthermore, from 1983 to shortly before he was hired by Trans-Colorado in 1986, he received five moving violation citations, two of which involved traffic accidents.

Flying under the influence of cocaine. According to the medical examiner's toxicology studies and pathology report on the captain, there were notable amounts of cocaine and benzoylecgonine (the principle metabolite of cocaine) in his system. More than a month after the initial autopsy, additional pathological samples were submitted to a private laboratory for a second toxicological analysis. Those results also indicated presence of cocaine and benzoylecgonine in the captain's body. The Board's best estimate was that the captain ingested cocaine between 12 and 18 hours before the accident.

Fellow Trans-Colorado pilots and other airline personnel who knew the captain all stated that they were unaware of his cocaine use, and that they never observed any unusual behavior. Likewise, the physician who performed the captain's most recent FAA medical examination described the process as routine, and that he showed no signs of being a drug user. Despite the captain's ability to hide his cocaine use from the workplace, it appeared that he might have been a user only on his days off and around people who were not associated with the airline.

Near the end of February 1988, a corporate pilot contacted the Board with information about the captain's cocaine use. He reported meeting a woman who said that she had been living with the captain, and that they were engaged to be married. According to the informant, she told him: "I'm sure glad that we were able to bury him right after the accident, because the night before we had done a bag of cocaine . . . and I was worried that the autopsy would say there were traces of this in his system before he died." She admitted to him that she and the captain had used cocaine periodically. Coincidentally, the informant was a former drug counselor in the military, and noticed that although she was not inebriated when she gave him this information, she did, however, appear to have a "burned-out look" typical of someone with a drug problem. The woman gave the informant her address and phone number, but when the Board tried to contact her, her attorney told them that she had no information that could help the investigation. The attorney added that the woman had not been with the captain the 24-hour period before the accident, and in her opinion, the captain was, ". . . not an habitual user of cocaine, alcohol or other similar drugs."

Another close acquaintance of the captain, however, stepped forward during the investigation and described the captain's dramatic personality change that took place over the course of just a year. She explained to the Board that she had seen the captain almost daily from early 1984 through mid-1986. For that period of time she characterized him as, "a very stable person . . . a nice guy . . . fun to be with." Although she kept in touch with him by phone, she did not see him again in person until the summer of 1987. It was during that visit that she noticed he had a much different demeanor.

> He wasn't himself anymore. I knew right off that there was some kind of drug problem. He acted very nervous like he was scared of something. He'd look over his shoulder a lot as if there was someone behind him when there wasn't. When I was over at his house, every time a car came through he'd jump up and look out that window. I thought he gained more weight than I had ever seen him gain before. And he was real jittery.

She made the comment to the captain that he had changed his phone number three times, and that he must be consuming "a lot" of drugs. He responded: "She's like a sickness, it's all a disease and there is no cure." The acquaintance believed that the captain's girlfriend and the use of cocaine were "combined together," and that their close relationship influenced his behavior. She also believed that because the captain was such a private person, it was probably easy for him to keep his drug habit from his coworkers.

Cocaine and its effect on human behavior. As part of the investigation, the Board read volumes of research dealing with the effects of cocaine on human behavior and performance. From the available literature on the subject, the Board soon discovered the similarities between the results of the studies and the reported behavioral patterns of the captain.

According to numerous clinical studies, cocaine use is difficult to detect, even by individuals who interact daily with the abuser. Since the method of ingestion (intranasal, or "snorting," reaches the bloodstream the fastest), and the person's tolerance level to the drug vary, personality changes might be subtle, and can therefore go unnoticed. That might explain why no one from Trans-Colorado ever remembered observing the captain's change in demeanor. But if any of his coworkers had witnessed unusual behavior, they might have viewed it as just an isolated case of the captain having a "bad day at the office."

The research also addressed the evidence that proves using cocaine, even in relatively small doses, can produce a strong physiological and psychological addiction, much like amphetamines. Shortly after cocaine ingestion, a person experiences an increase in heart rate and blood pressure, and an altering of brain waves. Since the drug is a psychomotor stimulant, the user can also feel a heightened sense of alertness and performance, especially if he or she is fatigued. Similarly, cocaine is a euphoriant which can enhance the mood of the user and produce feelings of friendliness, vigor, and elation. But once those effects wear off, the user resumes whatever physiological (i.e., fatigue) or psychological (i.e., depression) states they were in before they consumed the cocaine. According to the experts, it's this type of vicious cycle, known as a "cocaine crash" that hooks the user into becoming habitual with their drug intake.

The studies indicated that a tolerance to cocaine will develop after continued use. Therefore, the person might need more of the drug to initially get "high," while avoiding the inevitable "crash." As a result, this will produce a stimulant withdrawal syndrome that is marked by mental depression. Because the mind and body are in demand for the drug to relieve the depression, a greater amount of cocaine is considered by the user as a necessity to function. Paranoia and suspiciousness have been found to have a direct relation to the amount of cocaine ingested. According to the statements made to the Board by an acquaintance of the captain, she observed definite traits of paranoia and suspiciousness when she saw him in the summer of 1987. She also reported that his personality was very negative and different from how he had once been as recent as the year before.

Additional clinical reports describe that the adverse effects of cocaine are ultimately realized when the user no longer needs the drug for a "pleasurable" or euphoric experience, but rather, must have it to avoid the dreadful consequences associated with a "crash." One of the documents states, in part: ". . . [after repeated administration] cocaine can no longer evoke . . . euphoria. Instead, dysphoria [a state of anxiety, depression, and restlessness] dominates . . . Anhedonia, the inability to enjoy, can persist for weeks . . . If we were to deliberately design a chemical that would lock people into perpetual usage, it would probably resemble the neurophysiological properties of cocaine." This common feeling of hopelessness and despair that a habitual user experiences, might have been the basis for the captain's alleged remark he made to his close acquaintance, ". . . it's all a disease and there is no cure."

A deadly combination. According to the Safety Board, the evidence was clear as to the first officer's poor piloting skills. His performance records showed a history of failed proficiency checks, especially when it came to instrument flying. Even during his training at Trans-Colorado, he continued to struggle with the skill needed to fly instrument approaches. It was unlikely that he was able to conceal his poor performance with his fellow pilots, yet he was somehow allowed to remain as a crewmember on a passenger-carrying airplane.

Likewise, the Board believed that the captain's most recent cocaine use degraded his own ability to act as pilot-in-command. He was, therefore, unable to closely monitor the first officer's attempt at making the approach, and probably lost a sense of situational awareness in the cockpit. It was also suggested that since Flight 2286 was already behind schedule, and that the captain had a reputation of rushing, he might have initially allowed the rapid approach in order to save time. In the Board's opinion, the captain had a "cavalier attitude" when it came to following rules and procedures. They supported their belief from the large number of traffic convictions, the falsification of his driver's license application and an FAA airman medical certificate application, and repeated violations of routine company policies.

The Board also believed that Trans-Colorado did not complete an adequate preemployment background check on either crewmember. Company personnel informed investigators that they were unaware of deficiencies in the first officer's performance history—including his termination record from Pioneer Airways—before he joined their airline. Trans-Colorado's vice-president of operations also told the Board that the company did not know about the captain's involvement in an aircraft accident, or the driving convictions of either crewmember.

As mentioned earlier, the Board found that the combination of poor airmanship, illegal-drug use, and insufficient preemployment policies were, at varying degrees, contributory to this accident. Unlike many other accidents where the "error chain" could have been broken relatively late into the flight, the circumstances surrounding the crash of Flight 2286, however, are quite unique.

The captain's habitual cocaine use, and his decision to take the drug just hours before his duty day, ultimately became a link in the proverbial "deadly chain of events" that was irreversible. Because of the absence of a CVR, the Board was unable to determine the extent to which the cocaine use degraded the captain's performance, however, they were convinced that it was at a significant level. Likewise, the first officer's repeated demonstration of his poor flying skills during instrument approaches could also be considered a link that would have been difficult to break. In the Board's opinion, the first officer's lack of piloting ability was serious enough to warrant it as being part of the probable cause. As a result, the margin of safety for Flight 2286 was severely compromised, possibly irreparably, the moment the drug-impaired captain and non-proficient first officer were crewed together.

FAA oversight. A principle operations inspector (POI) at the FAA Flight Standards District Office (FSDO) in Denver, Colorado, had the primary responsibility of monitoring the operations of Trans-Colorado. Although he was type-rated in the Metro III and had observed the airline's pilot training program, he was not required by the FAA to fly every Trans-Colorado approach, and therefore, was not familiar with the VOR/DME runway 20 approach at Durango. He stated that after reviewing the airline's request to fly that particular approach, he gave his approval because it appeared to be similar to other VOR/DME approaches that the airline was using. The POI assumed that the pilots were flying the published instrument approach: direct to the Durango VOR, outbound on the 096 degree radial to intercept the 11 DME arc. However, unbeknown to him, an FAA "special approach" waiver had been attached to it years before, which allowed pilots to fly a steeper descent rate of 400 ft/nm between the 7.5 DME fix and the 5 DME fix. During the investigation, it was reported that the FAA did not require the POI to be informed about certain provisions for instrument procedures, which prompted him to tell the Board that had he been aware of the waiver, he would have examined the approach more closely.

The Safety Board had determined that there were no FAA policies or guidelines for a POI to follow when a request for a special instrument approach procedure was issued, or when an approach was transferred from one carrier to another. It was from this latter oversight that, interestingly enough, airline deregulation played a unique role. The original FAA approach inspection was conducted for Frontier Airlines in 1977. In 1985, People Express purchased Frontier, which was later acquired by Continental Airlines. Continental purchased Rocky Mountain Airways, and in 1987, Trans-Colorado entered into an agreement with Rocky Mountain to provide airplanes and crews under the Continental Express designation. As a result, the VOR/DME approach procedures for runway 20, including the special waiver, was transferred through five separate flight departments in 11 years.

Lessons learned and practical applications

This accident should be a wake-up call for every aviator who has ever flown with an incompetent pilot, or suspect that a crewmember abuses drugs or alcohol. Don't ignore the problem, or assume that the FAA will take care of it. Refer back to case study I-4. Take the appropriate actions to rectify the situation. Although it's unfortunate that one person often has to shoulder this responsibility, in the interest of safety it's absolutely vital.

Case study reference

National Transportation Safety Board. 4 February 1989. Aircraft Accident Report: Trans-Colorado Airlines, Inc., Flight 2286. Fairchild Metro III, SA227 AC, N68TC. Bayfield, Colorado. January 19, 1988. Washington, D.C.

CASE STUDY I-7: Runway Incursion: Northwest Flights 1482 and 299

Safety issues: CRM, cockpit discipline, pilot role reversal, communication, role of ATC, judgment and ADM

On 3 December 1990, a Northwest 727 and a Northwest DC-9 collided near the intersection of runways 09/27 and 03C/21C at the Detroit Metropolitan Airport, Michigan.

Probable cause

The NTSB determined that the probable cause of this accident was a lack of proper crew coordination, including a virtual reversal of roles by the DC-9 pilots. This led to their failure to stop taxiing their airplane and alert the ground controller of their positional uncertainty in a timely manner before and after intruding onto the active runway.

Contributing to the accident were problems with ATC, surface markings at the airport, and company training. The Safety Board specifically cited deficiencies in the air traffic control services provided by the Detroit tower. This included the failure of the ground controller to take timely action to alert the local controller to the possible runway incursion and inadequate visibility observations. The controller also did not provide progressive taxi instructions in low-visibility conditions. Rather, he issued inappropriate and confusing taxi instructions that were compounded by inadequate backup supervision for the level of experience of the staff on duty. The Board also found deficiencies in the surface markings, signage, and lighting at Detroit Metropolitan Airport. The FAA was faulted in not detecting or correcting any of these problems. And lastly, the Board cited failure of Northwest Airlines to provide adequate cockpit resource management training to their line aircrews.

History of flights

Flight 1482 was a regularly scheduled passenger flight from Detroit, Michigan, to Pittsburgh, Pennsylvania. The DC-9 had 40 passengers and 4 crewmembers onboard.

Flight 299 was a regularly scheduled passenger flight from Detroit to Memphis, Tennessee. The 727 had 146 passengers and 8 crewmembers onboard.

Pilot experience

The captain of Flight 1482 had 23,000 total flight hours, 4000 in the DC-9. He had been on a six-year medical leave which required him to complete the DC-9 Initial Pilot Training Course before resuming command. He passed his line check three days before the accident.

The first officer had 4685 total flight hours, 185 in the DC-9. He was a former U.S. Air Force B-52 and T-38 instructor pilot, and had been with Northwest just a little longer than six months.

Weather

About 25 minutes before the accident, the local controller stated that he had made a prevailing visibility observation of ¼ mile. The tower supervisor testified that she had confirmed the visibility within "minutes" prior to the runway incursion. Neither controller had referenced the Federal Meteorological Handbook, which provides guidelines for hazardous weather conditions, or the current NWS report.

Sometime in the 15 minutes before the accident, an off-duty controller made a visibility observation using the handbook reference chart. She had determined that the prevailing visibility was ⅛-mile. Following her observation, she asked the local controller whether he wanted to change the visibility reading, and he responded that the ¼-mile call was still current. The east ground controller stated that he had concurred with the ¼-mile call, however, he admitted that he was unable to see certain airport landmarks that were closer than a ¼-mile.

The accident

As Flight 299 left gate F11, the west controller cleared the crew to runway 3C via a right turn from the gate, and to hold short of Oscar 7, a taxiway just before the C concourse. (Refer to Fig. I-H to follow locations and sequence of events.) Although the current ATIS weather information reported ¾-mile visibility, the crew noted that the visibility began to deteriorate during their taxi.

The crew was instructed to contact the east ground controller when they neared Oscar 9, proceed to runway 3C by way of Oscar 6 and the Foxtrot taxiway. They were to report crossing runway 9/27. By this time, an updated ATIS was reporting ¼-mile visibility, which was the company takeoff minimum for runway 3C. As they taxied through the Oscar 6 area, the crew noticed Flight 1482 taxiing eastbound on the outer taxiway toward Oscar 4. The captain later testified that he had lost sight of the DC-9 as it taxied away and into seemingly thicker fog. Shortly thereafter, the crew remembered

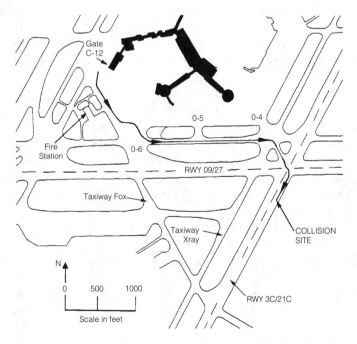

Fig. I-H. *Taxi route of Flight 1482 at Detroit Metropolitan Airport.* Adapted from NTSB

hearing a discussion on the east ground control frequency concerning an airplane that had missed the Oscar 6 intersection.

The crew of Flight 299 advised ATC when they had cleared runway 9/27, and turned onto taxiway X-ray. The captain noted that he could see about 1800 feet ahead of him, when the second officer commented on the rapidly deteriorating weather. The crew stopped at the hold line for runway 3C. At 1345:03, they began their takeoff roll. Five seconds later, the first officer remarked: "Definitely not a quarter mile, but . . . at least they're callin' [ATIS] it." The captain testified that he could still maintain the runway centerline, and because ATIS was reporting ¼-mile visibility, he believed that his decision to take off was correct.

When the airplane reached about 100 knots, the captain stated that a DC-9 suddenly appeared on the right side of the runway, directly in the path of the 727's right wing. He then shouted and moved his body to the left while pulling the yoke to the left and slightly aft. The aircraft struck the DC-9, and the captain rejected the takeoff and managed to stop at the end of the runway. The collision occurred 1 minute and 25 seconds after the tower had cleared Flight 299 for takeoff.

The taxi of Flight 1482. Flight 1482 left gate C18 and was cleared to a "right turn out of parking, taxi runway 3 center, exit ramp at Oscar 6, contact ground . . . ." The captain testified that he was able to follow the yellow taxiway centerline, but the first officer had commented, "it looks like it's goin' zero zero out here." Moments later, the ground controller requested Flight 1482's position, and the first officer replied that they were abeam the fire station. The controller then gave the crew an additional clearance of, "taxi Inner, Oscar 6, Fox, report making the right turn on X-ray." About 30 seconds later, the first officer stated: "Guess we turn left here." When the captain expressed some doubt about the left turn, the first officer replied: "Near as I can tell . . . I can't see . . . out here."

Shortly thereafter, the captain called for "flaps twenty and takeoff check when you get the time." The first six items on the checklist were completed when the first officer told the controller, "approaching the parallel runway on Oscar 6 . . . headed eastbound on Oscar 6." Seconds later, he added that they had missed Oscar 6 and were then following the "arrows to Oscar 5, think we're on Foxtrot now." The controller replied, ". . . you just approached Oscar 5 and you are . . . on the Outer?" The first officer answered, "yeah, that's right." Based on that information, the controller cleared Flight 1482 to "continue to Oscar 4, then turn right on X-ray."

The final three minutes of cockpit conversation was a revealing account of the confusion and role reversal that had taken place. Following are selected transcripts from the CVR tape, which represent the foundation for the lessons that will be discussed later.

The excerpts begin with the captain slowly taxiing the aircraft eastbound on the Outer taxiway and approaching the Oscar 4 intersection. The estimated visibility was 500 to 600 feet.

1342:00

Capt: This a right turn here . . .?

F/O: That's the runway.

Capt: (possibly a questioning tone) Okay, we're goin' right over here then.

F/O: Yeah, that way.

About 1342:30

F/O: Well, wait a minute. Oh . . . this . . . I think we're on . . . X-ray here now.

Capt: Give him [ground control] a call and tell him . . . .

F/O: . . . we're facing one six zero . . . cleared to cross it.

Capt: When I cross this, which way do I go? Right?

F/O: Yeah.

Capt: This is the active runway here, isn't it?

F/O: . . . should be nine and two-seven. It is. Yeah, this is nine-two seven.

Capt: Follow this . . . we're cleared to cross this thing. You sure?

F/O: That's what he said . . . Is there a taxiway over there?

At this point (1343:24), the captain set the parking brake. The crew of Flight 299 was about one-and-a-half minutes from starting their takeoff roll.

Capt: . . . I don't see one [taxiway]. Give him a call and tell him that . . . we can't see nothin' out here.

About 48 seconds later, the captain released the parking brake.

Capt: Now, what runway is this? This is a runway.

F/O: Yeah, turn left over there. Nah, that's a runway, too.

Capt: . . . tell him we're out here. We're stuck.

F/O: That's zero-nine [runway].

1344:47

The captain attempted to contact ground control, but because he was initially talking on some unknown frequency or over the intercom, he was unable to reach them until 11 seconds later, 5 seconds before the crew of Flight 299 added takeoff power.

Capt: [unsuccessful transmission] Hey, ground . . . we're out here . . . we can't see anything out here. Ground?

ATC: . . . just to verify you are proceeding southbound on X-ray now and you are across nine two-seven.

Capt: . . . we're not sure, it's so foggy out here, we're completely stuck here.

ATC: . . . are you on a taxiway or on a runway?

Capt: We're on a runway, we're right by . . . zero four.

ATC: . . . are you clear of runway 3 center?

Capt: . . . it looks like we're on 21 center here.

ATC:	. . . you are on 21 center?
Capt:	I believe we are, we're not sure.
F/O to Capt:	Yes, we are.
ATC:	. . . if you are on 21 center, exit that runway immediately, sir.

The two aircraft collided seven seconds later.

Impact and wreckage path

The 727's right wing initially sliced the DC-9 just under the first officer's right window. The cut continued in a straight line below the cabin windows along the right side of the airplane. Small pieces of debris from the wing tip of the 727, including shards of green glass from the right navigation-light lens, were found in the cockpit of the DC-9. The right engine was knocked off its pylon by the 727. The majority of the DC-9's fuselage was destroyed by fire from just aft of the cockpit to the aft bulkhead.

Approximately 13.5 feet of the 727's outboard wing had been sheared off at impact. Most of the leading-edge devices were broken off and found in and around the DC-9. The 727 sustained no other damage except several cuts on the right tires.

Accident survivability

None of the passengers or crew of Flight 299 were injured. However, seven passengers and one flight attendant on Flight 1482 sustained fatal injuries.

Following the collision, the captain of Flight 1482 shut off the fuel control levers, and ordered an evacuation. The two forward doors were opened, but the left evacuation slide had not been inflated prior to the crewmembers exiting the airplane. One passenger and a flight attendant attempted to deploy the tailcone exit, but due to a malfunctioning release mechanism, it would not open and both succumbed to smoke inhalation.

The investigation

The Safety Board focused the investigation on performances of ATC and the flightcrew.

ATC performance. The east-ground controller testified that the first time he realized that he was unsure of Flight 1482's whereabouts was when the crew told him that they were "completely stuck here." He soon became even more concerned when the crew added that they were "right by Oscar 4." Since he knew that Oscar 4 led onto runway 3C, he said that he loudly announced to the local controller, "I've got a lost aircraft out here, he might be on the runway." At that, he remembered the area supervisor stood up and told everybody to stop their traffic.

According to the local controller, he heard the east ground controller say that an aircraft was lost and possibly on the runway. He believed that Flight 299 was already airborne because he remembered hearing engine sounds, and that nearly a minute had elapsed from their takeoff clearance. Therefore, he made no attempt to contact the crew.

The area supervisor stated that she was seated at a desk doing paperwork when she heard the east-ground controller say something to the effect of "I think this guy's lost." She immediately directed the controllers to "Stop all traffic." When the east-ground controller added that the airplane might be on the runway, she loudly exclaimed, "I said stop everything." She did not remember hearing any engine noises of an aircraft taking off.

In a subsequent interview, investigators asked her why the local controller had not contacted Flight 299 in light of her order to stop traffic. She believed that he was the only person who knew where the airplane was, and therefore, the only one who could make the decision.

Although the local controller thought Flight 299 was already airborne, it was apparent to the Board that there was no basis for that assumption. The controller did not verify the aircraft's departure visually or on radar. In fact, he had cleared another airplane into position before the 727 began its takeoff roll. Therefore, the Board believed that the controller should have requested a "rolling" report from Flight 299. This would have enhanced his situational awareness, and might have prompted him to contact the crew when the order was issued to stop traffic.

At the time of the accident, controller workload was considered light, which freed the area supervisor to complete her paperwork. However, the Board believed that had she been monitoring the developing situation with Flight 1482, she might have detected the crew's confusion much earlier. The problem might have been corrected sooner, or at the very least, Flight 299 could have been warned before it was too late.

With regards to the visibility observation, the Board further suggested that the supervisor might have questioned the accuracy of a ¼-mile reading had she been more directly involved. It was noted that in the 30 minutes prior to the accident, there were patches of fog in which visibility ranged from ⅛-mile to ¾-mile. None of the controllers who believed the conditions were at ¼-mile looked at the observation reference guide, even though some at the time had visual indications that proved to the contrary. However, the off-duty controller did refer to the handbook and determined that the visibility was at ⅛-mile. Her query was quickly dismissed when she asked the local controller if he was going to change the official visibility report from ¼-mile. The Board believed that a more conscientious approach to gathering visibility information should have been maintained.

Flightcrew performance. The Safety Board believed that a nearly complete and unintentional reversal of command roles took place in the cockpit of Flight 1482. As a result, the captain became overly reliant on the first officer and essentially shifted the leadership position to him.

According to investigators, the breakdown began when the first officer implied to the captain that he was very familiar with the airport layout. The Board believed that the first officer did not want to appear inexperienced, but at the same time realized that the captain might need help. When the captain ask him for assistance with taxi clearances, he seemed to eagerly take on that responsibility. As the airplane left the gate, the first officer had already begun to dominate the decision-making in the cockpit.

The Board noted numerous examples of this domination in the early stages of the taxi:

- About 1322, the first officer explained to the captain the most accurate way to determine weight and balance.
- At 1325, the first officer told the captain that he had ejected from airplanes twice, and that he was a retired lieutenant colonel. The Board could find no basis for either statement in his military records.
- At 1331, the first officer explained to the captain the details concerning takeoff data for contaminated runways.
- At 1336, as they were initially looking for the yellow taxi line, the first officer said: "Just kinda stay on the ramp here." The captain replied: "Okay, until the yellow line, I guess?" The Board believed this conversation to be significant because the airplane was never positioned on the taxiway centerline that paralleled the ramp area and led to the open area at Oscar 6.
- About 1338, as the incorrect decision to turn left at the Oscar 6 sign was being made, the captain asked a series of questions about which way to turn. The first officer appeared to convince himself about their location, and then told the captain to turn left because they were on Oscar 6. The airplane was actually on the Outer taxiway.

There was no evidence to suggest that either pilot ever referred to his directional indicator to help determine their position. If they had checked the aircraft heading, they would have noticed that they were taxiing due east—Oscar 6 lies northwest/southeast. The Board believed that those cues should have been sufficient for the crew to orient themselves, and ask for ATC assistance. However, by the time the aircraft was on the Outer taxiway, the captain apparently felt that the first officer knew what he was doing and where the airplane was located.

In the Board's opinion, had the pilots admitted to themselves and to ground control that they were lost, this accident could have been prevented. Although the captain finally tried to assert his authority around 1344:47, and notify ATC, it was too late.

Lessons learned and practical applications

1. Ask for a progressive taxi. Get help especially if the visibility is poor, or you are not familiar with the airport, or both.
2. When in doubt, ask. Don't meander around an airport in hopes of finding the correct runway, taxiway, or gate. Swallow your pride, and ask ATC for help!
3. Maintain cockpit discipline. In this case, the nearly complete role reversal between the pilots was a direct result of the lack of cockpit discipline and CRM principles. There should always be a clear division of authority and responsibility.

4. Avoid role reversal. The term is "pilot-in-command," not "pilot-who-sits-in-left-seat." If you're new, then study the pubs, refresh your memory on airport layout and ATC frequencies, and review the departure/approach procedures *before* you get in the airplane.

5. Don't mislead crewmembers. Be careful not to leave the impression that you might know an important piece of information, when, in fact, you don't.

Case study reference

National Transportation Safety Board. 25 June 1991. Aircraft Accident Report: Northwest Airlines, Inc., Flights 1482 and 299. Runway incursion and collision. Detroit Metropolitan Wayne County Airport, Romulus, Michigan. December 3, 1990. Washington, D.C.

CASE STUDY I-8: United Airlines Flight 173

Safety issues: CRM, intimidation, distraction, cockpit discipline, fuel management

On 28 December 1978, a United DC-8 crashed into a wooded area near Portland, Oregon, after the aircraft experienced total fuel exhaustion.

Probable cause

The NTSB determined that the probable cause of this accident was the failure of the captain to properly monitor and respond to the aircraft's fuel state. He also failed to respond to the crewmembers' advisories regarding the fuel situation. Contributing to the accident was the failure of the two other crewmembers to either fully comprehend the criticality of the fuel state, or to successfully communicate their concerns to the captain.

History of flight

Flight 173 was a regularly scheduled flight from JFK International Airport, New York, to Portland, Oregon, with an intermediate stop in Denver, Colorado. The DC-8 departed Denver at 1647 Mountain Standard Time with 189 passengers and 8 crewmembers onboard.

Pilot experience

The captain had 27,638 total flight hours, 5517 as captain in the DC-8. He had been a United captain since 1959, and had successfully passed his last proficiency check three months before the accident.

The first officer had 8209 total flight hours, 247 as first officer in the DC-8. He was upgraded to that position six months earlier and had completed his last proficiency check in September 1978.

The second officer had 3895 total flight hours as a second officer, 2263 in the DC-8. He had been a flight engineer in that airplane for almost four years, and had passed his most recent en route proficiency check two weeks before the accident.

111

Weather

At the time of the accident, the weather was clear with 15 miles visibility.

The accident

Around 1709 PST, and as Flight 173 was descending through 8000 feet, the captain lowered the landing gear. The crew heard and felt unusually rapid and loud "thumps" as the gear extended. Although the nose-gear light illuminated, neither of the main-gear lights came on, and the aircraft yawed to the right. At 1712, the crew reported the problem to Portland Approach and requested to, ". . . stay at five [thousand] . . ." Interestingly enough, they had been cleared to only to 6000 feet, but by the time they had called approach, they had already busted their altitude clearance and were maintaining 5000 feet.

Portland approach began to vector Flight 173 in a holding pattern, and for the following 25 minutes the crew discussed and accomplished all of the emergency and precautionary actions to assure themselves that the landing gear was locked in the full-down position. Between 1738 and 1744, the crew talked with United's Maintenance Control Center to explain the situation and inform them that they were giving the flight attendants extra time to prepare the passengers and cabin.

For the next four minutes the crew discussed relatively minor issues, and expressed concern over the evacuation. However, the first officer did ask the captain, ". . . what's the fuel show now . . ." The captain replied, "Five [thousand pounds]."

In the DC-8, there are indicators on top of both wings that extend if the landing gear is down and locked. Shortly before 1750, the second officer determined from a visual inspection that the main gear, ". . . appeared to be down and locked . . ." This, however, did not motivate the captain to stop circling and head for the airport. Instead, the conversation lead to a brief discussion over figuring the fuel weight for another 15 minutes of flight time. They were estimating to land at 1805. The second officer informed the captain that, "Fifteen minutes is gonna . . . run us low on fuel . . ." The comment was, for the most part, passed over, and for the next six minutes the discussion was devoted to checklists.

The captain inquired about the remaining fuel at 1756, and was told there were 4000 pounds per engine left. Another six minutes elapsed when the second officer made the comment: "We got about three [thousand pounds] on the fuel . . . and that's it." There was no response to the remark. Shortly thereafter, the captain and first flight attendant discussed evacuation procedures. For a lengthy 57 minutes the crew had become so distracted over the gear malfunction and possible passenger evacuation, that they had devoted their attention to little else. That, however, changed at approximately 1806, when events took a turn for the worse.

Three-way miscommunication. *NOTE:* The following CVR transcript excerpts are taken from the final 27 minutes of Flight 173. Portions of the conversation have been condensed for clarity. Pay particular attention to the lack of direct interaction between the crewmembers. A side note of interest: The aircraft entered the holding pattern with 13,334 pounds of fuel. Its fuel burn was 13,209 pounds per hour, or 220 pounds per minute. Apply a little math, and you'll be able to calculate their actual fuel state.

1746:52

F/O: How much fuel we got . . . ?

S/O: Five thousand [pounds].

F/O: Okay.

1748:54

F/O: . . . what's the fuel show now . . . ?

Capt: Five [thousand pounds].

F/O: Five [thousand pounds].

Capt: That's about right. The fuel pumps are starting to blink.

1750:16

Capt: Give us a current [weight] . . . another fifteen minutes.

S/O: Fifteen minutes?

Capt: . . . give us three or four thousand pounds on top of zero fuel weight.

S/O: Not enough.

S/O: Fifteen minutes is gonna . . . really run us low on fuel . . .

?: Right.

1756:53

F/O: How much fuel you got now?

S/O: Four . . . thousand . . .

F/O: Okay.

Captain tells second officer to leave the cockpit and check on the cabin preparations. Second officer returns approximately four minutes later.

1806:34

Capt: . . . should be landing in . . . five minutes.

F/O: I think you just lost number four . . . .

F/A: . . . I'll go . . . I'm sitting down now.

F/O: Better get some cross feeds open . . . .

S/O: Okay.

1806:46

F/O: . . . goin' to lose an engine . . .

Capt: Why?

F/O: We're losing an engine.

Capt: Why?

F/O: Fuel.

806:55

S/O: Showing fumes.

F/O: It's flamed out.

S/O: We're going to lose number three in a minute, too.

S/O:	It's [number three engine] showing zero.
Capt:	. . . get some fuel in there [from crossfeeds].
Capt:	. . . watch one [engine] and two [engine].
Capt:	We're showing . . . zero or a thousand [pounds of fuel].
Capt:	. . . open all crossfeeds.
S/O:	. . . not very much more fuel.
S/O:	Number two is empty.

1809:45

Sound of engine spooling down.

Capt:	. . . see if we get gear lights.
Capt:	Yeah, the nose gear's down.
S/O:	. . . that fuel sure went . . . all of a sudden . . .
F/O:	Let's take the shortest route to the airport.

1813:21

S/O:	We've lost two engines . . .
S/O:	. . . just lost two engines, one and two.
Capt:	They're [the engines] all going.
F/O:	We can't make anything [airport or interstate].

1813:46

Capt:	Okay, declare a mayday.

1814:55

Sound of impact.

Impact and wreckage path

At close to 100 feet agl, the aircraft began to hit trees and two unoccupied houses along its 1555-foot wreckage path. The main landing gear and the nose section of the airplane struck a five-foot embankment next to a city street. As it traveled across the street, the vertical stabilizer snagged several high tension wires, and the aircraft finally came to rest in a wooded area and on top of the second unoccupied house.

The forward cockpit section separated from the fuselage and was found near the right wing. The first five rows of passenger seats were severely damaged from the force of impact, but the remaining interior of the cabin was left relatively intact.

Both main landing gear were fully extended but had been ripped away from their mounting structures. The nose gear was also fully extended and was still attached to the undercarriage. Investigators found no signs that indicated engine rotation at impact.

Accident survivability. The second officer, first flight attendant, and eight passengers sustained fatal injuries. They were all seated on the right side of the airplane, between the flight engineer's station and the fifth row of passenger seats. Most of the 23 seriously injured passengers and crew were located near those fatalities.

Although the inflatable slides quickly became torn from the debris, and five of the ten cabin exits were blocked by trees, the evacuation only took two minutes. Due to total fuel exhaustion, there was no post-impact fire.

The investigation

After close examination of the landing gear system, investigators determined that there was no evidence of a malfunction. Although the Safety Board understood the crew's concern over a possible gear problem, they believed that once the operations manual was checked and the company maintenance dispatch was contacted, there was nothing more the crew could do but attempt a precautionary landing.

The captain elected to circle at 5000 feet for one hour and two minutes, with the gear down and flaps set at 15 degrees. They had entered the holding pattern with about 13,334 pounds of fuel, but with the aircraft's configuration and altitude, the Board calculated their fuel burn to be approximately 13,209 pounds per hour—or 220 pounds per minute. According to investigators, routine procedures were not followed since there was no evidence that the crew ever discussed the fuel remaining in relation to time and distance to the airport during the final 30 minutes of flight.

The cockpit conversation revealed a lack of awareness on the part of the captain with regard to the aircraft's fuel-burn. Although the other two crewmembers made several comments about the fuel situation, they never spoke up in a clear and direct manner. Equally as puzzling to the Board was the crew's acceptance of a heading that took them away from the airport. There was no valid reason for such an action, and in the Board's opinion, "it was at this time that the crew's continuing preoccupation with the landing gear problem and landing preparation became crucial, and an accident became inevitable."

One distraction can lead to another. Oddly enough, the distraction caused by the possible gear malfunction was not what the crew devoted the majority of their time to. Instead, they concentrated a considerable amount of time and thought to evacuation procedures. According to the Board, there were numerous instances when the crew showed either a lack of concern or awareness over the fuel status. Just 17 minutes before the crash, with 4000 pounds of fuel remaining, the captain asked the first officer to check on the progress of the cabin preparations. Upon his return, four minutes later, the first officer told the captain the cabin would be ready in two or three minutes. But in the initial post-accident interview with the captain, he said he felt that the flight attendants wouldn't be in position for up to 15 minutes. Consequently, the fuel situation took a lessor priority, and at 1801, the crew accepted a vector that directed them away from the airport. Once the turn was completed, no one suggested heading back towards the airport. At that moment, because of the rate of fuel burn and the heading of the aircraft, it became physically impossible for Flight 173 to reach the airport.

Above all else, the Board believed that due to the crew's continuing preoccupation with the gear problem and landing preparations, their priority had long been shifted away from flying the airplane. An accident eventually became inevitable. It was just a matter of time. In this case, one hour and two minutes.

Tunnel ears. About seven minutes prior to the accident (refer to transcript excerpt), the captain was discussing the evacuation procedures with the first flight attendant. The first officer said, "I think you just lost number four . . ." The captain continued to speak with the flight attendant while the first and second officers tended to the emergency. It wasn't until the first officer repeated, ". . . goin' to lose an engine . . .", that the captain responded.

It was apparent that the captain had tuned-out the background conversation to the point that the words, ". . . lost number four. . . ," didn't even trigger a reaction. According to the investigation, the captain had not been monitoring the fuel situation, so when he heard the flameout was due to fuel exhaustion, it caught him totally by surprise.

Lessons learned and practical applications

1. Isolate distractions. Don't allow them to spread into other areas of the flight.

2. Maintain situational awareness. Watch out for "tunnel vision" and "tunnel ears."

3. Practice effective CRM. One of the primary points in this case was the intimidating tone that the captain had set in the cockpit. The other crewmembers' actions (or inactions) were driven by that tone. Remember that flying is a team effort. Encourage proactive participation from all crewmembers.

4. Speak up. When it became apparent to the first and second officers that the captain was not appropriately dealing with the fuel situation, they should have become much more assertive. In a loud, clear voice, communicate your concerns. And, don't take silence for an answer.

Case study reference

National Transportation Safety Board. 7 June 1979. Aircraft Accident Report: United Airlines, Inc., McDonnell Douglas DC-8-61, N8082U. Portland, Oregon. December 28, 1978. Washington, D.C.

PART II

Meteorology and atmospheric phenomena

According to information from the Massachusetts Institute of Technology, severe weather has been the leading cause of major airline accidents between 1975 and 1994. The two leading causes, thunderstorms and windshear activity, accounted for the most; four crashes resulting in 419 deaths. Ice buildup was a close third with three accidents resulting in 122 fatalities.

In many of these accidents, the flightcrews appeared to not have fully understood the implications of certain meteorological factors and atmospheric phenomena with regards to the overall safety of the flight. Therefore, this section includes chapters ranging from basic weather principles to more advanced studies on aircraft performance. The topics covered are air masses and fronts, cloud formation, thunderstorms, microbursts, icing, turbulence, and wake vortex turbulence.

REFERENCES

Barnett, Arnold. April 24, 1995. Massachusetts Institute of Technology. Annual Summaries of Air Crashes. *Newsweek*: 23.

5
Air masses and fronts

After completing this chapter, you should be able to:

1. Identify the four types of air mass classification.
2. Discuss frontal zones and their effect on the weather.
3. Explain the flying hazards associated with each type of front.

AIR MASSES

An air mass is a body of air extending over a large area of at least 1000 miles across. Properties of temperature and moisture are fairly constant throughout the air mass. Horizontal changes of these properties are usually very gradual. The terrain surface beneath the air mass is the primary factor in determining air mass characteristics.

Air mass classification

cP: Continental polar. Air stagnating over northern continental regions forms arctic air masses. They are cold and dry, and very stable.

mP: Maritime polar. The air masses form over northern oceanic areas. They are normally not as cold as cP air masses, especially in the winter, have a higher moisture content, and can be either stable or unstable.

mT: Maritime tropical. The air masses develop over warm oceanic areas nearer to the equator. They are very humid and generally are the most unstable of all.

cT: Continental tropical. These air masses originate from arid, continental regions that are hot, dry, and unstable. Due to the absence of water vapor, they produce very few rain showers.

FRONTS

Fronts are transition zones between air masses that have different densities. The density of air is primarily controlled by the temperature and humidity of the air. Therefore, fronts in the mid-latitudes usually form between tropical and polar air masses.

Frontal zones, which are normally many miles in width, are most easily detected when the air masses have vastly different properties. They are mostly determined by a change in temperature, moisture, and wind direction and velocity. Specific weather conditions precede and follow a front as it moves through an area. Furthermore, weather associated with one section of a front frequently is different from the weather in other sections of the same front.

Cold Fronts

The leading edge of an advancing cold air mass is called a cold front. Cold air is overtaking and replacing warmer air. Surface friction holds back the air in contact with the earth's surface, creating a bulge in the front. This explains why a cold front has a steep slope near its leading edge. Cold frontal slopes average about 1:80 miles, meaning that at 80 miles into the surface position of the front, the frontal boundary is about one mile above the ground.

Cold fronts might be accompanied by very marked weather changes and some of the most hazardous flying weather.

Fast-moving cold fronts. Air is forced upward in the area ahead of the surface position of a rapidly moving cold front. Therefore, most of the heavy cloudiness and precipitation is located just ahead of the front where the opposing air currents meet. This type of front usually causes very hazardous flying weather.

As mentioned earlier, friction often hinders the movement of the front near the surface. This causes its slope to steepen, and a narrower band of weather results. If the warm air mass is moist and unstable, scattered thunderstorms and showers are likely to develop ahead of the front. A squall line might also form, which is often characterized by a massive wall of turbulent clouds building to more than 40,000 feet. The squall line sometimes develops between 50 and 200 miles ahead of the front and is roughly parallel to it. Some of the most turbulent weather ever reported was associated with a squall line. The weather usually clears rapidly behind the fast-moving cold front, with colder temperatures and gusty surface winds following its passage.

Slow-moving cold fronts. When a cold front is moving at less than 15 knots, there is a general upsloping of the warm air over the frontal surface. This results in a rather broad cloud pattern in the warm air with the clouds extending well behind the surface position of the front.

If the warm air is stable, the clouds are stratiform. Cumuliform clouds and frequent thunderstorms develop if the warm air is moist and unstable. A slow-moving cold front or a cold front over a shallow, cold air mass might have a cross section more like that of a warm front.

Flying hazards

Potentially dangerous weather can be found in a cold front when cumuliform clouds develop along the front or when a squall line is ahead of the front. Those conditions are especially hazardous due to the presence of turbulence, strong and variable low-level winds, thunderstorms, lightning, heavy rain showers, hail, icing, and possibly tornadoes.

WARM FRONTS

The edge of an advancing warm air mass is called a warm front; warmer air is overtaking and replacing colder air. Since the cold air is more dense than the warm air, it tends to be slow at dissipating. This produces a gradual, warm frontal slope that usually average 1:200 miles.

If the advancing warm air is moist and stable, stratiform clouds will develop. Often the progression of cirrus, cirrostratus, altostratus, and nimbostratus clouds indicate such a front. Precipitation usually increases slowly with the approach of this type of warm front, and normally continues until it passes.

If the advancing warm air is moist and unstable, altocumulus and cumulonimbus clouds, including thunderstorms, will be imbedded in the cloud masses that normally accompany the warm front. The presence of these thunderstorms is often unknown to pilots until they fly into one. Precipitation in advance of the front is usually in the form of showers.

Flying hazards

One of the most serious hazards is the presence of low-level windshear that can linger for longer than six hours prior to the passage of a warm front.

The widespread precipitation area ahead of a warm front often causes low stratus and fog. When this occurs, the precipitation raises the humidity of the cold air to saturation. This can produce low ceilings and poor visibility over thousands of square miles. The frontal zone itself might have extremely low ceilings and near zero visibilities over a wide area.

If the cold air has subfreezing temperatures, the precipitation might take the form of freezing rain or ice pellets. In summer months, thunderstorm activity is quite likely.

STATIONARY FRONTS

Sometimes the opposing forces exerted by adjacent air masses of different densities are such that the frontal surface between them shows little or no movement. Surface winds tend to blow parallel to the front rather than away from it. Since neither air mass is replacing the other, the front is considered stationary.

Although there is no movement of the front's surface position, an upglide of air can occur along the frontal slope. The angle of this flow of air in relation to the surface position of the front, and the intensity of the upgliding wind, determines the inclination of the frontal slope.

The weather conditions associated with a stationary front are similar to those found with a warm front, but are usually less severe. Since the weather pattern is stationary, poor weather might persist and hamper flights in one area of the country for several days.

OCCLUSIONS

Occlusions are the result of one frontal system overtaking a previously-formed frontal system. Since they combine the weather of both a warm and cold front into one extensive system, a line of rain showers and thunderstorms typically develops as the cold front merges with the low ceilings of the warm front. Precipitation and low visibilities are widespread over a large area on either side of the surface position of the occlusion. Strong winds will occur around an intense low at the northern end of the occlusion. Therefore, pilots should be aware that weather conditions can change rapidly in occlusions, and that conditions are usually most severe during the initial stages of development.

CHAPTER REVIEW

Air masses

- Air masses extend over large areas
- Temperature and moisture are fairly constant
- **cP:** cold, dry, stable
- **mP:** higher moisture content, stable or unstable
- **mT:** humid, unstable
- **cT:** hot, dry, unstable

Fronts

- Density of air primarily controlled by temperature and humidity
- Determined by change in temperature, moisture, wind direction, and velocity

Cold fronts

- Leading edge of cold air mass
- Steep gradient
- 1:80 miles ratio
- Marked weather changes
- Hazardous flying weather

Fast-moving cold fronts

- Heavy clouds and precipitation
- Hazardous flying conditions
- Squall lines

Slow-moving cold fronts

- Stratiform clouds will form if air is warm and stable
- Cumuliform clouds and frequent thunderstorms will develop if air is warm and unstable

Flying hazards

- Potentially dangerous weather
- Turbulence, strong and variable low-level winds, thunderstorms, lightning, heavy rain showers, hail, icing, tornadoes

Warm fronts

- Slow to dissipate
- 1:200 miles ratio
- Stratiform clouds will develop if air is warm, moist, and stable
- Sign of front can be progression of cirrus, cirrostratus, altostratus, and nimbo-stratus clouds
- Precipitation slowly increases
- Altocumulus and cumulonimbus clouds develop if air is warm, moist, and unstable
- Imbedded thunderstorms

Flying hazards

- Low-level windshear
- Widespread precipitation
- Low stratus and fog
- Low ceilings and poor visibility
- Freezing rain or ice pellets
- Thunderstorm activity

Stationary fronts

- No movement between air masses
- Weather similar to warm front, but less severe
- Poor weather might persist for several days

Occlusions

- One frontal system overtakes a previously formed frontal system
- Warm and cold fronts combine into one extensive system

- Line of rain showers and thunderstorms might develop
- Widespread precipitation and low visibilities
- Strong winds might develop near an intense low at the northern end of occlusion
- Rapidly changing weather conditions
- Most severe during initial stages of development

CHAPTER REFERENCES

Department of the Air Force. 1 January 1982. *Weather for Aircrews*.
Williams, Jack. June 1995. "Jetstreams." *Flight Training*: 47–49.

6
Cloud formation

AFTER COMPLETING THIS CHAPTER, YOU SHOULD BE ABLE TO:

1. Identify the four basic cloud families.
2. Identify the types of weather associated with each cloud type.
3. Understand the flying hazards associated with each cloud family.

A basic understanding of cloud types is key to identifying approaching and existing weather conditions. Very simply, there are four "families" or categories of clouds: low, middle, high, and extensive vertical development. When clouds form, the degree of stability of the air helps determine what type they will be.

LOW CLOUDS

The group of low clouds consists of stratus, stratocumulus, and cumulus. The bases of these clouds range from near the surface to about 6500 feet agl. If low clouds form below 50 feet, they are reclassified as fog.

These kinds of clouds are made almost entirely of water, and depending on the temperature, can be filled with ice crystals or supercooled water droplets. Therefore, the potential for icing must always be considered as a threat when flying in or near these conditions.

Stratus

The slow lifting of a fog layer often results in the formation of stratus. Stable air rising over sloping terrain can also produce this low-hanging, uniform looking cloud. Stratus is usually associated with fog or precipitation, and contains little or no turbulence. When temperatures dip near freezing, hazardous icing conditions can be present.

Stratocumulus

Stratocumulus clouds can form when a layer of stable air that is being lifted is mixed by blowing wind over rough terrain. They can also develop from the breaking up of a stratus layer or from the spreading out of cumulus clouds. Higher ceilings and better visibilities usually are associated with these types of clouds. They often appear as large, dirty puffs of cotton.

Cumulus

Cumulus clouds form in convective currents caused by the heating of the ground. They can also develop as a cold air mass is warmed by passing over a relatively warm surface. These clouds are characterized by flat bases and dome-shaped tops. Fair weather cumulus indicate a shallow layer of instability. Some turbulence, but no significant icing or precipitation can be expected. Although cumulus do not show extensive vertical development, continued growth might lead to towering cumulus and cumulonimbus clouds.

MIDDLE CLOUDS

The middle clouds include: altostratus, altocumulus, and nimbostratus. The height of the bases of these clouds range from 6500 feet to about 20,000 feet agl. These clouds might be composed of ice crystals or supercooled water droplets. Therefore, middle clouds might contain significant icing conditions.

Altostratus

The altostratus kind of cloud is usually so dense that even through its thinner areas sunlight is seen as very dim. Altostratus are relatively uniform gray to blue sheets that cover the entire sky. It might gradually form into cirrostratus, and rarely contains turbulence. However, moderate icing might be present.

Altocumulus

These clouds often develop from dissolving altostratus, and appear as white or gray patches of solid mass. They are composed of water droplets and at very low temperatures ice crystals might form. Some turbulence and small amounts of icing will occur in altocumulus.

Nimbostratus

The massive gray and dark-layered nimbostratus is generally associated with continuous rain or snow. These clouds pose a serious icing problem if temperatures are near freezing. As conditions weaken, nimbostratus might merge into low stratus or stratocumulus-type clouds.

HIGH CLOUDS

High clouds in this family include cirrus, cirrostratus and cirrocumulus. The height of the bases ranges from about 16,000 feet to 45,000 feet in mid-latitudes. In the tropics, the upper limits of these clouds might reach 60,000 feet. A cirroform cloud is composed of ice crystals, and therefore, does not present a significant icing hazard.

Cirrus

Cirrus clouds often evolve from the upper part of thunderstorms or cumulonimbus. They can blow away from the main cloud, or the core of the cloud might evaporate, leaving only the ice crystal top portion. These feather-like clouds are spread in patches or narrow bands. If cirrus clouds are arranged in bands or connected with cirrostratus or altostratus, it might be a sign of approaching bad weather.

Cirrostratus

Cirrostratus clouds occur only in stable layers, therefore, no turbulence and little icing can be expected. However, any cirroform cloud might produce restricted visibility. They appear across the sky as a thin, whitish veil.

Cirrocumulus

Cirrocumulus clouds might develop from the lifting of a shallow, unstable layer of air. Heat loss by radiation occurs from the top of the cirrus layer, and the cooler air on top sinks into the cloud. As a result, shallow convective currents can be produced within the layer. Due to the lifting movements, some turbulence could be experienced. Cirrocumulus appear as small, white flakes of cotton.

EXTENSIVE VERTICAL DEVELOPMENT

Clouds in this family include towering cumulus and cumulonimbus. The height of their bases ranges from those in the low category all the way to the highest.

Towering cumulus

Although similar in nature to the cumulus, towering cumulus clouds have greater vertical development. They are often associated with the presence of thunderstorm formation, turbulence, and icing.

Cumulonimbus

Water droplets form the major portion of cumulonimbus clouds, but ice crystals usually appear in the upper limits. These clouds are synonymous with thunderstorms, and produce strong winds, lightning, and intermittent showers. The well-developed cumulonimbus might be the parent of the hailstorm and the tornado.

Flying hazards

Avoid flying near or directly under towering cumulus and cumulonimbus clouds. They can rapidly develop in groups or lines and become imbedded in stratiform clouds, resulting in hazardous instrument-flight conditions. Turbulence and icing might be severe enough to cause structural damage to the aircraft.

SPECIAL CLOUD TYPES

Because these types of clouds have unique characteristics, they do not fall into any formal classification.

Altocumulus standing lenticular

Altocumulus standing lenticular clouds develop on the crests of waves created by barriers in the wind flow. Condensation in the rising portion of the wave forms the clouds. In the descending section of the wave, the cloud evaporates. Thus, the cloud appears not to move, although the wind can be quite strong and dangerous blowing through it. These clouds must be avoided.

Rotor clouds

Rotor clouds form on the lee side of mountains. The rotor looks like a line of small cumulus clouds parallel to the mountain. Sometimes a person is able to see the rapid swirling motion of the rotor at ground level. Severe turbulence can be encountered in the vicinity of a rotor cloud. The threat of erroneous pressure instrument readings is also a possibility.

Virga

These streaks of water or ice particles, called virga, are usually seen hanging from altocumulus and altostratus clouds. Precipitation that falls from these high-based clouds, evaporates and cools the air, creating a downdraft. Do not fly near virga.

CHAPTER REVIEW

Low clouds

- Stratus, stratocumulus, cumulus
- Surface to 6500 feet
- Classified as fog when under 50 feet

Stratus

- Fog, precipitation, little or no turbulence
- Potential for icing

Stratocumulus

- Higher ceilings, better visibilities

Cumulus

- Formed by convective currents
- Some turbulence
- No significant icing or precipitation

Middle clouds

- Altostratus, altocumulus, nimbostratus
- 6500 feet to 20,000 feet

Altostratus

- Moderate icing
- No turbulence

Altocumulus

- Some turbulence
- Little chance of icing

Nimbostratus

- Serious icing conditions

High clouds

- Cirrus, cirrostratus, cirrocumulus
- 16,000 feet to 45,000 feet in mid-latitudes
- Upwards of 60,000 feet in tropics
- Cirroform clouds composed of ice crystals

Cirrus

- Evolve from upper part of thunderstorms or cumulonimbus
- If arranged in bands or connected with cirrostratus or altostratus, might be sign of bad weather

Cirrostratus

- No turbulence
- Little icing
- Restricted visibility

Cirrocumulus

- Some turbulence
- Extensive vertical development

Towering cumulus and cumulonimbus

- Base ranges from the low to high category of clouds

Towering cumulus

- Greatest vertical development
- Associated with presence of thunderstorm formation
- Turbulence and icing

Cumulonimbus

- Synonymous with thunderstorms
- Strong winds, lightning, intermittent showers
- Hailstorms and tornadoes

Flying hazards

- Avoid flying near towering cumulus and cumulonimbus
- Severe turbulence and icing

Special cloud types

Altocumulus standing lenticular

- Develop on the crests of mountain waves
- Strong and dangerous winds
- Avoid flying near these cloud formations

Rotor cloud

- Form on lee side of mountain
- Severe turbulence
- Possibility of erroneous pressure instrument readings
- Avoid flying near these cloud formations

Virga

- Precipitation that evaporates before it hits the ground
- Downdrafts
- Avoid flying near virga

CHAPTER REFERENCES

Battan, Louis J. 1979. *Fundamentals of Meteorology*. Prentice-Hall: Englewood Cliffs. 147–180.

Department of the Air Force. 1 January 1982. *Weather for Aircrews*: 7/1–9.

Federal Aviation Administration. 1975. *Aviation Weather*: 53–62.

Hapgood, Fred. April 1994. "Up in the Sky, There's a Good Time to be had Everywhere, Always, for Free." *Smithsonian*: 37–42.

Towers, Joseph F. January 1989. "Lenticular Clouds: Signs of Turbulence." *Air Line Pilot*. 22–23.

Williams, Jack. 1992. *The Weather Book*: Vintage Books: New York. 72–79, 113–167.

7
Thunderstorms

AFTER COMPLETING THIS CHAPTER, YOU SHOULD BE ABLE TO:

1. Discuss the three basic requirements for the formation of a thunderstorm.
2. Describe the life cycle of a thunderstorm cell.
3. Describe the types of thunderstorms.

There are three basic requirements necessary for the formation of a cumulonimbus (thunderstorm) cloud: unstable air, lifting action, and a high moisture content in the air.

UNSTABLE AIR

Air will become unstable when it reaches an altitude where it turns warmer than its local environment. Warm air will continue to rise until it has cooled to the temperature of the surrounding air.

LIFTING ACTION

Some type of external lifting action is needed to bring the warm air from near the surface to the point where it will continue to rise. That latter stage is called *free convection*. A strong lift is usually generated by mountainous terrain, fronts, low-level heating, or an atmospheric convergence. A convergence is when air coming from different directions merge, and the force that's created by the collision pushes the air upward in a swift, vertical motion.

MOISTURE

The mere lifting of warm air will not necessarily cause free convection. Clouds can still form when moisture condenses, but they will not grow significantly unless the air is lifted to the level of free convection. The higher the moisture content, the easier it is for the air to reach free convection. Once a cloud develops, the latent heat of condensation that is released by the change of state, vapor to liquid, tends to make the air even more unstable.

LIFE CYCLE OF A THUNDERSTORM CELL

A thunderstorm cell progresses through three stages during its life cycle: Cumulus, or growth, mature, and dissipating. Often a cluster of cells will be imbedded inside a thunderstorm. Since each cell might be in a different stage of the life cycle, the outward appearance of the cloud could be quite deceiving. The life cycle of a thunderstorm might last from 20 minutes to 3 hours, depending on the number of cells contained and their stage of development. It's virtually impossible for a person to visually detect the transition from one stage to another, so a pilot should never try to outguess the system and fly near a building thunderstorm.

Cumulus stage

Although most cumulus clouds don't become thunderstorms, the initial stage of a thunderstorm is always a cumulus cloud. The main feature of this stage is a rapidly moving updraft, that might originate near the ground and extend several thousand feet above the visible cloud top. The greatest updraft occurs at higher altitudes late in the cumulus stage, when it could reach speeds in excess of 3000 fpm.

As the cloud forms, water vapor changes to liquid and/or frozen particles. This results in a release of heat that provides a source of energy for the developing cloud. It is this release of heat that helps keep the cloud growing.

During this early stage, cloud droplets are very small, but turn into raindrops as the cloud builds upward. The raindrops might remain in a liquid state as the updraft pushes them well above the freezing level, which in the most intense storms could be more than 40,000 feet. There is usually no falling precipitation during this stage because the water drops and ice particles are still being carried aloft by the ascending air currents.

Mature stage

The beginning of falling rain or hail from a cloud indicates that a downdraft has developed, and the cell has entered the mature stage. The raindrops and ice particles in the cloud have grown too large for the updraft to continue to support them. By this time, the average cell has reached a height of about 25,000 feet, although at high latitudes, tops might be as low as 12,000 feet.

As the drops start to fall, the surrounding air begins a downward motion. Since the air is unstable, the cold air accelerates and forms a downdraft. The velocity of a downdraft can easily reach speeds of 2500 fpm, as it spreads outward near the ground. This produces a sharp decrease in temperature and strong, gusty surface winds. The leading edge of this wind is called the *gust front*.

Early in the mature stage, any remaining updrafts will continue to gain speed, and might exceed 6000 fpm. The violent flow of updrafts and downdrafts near each other create the vertical shears that can cause severe turbulence. All thunderstorms have reached their greatest intensity at this time.

Dissipating stage

Throughout the mature stage, the downdrafts continue to develop as the updrafts begin to weaken. As a result, the entire thunderstorm cell ultimately becomes an area of downdrafts. Since updrafts are necessary to produce condensation and latent heat energy, the thunderstorm begins to dissipate. Don't be fooled, though, into thinking a weakening thunderstorm can't still hold a punch or two. Strong, upper level winds will typically push the top of the cloud into an anvil shape. Although this can be *one* sign that the thunderstorm is dissipating, severe weather can still be present in many systems with a well-defined anvil.

The most intense thunderstorms occasionally do not dissipate in the manner just described. Although they are technically considered in the dissipating stage they behave more like a prolonged mature stage. If horizontal wind speeds dramatically increase with altitude, the storm clouds will shift to a tilted position. Precipitation will fall through a small portion of the rising air. This will be followed sometime later with it falling through the relatively calm air near the updraft, or even completely outside the cloud itself.

The updrafts produced by this prolonged mature stage can continue until their source of energy is exhausted. The thunderstorm, therefore, might dissipate without going through the normal process of tremendous downdraft activity. However, precipitation falling from the tilted cloud might cause downdrafts to form in the clear air, just outside the storm's boundary.

TYPES OF THUNDERSTORMS

There are several types of thunderstorms that can develop, each presenting unique and potentially dangerous flying conditions.

Warm front thunderstorms. Due to the shallow slope of a warm front, thunderstorms in those areas are usually the least severe of all frontal-type storms. Nevertheless, thunderstorms might still be found hidden within the stratiform clouds that are normally associated with a warm front. You might be able to spot them poking through the hazy layers if you're flying high enough, or become aware of their presence from sudden and loud spurts of static over the radio.

Cold front thunderstorms. These storms are quite severe, and usually form in a continuous line that is easy to recognize. These thunderstorms are generally most active during the warmer afternoon hours.

Stationary front thunderstorms. Occasionally, thunderstorms will develop in stationary fronts and become widely scattered.

Occluded front thunderstorms. These thunderstorms are particularly dangerous because they are not well-developed, but can become imbedded in stratiform clouds, making them difficult to detect.

Squall-line thunderstorms. A squall line is a nonfrontal, narrow band of active thunderstorms. It often develops 50 to 300 miles ahead of a rapidly moving cold front in moist, unstable air. The existence of a front, however, is not absolutely necessary for a squall-line to form.

The thunderstorms in a squall line are fast-building and are generally more violent than storms associated with a cold front. Severe weather, including heavy hail, destructive winds, and tornadoes are characteristic of a squall-line.

Air mass thunderstorms

Thunderstorms can form within warm, moist air that is not part of a front. This usually occurs from surface heating and convergence that takes place over land during the middle and late afternoon. Along coastal regions, air mass thunderstorms tend to reach their maximum intensity throughout the night and early morning when the cool air flowing off the land is heated by the warmer water surface. As a result, these thunderstorms often form a short distance offshore. Whether they develop over land or water, air mass thunderstorms are usually isolated or widely scattered over a large area.

Orographic thunderstorms

Thunderstorms will form on the windward side of a mountain if the wind forces moist, unstable air up the slope. The storm activity is usually scattered along the individual mountain peaks, but occasionally there will a long, unbroken line of thunderstorms. Stratus and stratocumulus clouds frequently enshroud the mountain peaks and obscure the storm.

METEOROLOGICAL OBSERVATIONS

Thunderstorms are categorized by levels of echo intensity and rainfall rate.

Level 1: Light precipitation. Light to moderate turbulence is possible with lightning.

Level 2: Moderate precipitation. Light to moderate turbulence is possible with lightning.

Level 3: Heavy precipitation. Severe turbulence is possible with lightning.

Level 4: Very heavy precipitation. Severe turbulence is likely with lightning.

Level 5: Intense precipitation. Severe turbulence is likely with lightning, organized wind gusts, and hail.

Level 6: Extreme precipitation. Severe to extreme turbulence is likely with large hail, lightning, and extensive wind gusts.

CASE STUDY REFERENCES

II-2, II-4, II-5

CHAPTER REVIEW

Unstable air

• Warm air rises until it has cooled to the surrounding temperature

Lifting action

• Free convection

Moisture

• High moisture content

Life cycle

• Cumulus, mature, and dissipating stages
• Might last 20 minutes to 3 hours

Cumulus stage

• Rapidly moving updraft
• Release of heat
• No falling precipitation

Mature stage

• Falling rain or hail
• Updrafts and downdrafts
• Decrease in temperature
• Strong, gusty winds
• Gust front
• Thunderstorm is at its greatest intensity

Dissipating stage

• Tremendous downdraft activity
• Severe weather can still be present when a thunderstorm has a well-defined anvil top
• Precipitation might fall near the updrafts or completely outside the cloud
• Tilted clouds might cause downdrafts to form in clear air

Warm front thunderstorms

• Least severe

Cold front thunderstorms

• Very severe

Stationary front thunderstorms

• Widely scattered

Occluded front thunderstorms

• Very dangerous
• Not well-developed
• Can be imbedded in stratiform clouds

Squall line thunderstorms

• Nonfrontal, narrow band of thunderstorms
• Fast-moving
• Most violent of storms
• Heavy hail, destructive winds, tornadoes

Air mass thunderstorms

• Often form a short distance offshore
• Isolated, widely scattered over a large area

Orographic thunderstorms

• Form on windward side of mountain
• Moist and unstable winds
• Stratus and stratocumulus clouds

Meteorological observations

• Six levels of echo intensity and rainfall rate.

CHAPTER REFERENCES

Battan, Louis J. 1979. *Fundamentals of Meteorology*. Englewood Cliffs: Prentice-Hall: 181.

Department of the Air Force. 1 January 1982. *Weather for Aircrews*. Washington, D.C.

Miller, Wally. June 1, 1995. "Killer Thunderstorms." *Aviation Safety*: 1–5.

Williams, Jack. 1992. *The Weather Book*. New York: Random House.

8
Microbursts and low-level windshear

AFTER COMPLETING THIS CHAPTER, YOU SHOULD BE ABLE TO:

1. Describe the difference between a microburst and macroburst.
2. Discuss the types of microburst classifications.
3. Identify the weather activity associated with each type of microbursts.
4. Discuss the types of parent clouds associated with microburst activity.
5. Identify the weather activity associated with low-level windshear.
6. Identify the cues present during windshear activity.

It seems the terms "microburst" and "windshear" have been in a pilot's vocabulary for many years, but, in fact, the current information we have today evolved from research conducted only a decade ago. In the 1970s, many meteorologists considered a downdraft as the most hazardous activity associated with a thunderstorm. The conventional theory was that a downdraft was strongest inside or directly beneath the cloud. It was thought to significantly weaken long before reaching the ground, and as a result there was little danger in an outflow of wind disturbance. This concept sounds undoubtedly archaic to us now, but that piece of erroneous guidance was all those pilots had to fly with.

The study and identification of microbursts began in 1974, when Professor Theodore Fujita, of the University of Chicago, began his world renown research on weather phenomenon. Scoffed at by his colleagues for his outlandish ideas, Fujita aggressively pursued the causes of aircraft accidents around thunderstorm activity. That year, the southeast United States experienced a "super outbreak" of 148 tornadoes and severe thunderstorms (a Southern Airways DC-9 crashed as a result of those storms). During an aerial survey of the aftermath, he found a strange pattern of tree damage. Unlike the swirling effect of fallen trees, commonly seen in the wake of tornadoes, hundreds of trees were blown down and outward, as in a starburst pattern. Trees near the center of these starbursts, were flattened or uprooted, and splattered with dirt. Nearly 15 percent of the almost 2600 miles of damage paths were attributed to starburst patterns.

A year later, an Eastern Airlines 727 (refer to case study II-5) crashed at Kennedy Airport, New York, while attempting to land in a thunderstorm. On the basis of his starburst airflow theory, Fujita hypothesized that the aircraft encountered such a system, causing it to be pushed earthward. He believed his idea was correct after examining the jet's FDR. Subsequently, he coined the term *downburst*. He explained the definition as: "a strong downdraft, which induces an outburst of highly diverging winds on or near the ground. The sizes of downbursts vary from less than one kilometer (km) to tens of kms." This concept was quite controversial for the day, and was not well-received by those in the meteorology discipline.

By 1978, Fujita had collected enough data to change the minds of the skeptics. The first field program on downbursts was conducted by the University of Chicago, and operated for 42 days. By using three strategically-placed Doppler radars and other measuring equipment, Fujita and his team discovered the phenomena of "macrobursts" and "microbursts." In the early 1980s, Fujita continued his studies through the University and with the National Center for Atmospheric Research.

According to Fujita, a downburst is subdivided into macrobursts and microbursts. The following are the official definitions:

Macroburst: A large downburst with its outburst winds extending in excess of four km (2.5 miles) in horizontal dimension. An intense macroburst often causes widespread, tornado-like damage. Damaging winds, lasting 5 to 30 minutes, could be as high as 60 meters/second, or 134 mph.

Microburst: A small downburst with its outburst, damaging winds extending only four km or less. In spite of its small horizontal scale, an intense microburst could induce damaging winds as high as 75 meters/second, or 168 mph.

MACROBURSTS

According to Fujita, due to its large horizontal scale, a macroburst is characterized by a big layer of cold air that is created by a succession of downdrafts beneath the parent rain cloud. Since a dome of cold air is heavier than the warm air surrounding it, the atmospheric pressure inside the dome is higher than its environment. The pressure gradient force, pointing outward from the dome area, pushes the cold air outward, inducing gusty winds behind the leading edge of the cold air outflow. The gust front denotes the leading edge of gusty winds which push the dome boundary away from the subcloud region. Those conditions produce a violent clash between the cool outflowing air and the warmer thunderstorm. This usually causes a windshift and drop in temperature that precedes a thunderstorm.

MICROBURSTS

Very strong outbursts of winds over an area less than 2.5 miles are the two basic characteristics of microbursts. Interestingly, while the core of a microburst is quite severe, its boundary propagates outward rather slowly. Photographic evidence has shown that a downflow of wind might be traveling at 40 knots. When a swirling horizontal vortex near the ground encircles the downflow center, it forms a vortex ring. The outbursting winds beneath this ring are accelerated and might reach 100 knots. This outburst continues to gain speed as the ring expands and stretches.

TYPES OF MICROBURSTS

Not all microbursts are alike; some are accompanied by heavy rain while others form beneath small virga. The first major classification is based on the amount of precipitation that reaches the ground. The second is distinguished by change in temperature. Refer to Fig. 8-1.

Wet microburst

The environment in which a wet microburst typically forms is marked by a deep, nearly saturated layer of air, a moist adiabatic lapse rate, and topped by an elevated dry

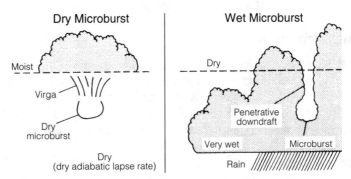

Fig. 8-1. *Environmental conditions associated with microbursts.*
Adapted from NOAA

141

layer of air. This mixture can produce enough negatively, buoyant potential energy to drive a severe downdraft. A descending wet microburst might first appear as a darkened mass of rain falling through an area of lighter rain. A strong downdraft pushes rain toward the surface at a much faster rate than it can fall at terminal velocity through still air. As the downdraft approaches the ground, it decelerates in the vertical, allowing a heavy load of water to accumulate just above the surface. The most pronounced visual indicator of a potential microburst, is a descending, high-density globular mass of rain, and a clearing out of precipitation in its wake.

Dry microburst

In the extremely dry environment, where moist convection is just barely possible, cumulus clouds with very high bases can form. Below the cloud layer is a deep and dry adiabatic lapse rate. The storms might not produce lightning, even though the cumulus have a fibrous appearance and a prominent anvil-shaped top. At first glance, the weather might appear good and nonthreatening, but conditions can change in a matter of minutes. A dry microburst becomes visible when the expanding ring of dust under a virga shaft descends from a high-based cumulonimbus. The precipitation mostly evaporates before reaching the surface, so rainfall is probably no more than a trace. As the microburst develops, the ring of dust spreads out over the ground.

Anvils of large, dry-line thunderstorms might also produce high-level virga, which can result in dry microbursts. These conditions could develop far away from the parent hail-storm and its associated heavy radar echo.

Traveling microburst

The traveling motion of a microburst distorts the airflow from a circular to elliptical shape. The front-side wind intensifies, while the back-side wind weakens, resulting in a crescent-shaped area of high winds.

Radial microburst

Radial streamlines are seen if microburst winds are not rotating.

Twisting microburst

When a microburst descends inside a cyclonic airflow at the surface, it curves and twists. This type has been observed during tornadoes that were spawned by supercell thunderstorms. Frequently, damage maps of tornadoes reveal a widening of the paths prior to their dissipation. When the end of a tornado's path becomes two to three miles wide, the flow pattern becomes similar to that of a twisting microburst.

Surface microburst

No microbursts form on or near the ground. But when one touches the ground it's called a *surface microburst*. When it spreads out over a large area, the downflow keeps supplying the mass until it sinks to the ground. After that, a microburst flattens and its expansion terminates.

Outflow microburst

A slow-traveling outflow microburst is often characterized by a vortex ring encircling it. The ring keeps stretching as a surface microburst gets older, until it reaches its limit. Thereafter, the vortex is cut into several pieces of roll vortices, each with a horizontal axis. It eventually turns into a rotor microburst.

Rotor microburst

When those vortex rolls "run away" from their source region, bands of high winds are produced. They can last two or three minutes. This type of microburst behaves like a tornado with a horizontal vortex axis. It usually creates a narrow, but severe, damage path. It's also frequently accompanied with a roaring sound, and is often misidentified as a tornado. In 1981, Fujita called this rotor microburst swath, a "burst swath."

TYPES OF PARENT CLOUDS

Anvil

Many microbursts form beneath virga descending from anvil-shaped clouds.

Supercell

These thunderstorms are likely to produce strong tornadoes. Microburst activity is almost always found in the vicinity of such storms.

Bow echo

Usually, high winds push out of a strong thunderstorm to form the shape of a bow (as in archery). During the mature stage of a bow echo, tornadoes and microbursts could occur simultaneously. The outburst winds tend to dissipate rather quickly, thereby drying the source region.

Isolated shower

An isolated shower, with or without thunder, can induce microbursts. The microbursts that caused the accidents described in case studies II-3 and II-6, were from isolated showers that were relatively small and short-lived.

Cumulus

Large cumulus and altocumulus, which produces rain or snow, can produce microbursts one of three ways.

Mushroom cloud. Due to the echo top being much higher than the inflow height, the mound-shaped top is not affected by the downflow. In turn, the shape of a mushroom remains during the microburst.

Sinkhole cloud. This is a cumulus cloud with a small vertical growth. The entire cloud is imbedded inside the inflow layer of an induced microburst. A sinkhole develops atop the cloud, and directly above the microburst downflow in mature and post-mature stages.

Giant anteater cloud. This is a cumulus or altocumulus cloud with a glaciated top. When a downflow forms on the upwind side of the cloud, the cloud base lowers, turning into the shape of giant anteater head. A microburst descends from the head section, which keeps lowering until reaching the ground. When a surface microburst forms, the entire head section descends to the ground, turning into a headless giant anteater. I think some scientists let their imaginations run wild with this one.

TEMPERATURE PARAMETERS

Most macrobursts are accompanied by a dome of high pressure, induced by rain-cooled air. Along the leading edge of the front of a macroburst, a pressure surge or jump takes place, as well as gusty winds and a temperature drop.

In microbursts, however, a strong downflow descends very close to the ground before spreading. The downflow air warms up dry-adiabatically, all the way to the ground, unless imbedded raindrops evaporate fast enough to maintain a moist-adiabatic descent. However, this is very unlikely in a strong microburst. The air temperature in microbursts can be either warmer or colder than their environment. The surface pressure can also be higher or lower than the environmental pressure, because the outburst winds lose their pressure head while being accelerated outward from the microburst center. Thus, the changes in the meteorological parameters in microbursts are very complicated.

LOW-LEVEL WINDSHEAR

Windshear is a change in wind speed, direction, or both over a short distance. There are several weather phenomena that produce such conditions. These include thunderstorms, fronts, radiation inversions, funneling winds, and mountain waves.

Thunderstorms

As discussed in the previous section, a downdraft exiting the base of a thunderstorm spreads outward in all directions. This forms a gust front that can extend 10 to 15 miles away from the source region. Extreme windshears of 10 knots per 100 feet of altitude have been measured immediately behind a gust front, while horizontal windshears of 40 knots per mile have been recorded across such an area. The most severe thunderstorms can produce directional shears of 90 to 180 degrees.

An aircraft passing through the gust front and downdraft would encounter not only a rapid change in the horizontal wind field but also a downward vertical motion. The latter motion, can add or subtract hundreds or even thousands of fpm to the descent or climb rate of the airplane.

Fronts

Winds can be significantly different in the two air masses that meet to form a front. Those most conducive to windshear are fast moving, 30 knots or more, have at least a 10 degree F (5 degree C) temperature differential, or both.

Windshear can occur with a cold front after the front passes. Because cold fronts have a greater slope and normally move faster than warm fronts, the duration of low-level windshear is usually less than two hours. However, windshear associated with a warm front is more dangerous due to the strong winds aloft. This might cause a rapid change in wind direction and speed where the warm air overrides the cold, dense air near the surface. Windshear might persist for more than six hours ahead of the warm front because of the front's shallow slope and slow movement.

Radiation inversion

This condition can start to form at sundown, and reaches its maximum intensity just before sunrise. It then dissipates, caused by daytime heating. The cooling of the earth creates a calm, stable dome of cold air that is 300 to 1000 feet thick, known as an *inversion layer.* Speeds of 30 knots are common above the top of this layer, and speeds in excess of 65 knots have been reported. Anytime a radiational inversion is present, the possibility of low-level windshear exists.

Funneling winds and mountain waves

When strong prevailing winds force a large mass of air through a narrow space, such as a canyon, it accelerates and spills out into a nearby valley or open space. These winds sometimes reach velocities of in excess of 80 knots. Refer to case study II-4 for an interesting analysis of this type of windshear.

INDICATORS OF WINDSHEAR

1. A recent cold frontal passage or an impending warm frontal passage might produce a windshear. The isobars around the frontal system at the surface yield a good approximation of the directional shear that will be encountered. For example: The surface winds at airport A are reported to be 320 degrees at 25 knots. The surface winds at airport B are reported to be 040 degrees at 20 knots. The direction of the isobars in the warm sector (from about 220 degrees) is an estimation of the wind direction above both the cold- and warm-frontal surfaces. Therefore, a pilot flying in or out of either airport can expect winds to be from about 220 degrees above the frontal surfaces. Shearing should be encountered from around the same direction, but below the frontal surfaces.

2. Abnormal power setting and rate of descent signal a possible windshear. Fluctuations in the indicated airspeed and the vertical velocity indicator always accompany windshear. Another determinant is a large difference between indicated airspeed and groundspeed. Any rapid changes in the relationship between the two represent a windshear.

3. Inertial navigation system (INS) comparisons are good indicators. Crews can compare the wind at the initial approach altitude with the reported runway surface wind to see if there is a windshear situation present. But remember, INS winds are in degrees true; tower winds are in degrees magnetic. This will make little difference at airfields where variation is only a few degrees, but it makes a considerable difference when variation is 20 degrees or greater.

FLYING THROUGH WINDSHEAR

Anticipating abnormal flight characteristics is key to successfully flying through a windshear encounter. For example, according to NASA studies, when large aircraft encounter low-level windshear, the oscillations near the jet's phugoid—caused when an aircraft is displaced along the longitudinal axis—frequency are significantly amplified. Therefore, longitudinal wind gusts might result in severe airspeed oscillations that could be difficult to recover from. Other examples are the more common expectations of airspeed fluctuations and rapid rates of descent.

Since this text is not designed to be a flight manual, actual flying techniques will not be discussed. However, I highly recommend that you study the current AIM and talk with a CFI for specific methods and practices.

CASE STUDY REFERENCES

II-2, II-3, II-5, II-6

CHAPTER REVIEW

Microburst research

- Fujita studies
- Downburst

Macroburst

- Large downburst
- Excess of 2.5 miles

Microburst

- Small downburst
- Less than 2.5 miles

Types of microbursts

- Wet
- Dry
- Traveling
- Radial
- Twisting
- Surface
- Outflow
- Rotor

Types of parent clouds

- Anvil
- Supercell
- Bow echo
- Isolated shower
- Cumulus
- Mushroom
- Sinkhole
- Giant anteater

Temperature parameters

- Macroburst: drop
- Microburst: rise or drop

Low-level windshear

- Thunderstorm
- Fronts
- Radiation inversion
- Funneling wind
- Mountain wave

Indicators

- Isobar shift
- Abnormal power setting/rate of descent
- INS

Flying techniques

- Anticipate abnormal flight characteristics
- Consult AIM and CFI

CHAPTER REFERENCES

Caracena, Fernando, Ronald Holle and Charles Doswell, III. February 1990. *Microbursts. A Handbook for Visual Identification*. National Oceanic and Atmospheric Administration.

Fujita, T. Theodore. 1985. *The Downburst: Microburst and Macroburst*. University of Chicago.

"How Safe Is This Flight?" April 24, 1995. *Newsweek*: 18–29.

Lansford, Henry. April 1987. "Avoiding Hazardous Weather." *Air Line Pilot*: 19–28.

Melvin, William W. November 1994. "Windshear Revisited." *Air Line Pilot*: 34–38.

Schiff, Barry. March 1982. "The Invisible Menace." *AOPA Pilot*: 113–116.

9
Icing conditions

AFTER COMPLETING THIS CHAPTER, YOU SHOULD BE ABLE TO:

1. Discuss the icing hazards associated with stratiform, cumuliform, and cirroform clouds.
2. Explain the flying hazards associated with clear, rime, and mixed ice.
3. Discuss the hazards of structural icing on fixed-wing and rotor-wing aircraft.
4. Discuss the hazards of induction icing.

STRATIFORM CLOUDS

Icing in middle and low-level stratiform clouds usually is confined to a layer between 3000 and 4000 feet thick. The intensity of the icing generally ranges from a trace to light rime or mixed, with the most significant amounts occurring in the upper portions of the cloud. The primary hazard lies in the great horizontal extent of some of these cloud decks. High-level stratiform clouds are composed mostly of ice crystals and produce little icing.

In thick stratified clouds, concentrations of water droplets are usually greatest where there are warmer temperatures. Therefore, the potential for significant icing can normally be found at or slightly above the freezing level. In layer-type clouds, continuous icing conditions are rarely present at more than 5000 feet above the freezing level, and are ordinarily 2000 or 3000 feet thick.

CUMULIFORM CLOUDS

Cumuliform clouds characteristically build vertically, therefore, icing conditions can be found at a greater range of altitude than with other cloud types. The most common forms are clear and mixed, and usually develop in the upper level of maturing cumulus. Depending on a particular cloud's stage of growth, the spectrum of icing intensities can be a trace in a small puffy cumulus to severe in a large towering cumulus or cumulonimbus.

The unstable conditions associated with cumuliform clouds can produce and maintain large water droplets that are conducive to the formation of clear ice. The powerful updraft can carry a sizable amount of liquid water well above the freezing level. There have been occasional reports of pilots encountering ice between 30,000 and 40,000 feet where the free-air temperature was colder than −40 degrees C.

CIRRIFORM CLOUDS

Aircraft icing rarely occurs in cirrus clouds even though some contain a small level of water droplets. However, light icing has been reported in the dense, cirrus, anvil-tops of cumulonimbus where updrafts might maintain considerable water at rather low temperatures.

STRUCTURAL ICING

Two conditions must be present for structural icing to form on an airplane in flight. First, the aircraft must be flying through visible liquid water. Clouds are the most common form of this moisture. Although icing can occur when the humidity level is very high, it most likely takes place when supercooled water droplets are present. Second, the free-air temperature and the aircraft surface temperature must be 0 degrees C or below.

Supercool conditions

A supercool water droplet forms in subfreezing temperatures and is considered to be in an unstable liquid state. When it hits the surface of an already cold airplane, part of the drop freezes instantly, thereby creating a rapid rate of ice accretion on exposed surfaces.

Research studies have shown that water droplets in the free-air do not freeze at 0 degrees C. Instead, their freezing level varies from −10 degrees to −40 degrees C. The smaller the droplet the lower the freezing point. Although a common line of reference is that severe icing conditions are rare in clouds with temperatures below −20 degrees C, pilots should never totally rely on such a general rule. As long as there is visible moisture and the temperature is below freezing, the potential for icing still remains.

Free-air temperature

Scientists have discovered during wind-tunnel experiments that when saturated air flows over a stationary object, ice might form on the object when the free-air temperature is as high as 4 degrees C. This happens because the temperature of the object is cooled by evaporation and pressure changes in the moving air currents. Conversely, the object is heated by the friction and impact of the water droplet. Therefore, when an aircraft travels at less than 400 knots TAS, the cooling and heating effects tend to neutralize, which causes the possibility of structural icing to occur at or below 0 degrees C.

Types of structural icing

There are four basic types of structural icing: clear, rime, mixed, and frost. How each type is formed depends upon the size of the water droplet and the temperature. When a supercooled water droplet freezes to a surface, the latent heat, the quantity of heat released by a substance undergoing a change of state, that was generated by the fusion process raises the temperature of the unfrozen portion of the droplet to the melting point. Aerodynamic effects, such as airspeed and wind velocity, might cause that unfrozen portion to freeze. The way in which that occurs determines the type of icing.

Clear ice. Clear ice is easily recognizable for its shiny, glazed appearance, and can freeze with either a smooth or rough texture. Formation is most likely to occur in the presence of large rain drops and when the temperature is between 0 degrees C and –10 degrees C. However, the temperature can be as cold as –25 degrees C if favorable conditions are encountered in any cumuliform cloud. Clear ice can be found in widespread, winter systems that produce altostratus and nimbostratus cover. The continuous rain associated with such conditions can spread over thousands of square miles, causing the long-term potential for ice formation.

After large, supercooled droplets make contact with the aircraft, they tend to flow out over the surface, gradually freezing as a smooth sheet of solid ice. Clear ice is considered to be the most serious type of structural ice because it sticks so firmly to an object as it spreads beyond the surfaces of the aircraft that are protected by deice/anti-ice systems. For clear ice to form rough and irregular edges on a surface, it must be mixed with either snow, ice pellets, or small hail. The whitish layer of deposits are usually shaped with blunt, uneven protrusions that bulge out against the airflow.

It should be noted that clear ice can occur on an aircraft still on the ground. Obviously, exposure to freezing precipitation is the most common means, however, water or slush that is splashed on an unprotected surface can just as easily turn into a coating of clear ice.

Rime ice. Rime is the most common type of icing and is usually encountered in lower-level stratus clouds. It forms into a milky, opaque, and granular consistency that leaves a rough surface. It develops when small, supercooled droplets, which are found in stratiform clouds, fog, or light drizzle, instantaneously freeze on impact. For the ice to accumulate, the temperature is generally between –15 degrees and –20 degrees C, although it can form when temperatures are as low as –40 degrees C in cumulus-type clouds and near thunderstorms.

Because the droplets retain much of their spherical shape and freeze so rapidly, large amounts of air become trapped, giving the ice a milky-opaque appearance. The trapped air accounts for the granular texture, which is quite brittle to the touch. The rough residue is a result of the droplets colliding into each other and freezing.

The particular design of the ice formation depends on the airflow over the aircraft's surface, and the length of time that the conditions are favorable for an appreciable accumulation. If there is a heavy build-up of ice, the distortion around the airfoil will change or deflect a droplet's impact pattern, which, in turn, will further distort the surface of the airfoil. A common occurrence is when the ice accumulates above and below the leading edge of the airfoil, causing a design similar to the horns on a ram.

Although rime ice significantly degrades the airflow over control surfaces, it rarely spreads like clear ice. Therefore, it tends to remain localized and relatively easy to remove by conventional methods. Nevertheless, the development of rime ice, as with all types of ice, should always be monitored closely and regarded as a serious threat to the safety of a flight.

Mixed ice. Mixed ice is a combination of clear and rime that can form rapidly when the supercooled droplets vary in size or when liquid drops merge with snow or ice particles. These ice particles can then become imbedded in a layer of clear ice, creating a very rough surface. The temperature range is usually between –10 degrees and –15 degrees C. Sometimes it takes on a unique mushroom shape on the leading edge of wings, or covers unprotected surfaces.

Frost

Frost is created by crystalline ice and usually leaves a thin layer of clear or whitish residue. It forms on aircraft when the temperature of an exposed surface is below freezing while the free-air temperature is slightly warmer. The ice usually forms during night radiational cooling which explains the scraping nightmare that sometimes awaits us when we arrive to our airplanes for an early morning flight.

A common misconception is that frost appears only at ground level. It can form in flight when a cold aircraft descends from subzero temperatures to a warmer and moist altitude below. The air is chilled suddenly to a subfreezing temperature by contact with the cold aircraft. Sublimation, the formation of ice crystals directly from water vapor, then occurs, possibly causing frost to develop over the windshield or other exposed surfaces.

Frost is truly a deceptive form of icing because we often disregard it as inconsequential to the safety of a flight. In reality, *any* surface contamination can affect the lift/drag ratio of an airplane. Never assume that it will blow off or melt on takeoff.

Cold Soaking

The effect of underwing frost on transport-type aircraft can seriously compromise an airfoil's lift capability. Since high-performance jets are exposed to below-freezing ambient temperatures at cruise altitudes, the fuel in the wing tanks will cool to very low temperatures. Under those conditions, the fuel is often referred to as *cold-soaked.*

After landing, the remaining fuel in the wing tanks can cool the lower surface of the wing to sub-freezing temperatures. The size of the affected wing area depends on the location of the tank, the amount of unused fuel, the fuel temperature, and the design of the wing structure and the tanks.

If the ground level humidity is high, moisture in the air will condense on the wings in the location of the cold fuel to create frost. Ice might form if the frost partially melts. As a result, the water then flows toward the fuselage, due to the wing dihedral, and refreezes. The frost usually occurs on the undersurface of the wing, often confined by the front and rear spars, and is thus termed *in-spar* frost.

Ice formed by cold-soaked fuel was causal in a 1989 accident. An Air Ontario Fokker 28 (F28) crashed on takeoff at Dryden, Ontario, resulting in 24 fatalities.

ICE FORMATION ON FIXED-WING AIRCRAFT

Propeller-driven airplanes and helicopters tend to be more susceptible to structural icing than jet aircraft due to their lower airspeeds, which result in less aerodynamic heating. Besides the physical characteristics of slower aircraft, they are subjected to poor weather conditions over longer periods of flight time, partly because they operate at altitudes that are more conducive to those hazards.

Wing surfaces

Ice build-up on a wing or tail surface disrupts the airflow around those airfoils. This results in a loss of lift, an increase in drag, and causes the aircraft to stall at a higher airspeed than normal. A heavy and rapid accumulation can add a lot of weight to the airplane, but the increase in weight *alone* most likely will not cause the aircraft to go down.

Research has shown that only one-half inch of ice on the leading edge of an airfoil can be enough for some aircraft to lose as much as 50 percent of its lifting power and increase its drag by the same amount. Under fairly common conditions, ice can accumulate to a dangerous level in less than two minutes. Consequently, the aircraft might stall much sooner than would be expected, putting the pilot in a very precarious and possibly unrecoverable position.

Horizontal tail surfaces

Tailplane icing can lead to a sudden stall, and must be seriously considered during winter operations. As of 1993, the FAA believed that the potential for icing-induced tailplane stalls was so great on certain turboprops, that the agency issued an Airworthiness Directive (AD) for the EMB-110, Saab SF-340, ATR-42, Jetstream 3101, and YS-11 commuter aircraft.

According to research presented at the FAA- and NASA-sponsored 1991 International Tailplane Icing Workshop, aviation author Dan Manningham compiled an informative analysis concerning this safety hazard. He stated that any time the weather is conducive to a rapid accumulation of ice, tail surfaces are likely to ice up before the wings, and at a faster rate, because the surfaces are much smaller. The size and shape of an airfoil are key to the susceptibility of ice accretion. Furthermore, many tailplanes are engulfed in a high-velocity propwash, causing the surfaces to have a lower ambient temperature than the wings.

Due to those two factors, tailplanes have been known to collect ice three to six times thicker than ice on the wings, and 50 percent thicker than on the windshield-wiper arms. Therefore, if a pilot sees one inch of ice on the wings before turning on the deicing system, there might be up to six inches of ice already on the tail.

Tailplanes generate lift in a downward direction. As a result, they respond to the AOA opposite the wings. The wings need a positive AOA to produce lift. Conversely, the tailplane requires a negative AOA to create its downward lift. Therefore, a loss of lift causes the tail to rise. When the negative AOA exceeds a certain limit, airflow separation occurs on the lower surface and the tailplane stalls.

Since the wing greatly influences the tailplane's AOA, any change in configuration can exacerbate the situation, most notably by flap extension. According to the research, the "net effect is that every increment of wing flap adds to the tailplane's negative AOA and moves it closer to a stall."

Airspeed affects negative AOA. Combined with the increased downwash of flap extension, the tail can come even closer to stalling. Not surprising then, is the fact that the highest negative AOA is usually produced during a landing approach.

Because tailplane stalls occur often, but not exclusively, at low altitudes, it's important to understand the warning signs. Most likely, there will be a buffet through the airframe, yoke, or both. Another indicator is a light feel in the elevator control.

For specific guidance with regards to tailplane icing on your particular type of aircraft, contact a representative from the manufacturer. And, get some tips from pilots who have experienced such an event. Although tailplane icing is not often talked about, except from those who have come through it, it's a very real and potentially fatal safety hazard. Be prepared.

Propellers

The accumulation of ice on the propeller hub and blades reduces the efficiency of the propeller, thereby causing a loss in airspeed. Depending on the severity of the icing, increasing the power setting will not necessarily produce enough thrust to maintain flight. It will, however, burn more fuel, not a good situation either.

The greatest danger associated with this type of icing is from propeller vibration, caused by the uneven distribution of ice on the blades. The propeller is meticulously balanced, and even a small amount of ice can create an imbalance. The resulting vibration, therefore, places stress on the engine mount as well as the propeller itself. Propellers operating with a low RPM setting are more susceptible to icing than those spinning at higher RPMs. Ice also usually forms faster on the hub of the propeller than on the blade because the differential velocity of the blade causes a temperature increase from the hub to the propeller tip.

Pitot tube and static pressure ports

When icing is observed on any part of the aircraft, the pilot should assume that the static ports are accumulating ice, possibly even faster than on the external surfaces. Icing of the pitot tube and other static pressure ports will often form into solid blocks of ice that will cause erroneous readouts of the altimeter, airspeed, vertical velocity, and certain engine instruments.

ICE FORMATION ON ROTARY-WING AIRCRAFT

Icing on rotary-wing aircraft is somewhat unique from that of fixed-wing airplanes due to the speed and cyclic pitch of the rotor blades. Dangerous levels of ice can accumulate faster on helicopters for the following reasons: lower rotational speed of the rotors, a smaller wing surface, the aerodynamic conditions in a hover, the variation of airspeed with rotor blade span, the cyclic pitch changing of the blades, and the cyclic variation of airspeed at any given point on the blade in conditions of forward flight.

Rotor systems

The relatively low rotational speed of the main- and tail-rotor blades, especially while hovering, tends to produce rapid ice accumulation. Consequently, ice formation on a helicopter's main rotor system or anti-torque rotor system might produce a severe vibration, loss of control or dangerously low engine RPMs.

Main- and tail-rotor blades

There are several factors that might reduce ice accretion on the main or tail rotor blades in forward flight, including the centrifugal force of rotation and the bending of the blades during rotation. However, a helicopter is extremely vulnerable to ice when it's hovering. Severe icing can quickly form on the center two-thirds of the main- or tail-rotor blades. And, as little as $\frac{3}{16}$-inch layer of ice is enough to prevent some rotorcraft from maintaining flight.

A severe vibration of the blades might be caused by rough or uneven accumulation of ice on the main rotor-head assembly or pieces of ice breaking away from any part of the rotor system.

Air intake screens

Due to inadequate cooling, ice can build up on the engine and transmission air intake screens at an alarming rate. Freezing water that passes through the screens can coat control cables, which might prohibit throttle movement.

FRONTAL ZONES

About 85 percent of all icing conditions reported occur in the vicinity of frontal zones. For severe icing to form above the frontal surface, the warm air must be lifted and cooled to saturation at temperatures below freezing, causing it to contain supercooled water. If the warm air is unstable, icing might be sporadic; if it is stable, icing might be continuous over an extended area. Icing could form in this manner over either a warm frontal or a shallow cold frontal surface. A line of showers or thunderstorms along a surface cold front might produce icing, but here the icing will be in a comparatively narrow band along the front.

Icing below a frontal surface outside of the clouds occurs most often in freezing rain or drizzle. Precipitation forms in the relatively warm air above the frontal surface at temperatures above freezing. It falls into the subfreezing cold air below the front, becomes supercooled, and subsequently freezes on impact with the aircraft. Freezing drizzle and rain occur with both warm fronts and shallow cold fronts. Icing in freezing precipitation is especially hazardous since it often extends horizontally over a broad area and can extend downward to the surface.

Terrain

Icing is more probable and more severe in mountainous regions than over other terrain. Mountain ranges cause upward air motions on their windward side, and these vertical currents support large water droplets that would fall as rain over level terrain. The movement of a frontal system across a mountain range combines the normal frontal lift with the upslope effect of the mountains to create extremely dangerous icing zones.

Seasons

Icing can occur during any season of the year, but in the temperate climates of the contiguous United States it is most frequent in the winter. The freezing level is nearer to the ground in the winter months than in summer, leaving a smaller low-level layer of airspace free of icing conditions. Frontal activity is more common in winter due to the widespread and extensive cloud systems.

Geographic regions at higher altitudes, such as Canada and Alaska, normally have the most severe icing conditions in spring and fall. During winter, the air is usually too cold in the polar regions to contain heavy concentrations of moisture necessary for icing. Furthermore, most cloud systems are stratiform and composed of ice crystals.

INDUCTION ICING

Ice frequently forms in the air intake of an engine, preventing a sufficient amount of air from entering to maintain combustion. Induction icing is particularly insidious because it doesn't need visible water droplets to develop, and therefore, can appear even on a clear and warm day. The range in temperature changes varies considerably with different types of engines (piston vs. jet), but generally speaking, if the free-air temperature is 10 degrees C or below, and the relative humidity is high, the potential for induction icing exists.

Carburetor icing

This type of ice forms during vaporization of fuel combined with the expansion of air as it passes through the carburetor. If the relative humidity of the outside air being drawn into the carburetor is high, ice can develop inside the carburetor when the temperature is as high as 22 degrees C (72 degrees F). The temperature drop in the carburetor is usually 20 degrees C or less, but can go as low as 40 degrees C.

Provided a certain level of moisture is present, ice will form in the carburetor passages if cooling is sufficient to bring the temperature inside the carburetor down to at least 0 degrees C. It is likely that ice will develop at the discharge nozzle, in the venturi, or around the butterfly valve.

The carburetor heater is an anti-icing device, not a de-icing system; therefore, its primary function is to prevent icing. When the heater is turned on, the air is heated before it reaches the carburetor and keeps the fuel/air mixture above the freezing point. It might be able to melt small amounts of ice and snow as they enter the intake. Because carburetor heating can adversely affect aircraft performance, use it only as outlined in your OPS manual.

Fuel system

Water easily mixes with jet fuel, therefore, the fuel absorbs considerable water when the air humidity is high. Occasionally, enough water is absorbed to create icing of the fuel system when fuel temperature is at or below the freezing temperature of water.

Induction system

There is a potential for ice to form in the induction system any time atmospheric conditions are favorable for structural icing, meaning visible liquid moisture and freezing temperatures. It can develop in clear air when the relative humidity is high and the free-air temperatures are around 10 degrees C or colder.

Air intake ducts

Duct icing can develop under conditions similar to carburetor icing; the presence of supercooled water droplets or humid and above freezing temperatures.

The air pressures going into the intake system are much lower when taxiing and during takeoff and climbout. Therefore, the temperature can drop to produce either condensation or sublimation. As a result, ice can form in the duct which decreases the radius of the opening and limits the air intake.

Inlet guide vanes

As supercooled water droplets freeze into ice on the inlet guide vanes, the air flow is reduced to the powerplant. This causes a decrease in engine thrust, and in extreme cases, the eventual failure of the engine. Once this icing condition develops, there's a real threat that chunks of ice ahead of the compressor inlet might be ingested into the engine, resulting in severe damage.

CASE STUDY REFERENCE

I-1

CHAPTER REVIEW

Stratiform clouds

- Trace to light rime, or mixed
- Significant amounts of ice in the upper portions of the cloud
- Continuous icing is rare at 5000 feet above freezing level

Cumuliform clouds

- Icing conditions can be found at a greater range of altitude
- Clear and mixed are most common

Cirroform clouds

- Icing can be found in the cirrus portion of cumulonimbus clouds

Structural icing

- Visible liquid water
- Supercooled water droplets
- Free-air temperature must be at least 0 degrees C

Supercooled conditions

- Forms in subfreezing temperatures
- Unstable liquid state
- Rapid rate of ice accretion
- Freezing level varies between −10 degrees C to −40 degrees C

Free-air temperature

- Ice might form at temperatures as high as 4 degrees C
- Object is cooled by evaporation and pressure changes in the moving air currents
- Object is heated by the friction and impact of water droplets
- Possibility of structural icing at or below 0 degrees C, and when traveling slower than 400 knots TAS

Types of structural icing

- Clear, rime, mixed, and frost
- Dependent upon the size of the water droplet and temperature

Clear ice

- Shiny, glazed appearance
- Large rain drops
- 0 degrees C and –10 degrees C
- Altostratus, nimbostratus, and cumuliform clouds
- Most serious type of icing

Rime ice

- Milky, opaque, and granular consistency
- Small, supercooled droplets
- Stratiform clouds, fog, light drizzle
- Develops between –15 degrees C and –20 degrees C
- Usually doesn't spread like clear ice

Mixed ice

- Combination of clear and rime ice
- Liquid drops merge with snow or ice particles
- Temperature ranges between –10 degrees and –15 degrees C

Frost

- Crystalline ice
- Thin layer of clear or whitish residue
- Forms during night radiational cooling
- Can form in flight and on the ground
- Sublimation
- Scrape off aircraft before flight

Cold-soaking

- Sub-freezing fuel temperatures
- Frost, ice, or both can form on undersurface of wings, over tanks

Ice formation on fixed-wing aircraft

- More susceptible at low airspeeds

Wing surfaces

- Disrupts airflow
- Loss of lift, increase drag
- Higher stall airspeed
- Added weight
- One-half-inch of ice on leading edge of an airfoil can reduce lifting power and increase drag by 50 percent

Horizontal tail surfaces

- Ice-induced tailplane stall
- ADs
- Forms faster than on wing
- 3 to 6 times thicker
- Tail requires negative AOA
- Airflow separation
- Flaps and speed affect negative AOA

Propellers

- Reduced efficiency
- Loss of airspeed
- Increase fuel burn
- Propeller vibration
- Low RPM settings are more susceptible
- Ice forms faster on propeller hub than blades

Pitot tube and static pressure ports

- Forms into solid blocks of ice
- Erroneous readout of the altimeter, airspeed, vertical velocity, and certain engine instruments

Ice formation on rotary-wing aircraft accumulates faster than fixed-wing airplanes because:

- low rotor speed
- smaller wing surface
- aerodynamic conditions in hover
- variation of airspeed with rotor blade span
- cyclic pitch changing of blades
- cyclic variation of airspeed

Rotor systems

- Rapid ice accumulation
- Severe vibrations
- Loss of control
- Dangerously low engine RPMs

Main- and tail-rotor systems

- Centrifugal force of rotation
- Bending of blades during rotation
- Severe icing on two-thirds of blades
- Small amount of ice can mean loss of lift
- Severe vibrations

Air intake screens

- Rapid build-up

Frontal zones

- 85 percent of all icing conditions occur near frontal zones
- Warm air lifts and cools
- Supercooled water
- Unstable air, icing might be sporadic
- Stable air, icing might be continuous and widespread
- Shallow cold frontal surface
- Line of thunderstorms might produce narrow band of icing
- Below frontal zone and outside of clouds, freezing rain or drizzle
- Most hazardous in freezing precipitation

Terrain

- Severe and probable in mountainous regions

Seasons

- Icing can occur in any season
- Most common in winter due to widespread and extensive cloud systems

Induction icing

- Can form on a clear and warm day
- Temperature range is usually 10 degrees C or below and high relative humidity

Carburetor icing

- Forms during vaporization of fuel, combined with the expansion of air as it passes through carburetor
- Can form with temperature as high as 72 degrees F
- Carburetor heater is anti-ice device, not a deice system

Fuel system

- Can form when fuel absorbs water and the humidity is high

Induction systems

- Similar conditions as structural icing
- Can develop in clear air when the humidity is high, and the free-air temperature is 10 degrees C or below

Air intake ducts

- Similar conditions as carburetor icing

Inlet guide vanes

- Decrease engine thrust
- Chunks of ice might be ingested into engine

CHAPTER REFERENCES

Bragg, M.B., D.C. Heinrich, W.O. Valerezo, and R.J. McGhee. November-December 1994. "Effect of Underwing Frost on a Transport Aircraft Airfoil at Flight Reynolds Number." *AIAA Journal of Aircraft.*

Hansman, R. John, Kenneth S. Breuer, Didier Hazan, Andrew Reehorst, and Mario Vargas. January 11, 1993. *Close-up Analysis of Aircraft Ice Accretion.* NASA Technical Memorandum 105952. 31st Aerospace Sciences Meeting and Exhibit, American Institute of Aeronautics and Astronautics. Washington, D.C.

Manningham, Dan. December 1993. "Watch Your Tail." *Business & Commercial Aviation*: 65–68.

National Transportation Safety Board. *Safety Report: Aircraft Icing Avoidance and Protection.* September 9, 1981. Washington, D.C.

10
Turbulence

A<small>FTER</small> COMPLETING THIS CHAPTER, YOU SHOULD BE ABLE TO:

1. Discuss convective turbulence.
2. Discuss mechanical turbulence.
3. Discuss mountain range turbulence.
4. Discuss clear-air turbulence.
5. Explain the flying hazards associated with each type of turbulence.

Turbulence is created from a change in the flow of air currents over a short distance. A knowledge of the location and causes of turbulence is helpful in minimizing its effects or avoiding it altogether.

CONVECTIVE TURBULENCE

Convective currents are a common cause of turbulence, especially at low altitudes. These currents are localized, vertical air movements in ascending and descending motion. For every rising current, there is a compensating downward current. Those downward currents frequently occur over a wide area, and therefore, usually have a slower vertical speed than the rising currents.

Distinct and well-formed convective currents are most active on warm, summer afternoons when the winds are light. Those currents usually dissipate, however, when there are strong gusts. Heated air at the surface creates a shallow, unstable layer, and the warm air rises. Convection increases in strength and to greater heights as surface heating increases. Barren surfaces, such as sandy or rocky wastelands and plowed fields become hotter than open water or ground covered with vegetation. Thus, air at or near the surface heats unevenly. Because of this, the strength of the convective currents can vary considerably within short distances.

As air moves upward it cools by expansion. A convective current continues upward until it reaches a level where its temperature cools to the same as that of the surrounding air. If it cools to saturation, a cloud forms. Pilots should, therefore, associate thermal turbulence with cumulus and cumulonimbus clouds. As a general rule, turbulence might be severe beneath or in the clouds, while the air above the clouds is usually smoother. Remember too, that dry air can also produce convective currents even though moist conditions do not exist for the presence of cumulus clouds. Pilots will have little indication of those currents until they encounter the turbulence.

MECHANICAL TURBULENCE

Turbulence can occur when air near the surface flows over rough terrain or other obstructions. The higher the wind speed or the rougher the ground surface, the greater the turbulence intensity. Unstable air allows larger eddies to form than those in stable air, but the instability breaks up the eddies quickly, while in stable air they dissipate slowly.

Variability of wind near the surface is an extremely important consideration during takeoff and landing. If the wind is light, eddies tend to remain as rotating pockets of air near the windward and leeward sides of nearby buildings. But, if the wind speed exceeds about 20 knots, the flow might be broken up into irregular eddies, which are carried a sufficient distance downstream to create a hazard in the landing area.

MOUNTAIN RANGE TURBULENCE

When winds blow across rugged hills or mountains, the resulting turbulence might increase as the wind speed increases. Extreme caution is necessary when crossing mountain ranges under strong wind conditions. Severe downdrafts can be expected on the lee side. Pilots should allow for this possibility when approaching mountain ridges against the wind. If the wind is strong, and the ridge line is sharp, pilots should climb their aircraft to a crossing altitude several thousand feet higher than the highest obstruction.

It is important to climb well before reaching the mountains to avoid having to climb, or not climb at all, in a menacing downdraft. Attempting to cross at a lower altitude will also subject the aircraft to much greater turbulence and sudden crosswinds caused by winds blowing suddenly parallel to the valley instead of in the prevailing direction.

When the wind blows across a valley or canyon, a downdraft will occur on the lee side, while an updraft will be present on the windward side. The mountains funnel winds into valleys, thus increasing wind speed and intensifying turbulence. Although canyon flying should never be attempted, if you do find yourself in such a precarious situation the safest flight path is along the windward side.

If the wind blows across a narrow canyon or gorge, it will veer down into the canyon. Turbulence will be found near the middle and downwind side of the canyon. Pilots must avoid the downwind side of narrow canyons, because they could encounter an unrecoverable rate-of-descent.

MOUNTAIN WAVE TURBULENCE

When stable air blows across a mountain range, a phenomenon known as a *mountain wave* might occur. It usually develops when the wind component flowing perpendicular to the top of the mountain exceeds 25 knots. The waves, which resemble ripples, remain almost stationary while the wind blows through them. Although they are usually associated with high mountain ranges, particularly the Colorado and Canadian Rockies in North America, the development of mountain waves can be found above any mountain with a crest of at least 300 feet.

The most dangerous features of a mountain wave is the extreme turbulence and high velocity updrafts and downdrafts found on the lee side of a mountain range. Researchers have documented areas of updrafts and downdrafts that have extended over 70,000 feet and as far as 300 miles downwind from the mountain range. The velocity and intensity of the waves, however, decrease the further they are from the primary source region.

Three kinds of cloud formations are associated with the presence of a mountain wave: cap, rotor, and lenticular.

Cap cloud

Cap cloud is a low hanging cloud with its base near the peak of a mountain. Although most of the cloud is visible on the windward side, it appears to have fingers pointing downward when viewed from the leeward side of the mountain.

Rotor cloud

A rotor cloud looks like a line of cumulus clouds floating parallel to the ridge line of a mountain. Depending on your vantage point, you can actually see the air currents rotating in a swirl of turbulence. The cloud is usually stationary, but is constantly forming updrafts and downdrafts of up to 5000 feet per minute. Downdrafts just to the lee of the mountain ridges and to the lee of the rotor cloud itself are the areas with the most severe turbulence.

Lenticular cloud

Many pilots associate the lens-shaped lenticular cloud with a mountain wave. Like the mountain wave, lenticular clouds are stationary and constantly forming in bands parallel to the mountain. Lenticulars tend to form at fairly regularly spaced intervals, horizontally or vertically, on the leeward side. These clouds are normally found above 20,000 feet, and are turbulent whether they are smooth or ragged in appearance.

Besides the obvious hazards of severe turbulent conditions, significant pressure altimeter errors are also associated with mountain waves. Due to the venturi effect of high winds over a mountain range, barometric pressures are considerably lower in a wave. Therefore, if a pilot is encountering a strong wave, the altimeter readout might be as much as 2500 feet higher than the actual altitude.

FLYING GUIDELINES AROUND A MOUNTAIN WAVE

First rule: Avoid any area that you even suspect is a mountain wave. But if you must fly, then the following guidelines are food for thought:

1. Avoid flying in the vicinity of cap, rotor, or lenticular clouds. It might feel like an "E" ticket at Disney World, but you'll be too scared to enjoy it.

2. As a minimum, fly at an altitude at least 50 percent higher than the height of the mountain range. This might keep the aircraft out of the worst turbulence, and could provide a margin of safety if a strong downdraft is encountered.

3. Approach the mountain range at a 45 degree angle. This enables you to make a quick getaway if you encounter a strong downdraft.

4. Be aware of possible pressure altimeter errors.

5. Penetrate turbulent areas at airspeeds recommended for your aircraft.

CLEAR-AIR TURBULENCE

The rough, bumpy air that sometimes buffets an airplane in a cloudless sky is referred to as *clear-air turbulence* (CAT). Contrary to the name, studies have shown that only 75 percent of all CAT encounters are in clear weather. Pilots can also experience this turbulence in cirrus clouds and haze layers, but there are usually no visual signs indicating such activity.

Clear-air turbulence is usually found above 15,000 feet, and in association with a drastic change in wind speed, most notably horizontal or vertical windshears. The presence of CAT is often in the vicinity of the jet stream and can occur in patches averaging about 2000 feet deep, 20 miles wide, and 50 miles long. Although the most severe cases of CAT are usually reported along the jet stream or near an upper level trough, there is always the potential for less dramatic encounters.

Flying guidelines for CAT

Over the years, the results of CAT research have provided a few guidelines for pilots to follow when experiencing turbulence near the jet stream. Most pilots try to immediately find smoother air, but without the help of recent inflight reports, a decision to climb or descend might be delayed. Here's an easy tip: Watch your temperature gauge for one to two minutes. If the temperature is rising, then climb; if it's falling, then descend. If the temperature remains steady, you can either climb or descend.

CHAPTER REVIEW

Convective turbulence

- Low altitudes
- Updrafts and downdrafts
- Common on warm, summer afternoons
- Increases with strength as surface heating rises
- Currents over short distances
- Cumulus and cumulonimbus clouds
- Possible severe turbulence underneath clouds
- Dry air can also produce convective currents

Mechanical turbulence

- Over rough terrain and other obstructions
- Can form in unstable and stable air
- Variability of wind
- Landing hazards

Mountain range turbulence

- Strong wind conditions
- Flying hazards
- Downdrafts on lee side
- Updrafts on windward side
- Can be found near the middle and downwind side of a canyon

Mountain wave turbulence

- Stable air over a mountain range
- Extreme turbulence
- High velocity updrafts and downdrafts

Cap cloud

- Low hanging near mountain peak
- Mostly visible on the windward side

Rotor cloud

- Constantly forming updrafts and downdrafts
- Severe turbulence

Lenticular cloud

- Stationary and constantly forming
- Severe turbulence

Flying hazards

- Erroneous pressure instrument readouts
- Severe turbulence
- Study flying guidelines

Clear-air turbulence

- Clear air, cirrus clouds, and haze layers
- Found above 15,000 feet
- Drastic change in wind speed
- Notable horizontal and vertical windshears
- Near jetstream
- Study flying guidelines

CHAPTER REFERENCES

Chebbi, B. and S. Tavoularis. June 1995. "Interaction of a Ribbon's Wake with a Turbulent Shear Flow." *AIAA Journal.*

Fuller, J.R. March-April 1995. "Evolution of Airplane Gust Loads Design Requirements." *AIAA Journal of Aircraft.*

Keller, John L. December 1978. *Prediction and Monitoring of Clear-air Turbulence: An Evaluation of the Applicability of the Rawinsconde System.* NASA Contractor Report 3072. Washington, D.C.

Kim, J.M. and N.M. Komerath. March 1995. "Summary of the Interaction of a Rotor Wake with a Circular Cylinder." *AIAA Journal.*

Department of the Air Force. 1 January 1982. *Weather for Aircrews.* Washington, D.C.

11
Wake-vortex turbulence

AFTER COMPLETING THIS CHAPTER, YOU SHOULD BE ABLE TO:

1. Understand the causes of wake-vortex turbulence.
2. Apply various flying techniques to avoid wake turbulence encounters.
3. Discuss recent wake-vortex turbulence encounters that involved Boeing 757 aircraft.

Every aircraft in flight generates a certain level of wake-vortex turbulence, produced by counter-rotating vortices trailing from each wing tip. The intensity of the vortex is dependent upon the aircraft's weight, speed, wing span and shape, angle of attack, and certain atmospheric conditions. The strongest vortex usually occurs when the generating aircraft is heavy, clean (gear, flaps, leading-edge devices, etc. are retracted), and slow. These vortices are extremely dangerous since they can cause a following aircraft to roll out of control. Specific instances will be discussed later in the chapter.

WAKE-VORTEX MOTION

According to the results of studies presented at a 1991 wake-vortex conference and the notable accomplishments of researchers Veillette and Decker, a vortex develops in five generally recognized stages. The first, is when a vortex initially forms over the wing as a series of vortices. A dominating pair of vortices absorb the weaker ones, thereby curling up into a "trailing edge vortex sheet." This roll-up occurs within two to four wing spans behind the aircraft. The vortices are not centered at the wingtips because the core of a vortex varies depending on wing design and flap configuration. A trailing vortex wake is created the moment the nose wheel lifts off the ground during rotation, and ends when it touches down on landing.

The second stage pertains to how each vortex affects the other. The vortex from the left wing causes the right wing's vortex to fall, and vice versa. This is due to the generation of lift on the wing that results in an equal and opposite reaction on the airflow, and in turn, induces a downward motion. The actual rate at which the vortices settle will vary. A joint U.S. Air Force, FAA, and NASA flight test study, established that a heavy jet, 300,000 pounds maximum gross weight, 140-foot wing span, 150-knot approach speed, created vortices that dropped at 350 fpm. Whereas, a light transport, 35,000 pounds, 95-foot wing span, 100-knot approach speed, produced vortices that settled at 150 fpm.

A study conducted by NOAA, found that sink speeds of vortices generated by 727, 757, and 767 aircraft depended on aircraft configuration, atmospheric conditions, and airplane/vortex proximity to the ground. For example:

Aircraft	Ground effect	Altitude (feet)	Sink rate (fpm)
727	Dirty	65	642
727	Clean	42	492
757	Dirty	74	594
757	Clean	49	444
767	Dirty	82	558
767	Clean	61	372

The third stage of vortex development depends on the growth of the vortex rotational axis. Although turbulence can be a significant factor in the breakup of a vortex, it is in no way a guarantee that hazardous vortices have dissipated. The core, the airflow diameter around the wing tip, can range in distance from 25 to 50 feet, and have a velocity of more than 90 knots. These currents usually remain close together until the vortex dissipates.

In the fourth stage, the core of the vortex intensifies and becomes well-defined. Before this stage is over, however, the tangential speed is decreased. Vortices will often change orientation and become distorted in the final moments of this stage. The exact cause of this event is unknown at this time, but research is aggressively ongoing.

The fifth stage is the creation of the distorted vortex rings. This is considered to be the least hazardous part of wake-vortex turbulence.

ATMOSPHERIC FACTORS

It has been documented that as vortices reach ground effect at a height of about half that of the aircraft's wingspan, they will push outward. The outflow and downward speed will then be nearly equal. A vortex also tends to bounce up when it touches the ground. In many cases, it will "jump" well over a height of two wingspans of the generating aircraft. Scientists have also determined that if the wind gradient, windshear, or both are weak, the downwind vortex rebounds higher. In strong vertical shears, the upwind vortex bounces even higher. German researchers of this phenomenon believe that this is more or less common, and not a rarity. Therefore, pilots can get caught in a bouncing vortex if they're not anticipating such an event.

Further studies have found that vortices formed in ground effect do not tend to sink. While some experts contend that these vortices might dissipate faster, others have presented data demonstrating that the very complex nature of vortex movement depends on additional factors, not just origin.

Winds significantly affect the motion of vortices. NOAA researchers have learned that the longest-lived vortices are upwind, which usually linger at the approach end of an active runway, and during a crosswind. This is caused because the wind increases the vortex's rotational energy. When the ambient wind speed is greater at the top of the vortex than at the bottom, it supports the upwind vortex rotation. Conversely, the rotation of the downwind vortex is opposite to the wind gradient, thus diminishing the strength of such a vortex.

Researchers also found that downwind vortices had a tendency to climb while moving, which would place a vortex at a higher altitude than most pilots would anticipate, and possibly into the flight path of an aircraft thought to be high enough to avoid a wake vortex. Pilots should be particularly cognizant of a wake environment near closely spaced parallel runways.

A series of NOAA flight tests revealed the correlation between vortex presence and vortex lateral movement. In the first test, scientists used C-130, C-141, and C-5 military aircraft to determine that vortices were most persistent when winds were three to ten knots. Vortices that lasted for as long as 60 seconds were generated in these low-velocity winds. A second test using 727, 757, and 767 aircraft corroborated the previous data. All of the vortices that remained for over 85 seconds were generated when the wind speed was less than five knots. When the wind speed was between five and ten knots, all of the vortices hung for more than 35 seconds. The tangential velocities for those vortices were greater than 200 fps.

Temperature also affects the life of a vortex. Scientists at the Idaho Nuclear Engineering Laboratory determined that vortices that were long-lived, of higher intensity, or both were generated under stable atmospheric conditions. This is most common during the dissipation of a temperature inversion. As the sun begins to warm the layer of air next to the ground, that layer will begin to convect heat away from the surface and into the surrounding layer of air. All of the NOAA flight test data that contained a vortex life cycle greater than 100 seconds, and tangential velocities of more than 240 fps, were found under stable atmospheric conditions.

FLYING GUIDELINES

NOTE: Refer to the current Airman's Information Manual for the latest updates on wake-vortex turbulence avoidance guidelines. A few basic points are discussed below.

Vortices tend to sink immediately below the flight path at a rate of 400 to 500 fpm and can normally peak at 800 to 900 fpm. Therefore, a general rule is that pilots should fly at or above the preceding aircraft's altitude. When vortices sink into ground effect, they move laterally over the ground at a speed of about five knots. A crosswind will influence this lateral movement often to the point where the downwind vortices might gain speed, whereas the upwind vortices remain at the slower speed. On the other hand, a gusty and stronger crosswind might blow the vortex movement across a parallel taxiway or runway. Therefore, it's important to be aware of the type of departing traffic on different runways so you can plan your takeoff or landing accordingly.

WAKE-VORTEX RESEARCH

Wake-vortex turbulence continues to be a serious threat to the safety of all flights, regardless of the type of aircraft involved. As part of a recent study, the NTSB noted that between 1983 and 1993 there were at least 51 accidents and incidents in the United States that resulted from "probable encounters" with wake vortices. The casualty rate included 27 fatalities and 8 seriously injured. Forty airplanes were substantially damaged or destroyed.

Authorities from the United Kingdom have perhaps the most comprehensive data base on wake turbulence. Between 1982 and 1990, there were 515 recorded incidents at London's Heathrow Airport. Researchers have since discovered that there were two separate blocks of altitude, in which the majority of these encounters occurred. One block was between 100 and 200 feet above the runway threshold, and the second was between 2000 and 4000 feet. The latter was attributed to crews leveling-off to intercept the localizer.

The British data also indicated that 747s and 757s as the generating aircraft produced significantly higher incident rates. Following aircraft most affected were DC-9s, 737s and BAC-111s. The authors of the study also concluded that, "the main effect of increasing wake separation distances is to decrease the risk of an incident, rather than its severity."

In a French government study, researchers determined the following facts common to wake turbulence incidents in that country:

1. The surface winds did not exceed eight knots in all wake-vortex encounter incidents.
2. The trailing aircraft was VFR and the pilot knew of the presence of the preceding aircraft.
3. In 80 percent of the cases, ATC informed the pilot of the preceding aircraft.
4. All of the pilots involved with wake encounters held commercial or ATP licenses.

BOEING 757 SAFETY CONCERN

According to the NTSB, a unique and potential safety hazard has developed concerning the wake vortices generated by Boeing 757 aircraft. From December 1992 to March 1994 there have been several accidents and incidents in which an airplane on approach encountered severe wake-vortex turbulence while flying behind a 757. Thirteen occupants died in two of the three accidents. In each mishap, the velocity of the core vortices of the 757 were so strong and violent that they were able to force the following airplanes into an unrecoverable loss of control. In two additional and separate instances, the wake vortex of 757s threw an MD-88 and a 737 into a severe, induced roll. The crews were able to successfully recover, but not before the aircraft dropped dangerously close to the ground.

In light of the recent accidents and incidents, the Safety Board conducted a special investigation to examine the circumstances associated with 757 vortex wake turbulence. The purpose of the report was to determine what improvements might be needed in existing procedures to reduce the likelihood of wake-vortex encounters.

WAKE-VORTEX ACCIDENTS

A common misconception among pilots is that wingtip vortices mostly affect light general aviation aircraft, and leave the big jets alone with not much more than a few bounces over the threshold. But, as the following accidents and incidents illustrate, no aircraft is completely safe from the powerful, and even deadly force, of wake-vortex turbulence.

NOTE: The FAA classifies airplanes as small, large, and heavy based on their maximum takeoff weight. Small airplanes are those that weigh up to 12,500 pounds. Large airplanes weigh between 12,500 and 300,000 pounds. Heavy airplanes weigh more than 300,000 pounds. The NTSB refers to these classifications in the following wake-vortex accident and incident reports.

CASE STUDY 11-1

18 December 1992: a Cessna Citation 550 crashed while on a visual approach to runway 27R at the Billings Logan International Airport, Montana.

Air traffic control provided to the Citation pilot the standard IFR separation of greater than 3 nm from his traffic, a 757 also landing at Billings. About four-and-a-half minutes prior to the accident, and at a distance of 4.2 miles from the 757, the Citation pilot received his clearance for a visual approach. He increased his speed while the crew of the 757 decreased their speed in preparation for landing. The controller advised the pilot of the Citation that the 757 was slowing and gave him the option to make a right turn to increase separation. Although the pilot never asked the controller about his distance from the 757, he apparently recognized how close he was getting to his traffic from a comment taped on the CVR, "almost ran over a seven fifty-seven," about 40 seconds prior to the wake-vortex encounter.

The Citation suddenly and violently entered an uncontrolled left roll about 2.78 nm (74 seconds) behind the 757. Witnesses reported seeing the airplane roll and hit the ground in a near vertical dive. The two crewmembers and six passengers were killed.

According to the investigation, the Citation's path was at least 300 feet below that of the 757 during the last four miles of the approach. The only clue available to the Citation pilot to determine his flight path relative to that of the 757 would have been his visual alignment of the airliner and objects on the ground. However, there could still have been a discrepancy depending on how the 757 pilot was lined up with the runway. For example, if the 757 was aimed at the touchdown zone, then the Citation would have had a similar flight path. If the 757 was aligned with the far end of the runway, the flight path of the Citation would have been below the larger jet. Or, if the 757 was positioned with the approach lights, then the flight path of the Citation would have been above the airliner.

The data revealed that the induced roll started when the Citation was just less than 3 nm from the 757. Investigators determined that had the Citation been exactly 3 nm, or an additional six seconds farther from the 757, the strength of the vortex would not have diminished enough to prevent the wake turbulence. As a result, the Safety Board concluded that lighter weight airplanes in the large category, such as the Citation, require a greater separation distance than three nm when following heavier airplanes in the same classification.

The Safety Board believed that the failure of the Citation pilot to ensure an adequate separation distance from the 757 strongly suggested that he did not realize the potential danger of severe wake vortices. Although the AIM recommends that the pilot of the following airplane should remain above the flight path of the preceding aircraft, the Safety Board noted that their agency was unaware of existing training material that discussed techniques for determining the relative flight paths of airplanes on approach.

CASE STUDY 11-2

1 March 1993: A Delta Airlines MD-88 suddenly encountered an uncommanded 13-degree roll while on a visual approach to runway 18R at Orlando International Airport, Florida.

The Delta jet was more than 4 miles behind a 757 when the crew received their visual approach clearance. They quickly increased their speed and closed the separation to 2½ miles. Shortly thereafter, the Delta crew reported a strong roll to the right. Data collected from the FDR indicated that at about 110 feet agl, the roll angle reached 13 degrees right wing down. The pilot made a rapid correction which deflected the ailerons about 10 degrees, and the rudder at 23 degrees. The crew regained control and successfully landed.

The recorded radar data showed that at the point of upset, the MD-88 was trailing the 757 by 65 seconds and slightly below its flight path. Both aircraft were descending at a three degree angle. The Safety Board determined that had the Delta crew maintained a distance of three nm, an additional 13 seconds of separation would have existed between the two airplanes. Nevertheless, the Delta jet was still below the flight path of the 757, so even with a small increase in distance, they still might have encountered a significant wake vortex.

CASE STUDY 11-3

24 April 1993: A United Airlines 737 encountered an uncommanded roll of 23 degrees while on a visual approach to runway 26L at Denver-Stapleton International Airport, Colorado.

The flightcrew reported that at about 1000 agl, the airplane rolled violently to the left with no yaw, the pitch decreased five degrees, and they lost 200 feet in altitude. They promptly initiated a go-around, and they successfully landed without further incident.

According to the FDR, the pilot rapidly corrected the induced roll with 60 degrees of aileron and 7 degrees of rudder. The collected evidence also showed that at the point of upset, the airplane was at 900 feet agl, and in two seconds its roll angle reached 23 degrees left wing down.

The recorded radar data indicated that the flight path of the United jet was about 100 feet below and 1.35 nm, 32 seconds, behind that of a 757 landing on a parallel runway. The wind was blowing from the north at around 10 knots gusting to 16 knots. Both airplanes had a flight path angle of three degrees.

Runway 26L is displaced 900 feet south of runway 26R. The threshold of 26L is offset about 1300 feet to the east of the threshold of 26R, resulting in a flight path to 26R that is about 70 feet higher than the flight path to 26L. Under the existing wind conditions at the time of the upset, a wake vortex from the 757 would sink and move to the south, toward a standard flight path to runway 26L.

Air traffic controllers are required to provide a 3 nm separation to IFR flights that are approaching 26L and 26R because the runways are divided by less than 2500 feet. However, the United crew accepted a visual approach. Therefore, within 12 nm from the runway both airplanes were on converging courses, prompting ATC to issue a series of S-turns to the United crew for spacing. After completing those maneuvers the separation distances were still not ideal. Laterally, the two airplanes were 4.55 miles apart, but longitudinally they were only .65 nm. Furthermore, the distance between both final approach paths had been reduced to .15 nm. As each aircraft converged to its respective runway alignments, the longitudinal component increased to an in-trail separation of 1.35 nm.

The Safety Board believed that the controller should have recognized the problems associated with tight spacing between parallel runways, and have issued additional S-turn maneuvers to the 737 crew in order to maintain a more acceptable three nm in-trail separation.

CASE STUDY 11-4

10 November 1993: A Cessna 182 encountered a sudden, uncommanded 90-degree roll and pitch up while on a visual approach to runway 32 at Salt Lake City International Airport, Utah.

The pilot of the Cessna reported that he was instructed by ATC to proceed "direct to the numbers" of runway 32 and pass behind a "Boeing" that was on final approach to runway 35. There was no evidence to suggest that the pilot was advised that his traffic was a 757.

The pilot of the Cessna further stated that while on final approach the airplane experienced a "burble" and then the nose pitched up. The aircraft suddenly rolled 90 degrees to the right, as the pilot immediately applied a full-left deflection of the rudder and aileron and full-down elevator in an attempt to level out. As the airplane began to recover, the pilot realized that he was near the ground and pulled the yoke back into his lap. He crashed short of the threshold to runway 32, veered to the northeast, and came to a rest on the approach end of runway 35. The pilot and the two passengers suffered minor injuries, and the airplane was destroyed.

The approach ends of runways 32 and 35 are about 560 feet apart. The recorded radar data indicated that the Cessna was less than 100 feet agl when it crossed the flight path of the 757. The airliner had passed that point just 38 seconds prior to the Cessna. Although the exact position of the upset was not determined, radar evidence suggested that the 182 was flying slightly above the flight path of the 757.

As mentioned earlier in the chapter, when wake vortices get caught in a ground effect, they tend to spread outward at a speed of three to five knots, plus the wind component. In this case, the left vortex of the 757 typically would have spread 200 to 300 feet to the west. The core might have been located about 75 feet above the ground, although researchers have noted the vortex has the potential to "bounce" twice as high as the steady state height. Furthermore, the diameter of the vortex's flow field is usually about equal to the wing span of the generating airplane. Therefore, the 182 could have been affected by the vortex at any altitude between ground level and 200 feet agl.

Although the Cessna's flight path was above that of the 757, the Safety Board believed that the pilot did not adequately compensate for the height of the vortex.

CASE STUDY 11-5

15 December 1993: A Westwind corporate jet encountered a sudden, uncommanded roll and pitch down while on a night, visual approach to runway 19R at the John Wayne Airport, Santa Ana, California.

According to the recorded radar data, the Westwind suddenly rolled and pitched down at a 45-degree angle just prior to impact. The two crewmembers and three passengers were killed. At the point of upset, the aircraft was at 1200 feet msl and 3.5 nm from the end of runway 19R. A 757 was on a final approach for the same runway, 2.1 nm and 60 seconds ahead of the Westwind, and on a flight path that was 400 feet above the smaller jet.

In reference to the CVR tape, the Westwind pilots were aware they were close to a higher-flying Boeing aircraft, and after experiencing a little buffet from its wake decided to fly their own approach at 3.1 degrees of glide slope instead of the standard three degrees. There was no evidence that the crew was advised specifically that they were following a 757.

Since both aircraft were flying in an easterly direction when the crews received vectors to the airport, each airplane had to make a converging right turn in order to set up for the 19R approach. Radar data and ATC voice transcripts showed that the Westwind was 3.8 nm northeast of the 757 when cleared for a visual approach. The Westwind started its right turn from a ground track of 120 degrees while the 757's ground track remained at about 90 degrees. As a result, the closure angles started at 30 degrees and became greater as the Westwind continued its turn.

About 23 seconds later, the 757 was cleared for the visual approach. The average ground speeds of the Westwind and the airliner were about 200 and 150 knots, respectively. The Westwind was established on course 37 seconds prior to the 757. Although the combination of the closure angle and the faster speed of the Westwind reduced the separation distance from about 3.8 nm to around 2.1 nm in 46 seconds, the primary factor in the decreased separation was the converging ground tracks. The only way the pilot of the Westwind could have maintained adequate separation was to execute major aerial maneuvers.

Based on radar data, at the time the visual approach clearance was issued, the separation distance was rapidly approaching the 3 nm required for IFR separation. To prevent a violation, the controller would have had to change the Westwind's track, or ensure that the pilot accept the visual approach within 29 seconds.

The Safety Board's analysis

After a thorough study and analysis of wake-vortex encounters with 757 aircraft, the Board found little technical evidence to support the theory that these vortices are significantly stronger than indicated by its weight. The Safety Board noted that the FAA's aircraft weight class system was established in 1970 based on jet aircraft existing at the time. Except for an update in 1975, no other modifications have been made. Obviously, many transport category turbojet airplanes have been introduced into service since those guidelines were written. Therefore, the Board expressed concern over the following three areas pertaining to wake-vortex upsets: (1) the outdated weight classification, (2) ATC procedures, and (3) pilot knowledge.

Weight criteria

In 1992, NOAA conducted flight tests to determine the characteristics of wake vortices produced by the 757. The results indicated that the 757 generated the highest vortex tangential velocity, 326 fps, of any tested aircraft, including the 747, 767, and C-5A. The most common theory to explain the occurrence was that the 757 wing flap design is different from the other aircraft. Most larger transport category airplanes have gaps between the trailing edge flaps that disrupt the uniform development of the vortex. Whereas, the 757's flaps are continuous from the fuselage to the ailerons, a design that is believed to be more conducive to the formation of a wake vortex.

ATC procedures

With regards to wake-vortex considerations, the FAA provides less radar separation for IFR airplanes under positive air traffic control than recommended by the International Civil Aviation Organization (ICAO) and required by the Civil Aviation Authority (CAA) of Great Britain. For example, a Citation or Westwind following an aircraft like the 757 would require a 5 nm separation under ICAO and 6 nm separation based on CAA standards. The FAA requires a 3 nm separation.

The Safety Board believes that the FAA should prohibit controllers from issuing a visual approach clearance to an IFR airplane operating behind a heavy-category aircraft until they are positive a 3 nm in-trail separation exists. Airspeed changes or S-turns are possible solutions that a controller is recommended to use.

Pilot knowledge

According to the Safety Board, the accident and incident data suggest that a combination of pilots' lack of understanding of the hazards of wake vortices and the difficulty of knowing the movements of these vortices are major contributors to wake-vortex encounters. It is apparent that a pilot's visual estimate of range is not sufficient to accurately judge a safe separation distance. Therefore, the Board encourages the FAA to develop a comprehensive training program related to wake turbulence avoidance, and to make it readily available to the flying community.

CHAPTER REVIEW

- Aircraft weight, speed, wing span and shape, AOA, atmospheric conditions
- Strongest vortex is generated by heavy, clean, slow aircraft.

Wake-vortex motion

- Vortex formation
- Trailing-edge vortex sheet
- Vortex effects
- Sink speeds
- Growth of vortex rotational axis
- Core of vortex
- Creation of distorted vortex rings
- Atmospheric factors
- Ground effect
- Wind gradient
- Temperature

Flying guidelines

- Refer to current AIM
- Vortices sink below flight path between 400 and 900 fpm
- In ground effect, vortices move laterally at five knots
- Crosswind will influence lateral movement
- Strong crosswind might blow vortex across parallel taxiway or runway

Wake-vortex research

- U.S. wake-turbulence encounters
- British data base

- 100 to 200 feet agl and 2000 feet to 4000 feet
- French study

Boeing 757 safety concern

- Case studies
- Weight classification controversy
- ATC procedures
- Pilot knowledge

CHAPTER REFERENCES

National Transportation Safety Board. February 1994. *Special Investigation Report. Safety Issues Related to Wake-vortex Encounters During Visual Approach to Landing*. Washington, D.C.

Subcommittee on Technology, Environment and Aviation. July 28, 1994. *Application of FAA Wake-vortex Research to Safety*. Washington, D.C.

U.S. Department of Transportation. Federal Aviation Administration. March 30, 1995. *Airman's Information Manual*. Washington, D.C.

Veillette, Patrick and Rand Decker, Ph.D. January 1994. "Beware of the Horizontal Tornado. Wake-vortex, Part I. *Air Line Pilot*: 26–31.

Veillette, Patrick and Rand Decker, Ph.D. February 1994. "Beware of the Horizontal Tornado. Wake-vortex, Part II. *Air Line Pilot*: 24–28.

CASE STUDY II-1: USAir Flight 405

Safety issues: icing conditions, procedural deviation, aircraft performance, CRM

On 22 March 1992, Flight 405 crashed into Flushing Bay during takeoff from runway 13 at La Guardia Airport, New York.

Probable cause

The NTSB determined that the probable cause of this accident was the failure of the airline industry and the FAA to provide to flightcrews procedures, requirements, and criteria compatible with departure delays in conditions conducive to airframe icing; and, the decision by the flightcrew to take off without positive assurance that the airplane's wings were free of ice accumulation after 35 minutes of exposure to inclement weather following deicing. Contributing to the cause of the accident were the inappropriate procedures used by, and inadequate coordination between, the flightcrew that led to a takeoff rotation at a lower than prescribed airspeed.

History of flight

Flight 405 was a regularly scheduled passenger flight from Jacksonville, Florida, to Cleveland, Ohio, with an intermediate stop at La Guardia. The Fokker F-28 jet departed La Guardia at 2134 Eastern Standard Time with 47 passengers and 4 crewmembers onboard.

Pilot experience

The captain had 9820 total flight hours, 2200 in the F-28. He had logged 1400 flight hours as an F-28 captain. He received his last proficiency check three months before the accident. USAir also required pilots to complete an annual 9-hour home study course on winterization, followed by a closed book examination. The captain passed the program in November 1991.

The first officer had 4507 total flight hours, 29 in the F-28. He had been a 727 second officer for Piedmont Airlines/USAir from 1989 to 1 February 1992, when he was upgraded to first officer on the F-28. His last line check was completed during his initial operating experience three weeks before the accident. He passed the winterization program in November 1991.

Weather

At 2000, the terminal forecast for the La Guardia area was: A ceiling of 500 feet overcast, visibility ¾ mile with light snow and fog; winds 070 degrees at 10 knots; occasional ceiling 300 feet obscured; visibility ½ mile with moderate snow and fog; chance of ceiling at 1100 feet overcast, visibility 2 miles with light snow and fog.

The following ATIS information was issued at 2124, and was the most current at the time of the accident: Indefinite ceiling, 700, sky obscured, ¾ mile visibility with light snow and fog. Temperature 31 degrees, dewpoint 30 degrees. Wind 110 degrees at 12 knots . . . Braking action advisories in effect. Runway 4/22 closed for snow removal. NOTAM: Runway 13/31 plowed 40 foot either side of centerline. Thin layer of wet snow on surface of runway has been sanded.

The accident

Flight 405 arrived at La Guardia a little more than one hour behind schedule. After the crew took a break and prepared for the next leg, they met back in the cockpit. Neither pilot performed a walkaround inspection of the aircraft, nor were they required to do so by USAir procedures. The first officer described the snowfall as "not heavy, no large flakes." He later told investigators that the windshield heat was on low and snow was sliding off the airplane. He also noticed the aircraft's nose had a watery layer as far as his arm could reach out the window.

USAir records showed that the jet was deiced with Type I fluid with a 50/50 water/glycol mixture at about 2026. One of the two deicing trucks broke down behind the airplane, resulting in a pushback delay of 20 minutes. The captain then requested a second deicing of the aircraft, which was completed around 2100.

At 2105, the flight was cleared to taxi to runway 13. The first officer, who was the non-flying pilot, recalled selecting engine anti-ice for both engines and that there were no visual or directional control problems. The captain announced that the flaps would remain up during the taxi, and he placed an empty coffee cup on the flap handle as a reminder. He later told the first officer that they would use an 18-degree flap setting and a reduced V1 speed of 110 knots, in accordance with the company's procedures for takeoff on a contaminated runway.

As the crew slowly taxied towards the active, the first officer remembered using the windshield wipers "a couple of times," and turning on the right wing inspection light "maybe ten times, but at least three." He told investigators that he looked at the wing, checked the upper surface and the black strip on the leading edge for ice buildup. Based on his observation, there was no contamination, nor had there been a heavy snowfall, so he didn't feel a need for a third deicing. He also recalled that as the airplane approached the number one spot for takeoff, the crew looked back at the wings several times.

At 2134:51, Flight 405 was cleared for takeoff, and about 26 seconds later the first officer made a callout of 80 knots. Shortly thereafter, the first officer called V1. The captain maintained a smooth, gradual rotation to 15 degrees at a normal rate. About seven seconds after VR (124 KIAS), the stickshaker activated, instantaneously followed by several stall warning beeps. The first officer said that he was aware the main landing gear came off the runway, but just as they were entering ground effect he felt a strong buffet develop in the airframe. The aircraft began rolling to the left, "just like we lost lift." As the captain leveled the wings and they headed towards the water, the first officer joined the captain on the controls. He testified that there were "no heavy control inputs," and that they used the right rudder to maneuver the airplane back toward the ground and to avoid the water. The first officer further recalled that they tried to hold the nose up so as to lessen the impact, and he did not touch the throttles. The last thing he remembered was an orange and white building that disappeared under the nose. At 2135, there was a sudden flash and a hard jolt before the aircraft abruptly came to rest partially inverted at the edge of Flushing Bay.

Impact and wreckage path

The initial ground scrape marks from the airplane were approximately 36 feet left of the runway centerline, and ranged from 5 feet to 65 feet in length. Pieces of the left wing were found about 200 feet farther down the runway,

A portion of the aircraft from the nose to just aft of the fourth-passenger row was found upside down and submerged under water. The left side and bottom of the forward section was crushed by the impact, and a hole was in the fuselage skin to the left of the captain's seat.

The section of fuselage between rows 4 and 11, was found floating in the water. The floor was torn and showed signs of fire damage. The remaining part of the cabin was under water, and portions of it had been burned or crushed. The vertical stabilizer and rudder assembly stayed attached to the empennage. However, various parts of the tail section were damaged by either fire or the impact.

Accident survivability. The captain, 1 flight attendant, and 25 passengers sustained fatal injuries. The leading causes of death were drowning and blunt-force trauma. All of the deceased passengers were seated between rows 4 and 11, near the overwing exits, and at row 13.

Nineteen passenger seats had separated from their floor attachments and were found scattered throughout the wreckage, some of which were damaged by fire. Nine seats were never recovered. The seats that were near the front of the cabin appeared to have been less damaged than those in the rear. Many of the survivors became disoriented in the darkened water, and found it difficult to release their seatbelts and make their way to the surface. After the crash, passengers remembered seeing fires in the left forward and aft portions of the aircraft, as well as several small fires on the water. A number of survivors, including the lead flight attendant and first officer, escaped through a hole in the cabin floor.

Due to the snowy and foggy conditions, airport rescue units arrived near the scene about four minutes after the call. Because the aircraft was below a constructed dike, fire crews were initially unable to see the wreckage until they climbed to the top and looked down into the water. Divers from the New York Police Department were also notified, but by the time they were able to conduct an underwater search, the remaining survivors had already been found.

The investigation

After thorough examination of the evidence, the Safety Board determined that the accident was not caused by an improper wing configuration or speedbrake deployment, or mechanical defects. Therefore, the focus of the investigation was on weather-related issues and flightcrew performance.

Aerodynamic performance of the F-28. The Safety Board evaluated simulation data provided by Fokker in order to determine the aircraft's rotation speeds and angle of attack (AOA) during takeoff. An F-28, without wing contamination should lift off about two seconds after the start of rotation, and accelerate about seven knots. Therefore, at a normal pitch attitude of one degree and airspeed of 124 knots, the airplane should lift off as it reached 131 knots and 5 degrees of pitch. The data also showed that the AOA would peak at 9 degrees as the aircraft transitioned to the initial climb. With a stall AOA of 12 degrees in ground effect, an F-28 without wing contamination, should have at least a 3-degree-AOA stall margin during the transition to climb. The margin would naturally increase as the airplane continues to accelerate and establishes a steady climb rate.

In the case of Flight 405, investigators heard sounds on the CVR that correlate with the extension of the main landing gear struts and the nose gear strut. They compared those sounds with the recorded timing data, and were able to determine an accurate airspeed and AOA analysis. The Safety Board concluded that the captain initiated a takeoff rotation when the airplane reached 119 knots, about 5 knots slower than the proper rotation speed. Their analysis showed that the jet would have lifted off at about 128 knots with an AOA of 5.5 degrees. Under those conditions, the AOA probably exceeded 9 degrees as the aircraft transitioned to a normal climb.

According to Fokker wind-tunnel data, a wing upper surface roughness caused by particles of only 1 to 2 millimeters in diameter, and at a density of 1 particle per square centimeter, can result in a loss of lift on the F-28. This can occur 22 percent of the time when the aircraft is in ground effect. Furthermore, when the aerodynamic characteristics of the wing were significantly degraded from contamination, the stall AOA in ground effect was reduced from 12 degrees to 9 degrees. Therefore, the Safety Board believed it was probable that during the transition to climb, immediately after liftoff, Flight 405 reached an AOA beyond the stall limits, which produced a loss of both lift and lateral control effectiveness. The abrupt roll that occurred during takeoff was consistent with this analysis.

F-28 wing contamination. Most wings are designed so the inboard sections will stall before the outboard portions. This design ensures that roll control can be maintained through use of the ailerons on the outboard wing sections. However, the variable disbursement of ice particles over shorter chord lengths of a wing can create an irregular stall distribution across the wing. A premature stall of the outboard sections usually occurs first, followed by a loss of lateral control. Although it is characteristic for a swept wing aircraft to have a significant nose-up pitching moment after its outboard wing stalls from contamination, wind tunnel tests have proven otherwise for the F-28. The airplane's sweep angle is only 16 degrees, therefore, it is most likely that the aircraft would experience a nose-down pitch attitude.

In any event, it was apparent to investigators that Flight 405 was unable to transition to a positive climb angle during the 11 seconds that it was airborne. The maximum airspeed that was recorded was 134 knots, just as the stickshaker activated. The airspeed then fluctuated between 128 and 130 knots until impact. According to the manufacturer's data, the aircraft should have been able to maintain a 3-degree AOA stall margin, *unless* it did not have clean wings. Since the airplane exhibited an abnormal flight performance, the Board considered that to be conclusive evidence that the wing's aerodynamic lift capability was "significantly degraded by an accumulation of frozen contaminant."

Fokker tests. As a result of this accident, Fokker conducted further tests to determine if changes in the F-28 operating procedures could enable the aircraft to successfully take off with wing contamination. They first looked at modifying the rotation speed. Researchers found that if the rotation speed was increased by 10 knots, the peak AOA would decrease approximately 3 degrees, from 12 degrees to 9 degrees. When a relatively slow rotation rate of 2-degrees-per-second was used, the peak AOA decreased from 12 degrees to 8 degrees. Concerned that a pilot should not be expected to control the rotation to such minute precision, Fokker believed that a change in the rotation rate alone might not be adequate.

The second consideration was to modify the target pitch attitude on takeoff. The data revealed that when pitch attitude was lowered from 15 degrees to 10 degrees, the peak AOA decreased approximately 5 degrees, from 12 degrees to 7 degrees. Fokker believed that this procedure proved more effective at lowering the wing AOA than would a slower rotation rate, or increased rotation speed, without imposing associated runway length or takeoff weight performance penalties.

A 10-degree pitch attitude was already approved for the F-28 engine-out procedure. Therefore, with both engines operating, researchers believed that the airplane can successfully climb out of ground effect. Once that occurs, the jet should be able to maintain a 15-degree rotation rate and establish a positive climbout.

Based upon this data, the Safety Board would like to see Fokker conduct additional flight dynamics' studies on pitch attitude, rotation rate, AOA, and wing contamination. The Safety Board's primary concern is how to structure the takeoff maneuver to prevent pilots from stalling the airplane, especially when it has just lifted off and is still in ground effect.

De-icing operations. The Safety Board found that the aircraft had been properly cleared of ice and snow during the two deicing procedures at the gate. However, approximately 35 minutes elapsed between the second deicing and the takeoff roll, during which the jet was exposed to continuous precipitation in below freezing temperatures. Although investigators were unable to determine the exact amount of contamination on Flight 405, they did believe that some accumulation had to have occurred in the span of 35 minutes. Therefore, they concluded that ice contamination led to the control difficulty shortly after rotation.

Flightcrew performance. Although the crew did not perform a walkaround inspection, the aircraft was checked by ground personnel after the first deicing. The Board believed that since the captain requested a second deicing after a 20-minute delay, showed his level of concern and prudence in appropriately dealing with the situation. Following that deicing, the crew most likely was satisfied that the airplane was free of contamination.

When a delay exceeds 20 minutes, USAir procedures dictate a careful examination of the airplane's surfaces. The first officer stated—and passengers confirmed—after the accident that he had turned on the wing inspection light several times. However, the only related comment recorded on the CVR was nearly 30 minutes after departing the gate and about 5 minutes before takeoff, when the first officer said, "looks pretty good to me from what I can see." The observation was made through the wet, closed cockpit window, 30 to 40-feet from the wing. Therefore, the Safety Board believed that did not constitute a careful examination.

The Board recognizes the dilemma all crews face when confronted with the decision to either return to the gate and incur further delays or even cancellation, or proceed with the takeoff and accept the risks. Nonetheless, they believed that the crew of Flight 405 should have taken more assertive steps to assure a contamination-free wing, such as entering the cabin to look at the wing from a closer range. Although the detection of minimal amounts of icing might not have been possible even from that vantage point, it might have afforded the crew additional information that might have prompted them to return to the gate. The Safety Board concluded that the crew's failure to take such precautions, and the decision to attempt the takeoff when they were not positive of the icing conditions, led to this accident.

Furthermore, the Safety Board determined that a V1 speed of 110 knots was not authorized for takeoff. The first officer could not explain why the captain chose that speed, but assumed he was concerned about the airplane's stopping ability on a slick runway. As a result, the selection of a low V1 speed caused the first officer to call VR prematurely. He stated that because V1 and VR are normally the same speed, he inadvertently followed his standard procedure of calling VR immediately after V1.

Data obtained from the CVR and FDR, showed that the VR call made by the first officer occurred at about 113 knots, approximately 11 knots below the correct rotation speed of 124 knots. The first officer noted that notwithstanding the premature VR call, the captain did not rotate the airplane until the appropriate speed. However, the analysis of the data revealed that the captain began the takeoff rotation 5 knots below the correct VR speed. The airspeed indicator bug was properly set for a VR of 124 knots, so the Safety Board speculated that the captain might have been reacting to the first officer's early VR callout without cross checking his own airspeed indicator. As a result of the early rotation, the airplane lifted off at an AOA that was about 0.5 degrees higher than normal. Combined with the wing contamination, sufficient lift could not be maintained.

The first officer also stated that following the stickshaker activation and control problems, both he and the captain knew that the airplane was not going to fly. They then focused their efforts to stay over land and remain upright. Other than initially applying rudder, there were no corrective actions taken by the crew. According to the first officer, they used the yoke to "hold on" to the aircraft.

The Safety Board was unable to determine whether any specific actions could have been taken by the crew to have prevented the type of impact or level of severity. However, based on the corroborating evidence from the CVR and FDR, they concluded that seconds after liftoff, the airplane was in a stall regime from which recovery was not possible.

Lessons learned and practical applications

1. Maintain procedural discipline. From his own admission, the first officer thought he might have been thrown off by the atypical V1 and VR speeds requested by the captain. It's much more difficult to be a coordinated, unified crew when one deviates from standard procedures.

2. Ensure your airplane is free of contamination. Even if it adds to a delay, the Board believes the consequences are too serious to ignore.

Case study reference

National Transportation Safety Board. 17 February 1993. Aircraft Accident Report: Takeoff Stall in Icing Conditions. USAir Flight 405, Fokker F-28, N485US. La Guardia Airport, Flushing, New York. March 22, 1992. Washington, D.C.

CASE STUDY II-2: Delta Airlines Flight 191

Safety issues: microburst, low-level windshear, weather dissemination, ATC factors, PIREPs, ADM, CRM

On 2 August 1985, a Delta L-1011 crashed while attempting to land during a thunderstorm at Dallas/Ft. Worth International Airport (DFW), Texas.

Probable cause

The NTSB determined that the probable causes of this accident were: (1) the flightcrew's decision to initiate and continue the approach into a cumulonimbus cloud, which they observed to contain visible lightning, (2) the lack of specific guidelines, procedures, and training for avoiding and escaping from low-altitude windshear, and (3) the lack of definitive, real-time windshear hazard information. This resulted in the aircraft's encounter, at low altitude, with a microburst-induced, severe windshear from a rapidly developing thunderstorm located on the final approach course.

History of flight

Flight 191 was a regularly scheduled passenger flight between Fort Lauderdale, Florida, and Los Angeles, California, with an intermediate stop at DFW. The L-1011 departed Fort Lauderdale at 1510 Eastern Daylight Time with 152 passengers and 11 crewmembers onboard.

Pilot experience

The captain had logged 29,300 total flight hours, 3000 in the L-1011. He had been with Delta since 1954, and had completed his last recurrency training and line check 11 months before the accident.

The first officer had accumulated 6500 total flight hours, 1200 in the L-1011. He had been employed by Delta since 1970, and had passed his last proficiency check and recurrency training about four months prior to the accident.

The second officer had 6500 total flight hours, 4500 in the L-1011. He had been a flight engineer with Delta since 1976, and had passed his last proficiency check and recurrency training about five months before the accident.

Weather

The NWS terminal forecast for the DFW Airport indicated a slight chance of a thunderstorm with a moderate rain shower. The NWS area forecast called for isolated thunderstorms with moderate rain showers for northern and eastern portions of Texas. There were no SIGMETs or severe watches or warnings in effect for the time and area of the accident.

The Delta dispatch and meteorology departments provided similar information to the flightcrew prior to the flight's departure.

Weather radar photographs taken by the NWS station in Stephenville, Texas, about 72 nm from the approach end of DFW's runway 17L, revealed the following: At 1748, a Level 2 cell (referred as cell "A") developed about 6 nm northeast of the end of the runway. By 1752, the cell had intensified to a Level 3. At the same time, a Level 1 cell (referred as cell "B") had developed 2 nm northeast of the end of the runway. In about four minutes, cell "B" had become a Level 3 storm and had moved to just north of the runway threshold. Cell "A" was stationary.

At 1800, cell "B" appeared to be the dominant echo and was still located near the end of the runway. By 1804, cell "B" had intensified to a Level 4 storm, and cell "A" was no longer displayed on the radar scope.

The accident

The flight was uneventful until it passed New Orleans, Louisiana, when the crew noticed that a line of weather along the Texas-Louisiana gulf coast had intensified. They elected to change their route of flight to the north, to avoid the developing storms in the south. Consequently, they were issued a 10- to 15-minute hold at the Texarkana, Arkansas, VORTAC for arrival sequencing at DFW.

At 1735, the crew received the ATIS information that stated, ". . . 6000 scattered, 21,000 scattered, visibility 10, temperature 101, dew point 67, wind calm . . . visual approaches in progress . . . ." By 1743, the crew had been cleared to descend to 10,000 feet and to change their heading to 250 degrees. The controller also told them that, "we have a good area to go through." The captain replied that he was looking at a "pretty good size" weather cell "at a heading of 255 . . . and I'd rather not go through it. I'd rather go around it one way or the other." The controller obliged and a couple of minutes later the crew was given another clearance.

Shortly thereafter, the captain told the first officer: "You're in good shape. I'm glad we didn't have to go through that mess. I thought sure he was going to send us through it." The second officer remarked, "Looks like it's raining over Fort Worth." The flight was then instructed to contact approach control.

Moments later, the east-approach controller transmitted the following: "Attention, all aircraft listening . . . there's a little rain shower just north of the airport and they're starting to make ILS approaches . . . tune up [frequency] for 17 left [runway]." The crew made a lighthearted comment about the rain as they informed the controller that they were at 5000 feet. At 1800, the controller asked the crew of an American Airlines flight, which was two airplanes ahead of the Delta, if they were able to see the airport. The reply was, "As soon as we break out of this rain shower, we will." Less than a minute later, the crew of Flight 191 was given further instructions for the approach, and was sequenced behind a Lear 25 jet.

At 1802, the crew was advised that they were six miles from the outer marker and "cleared for ILS 17 left approach." The crew made a normal reduction in airspeed and was then informed by the controller, ". . . we're getting some variable winds out there due to a shower . . . north end of DFW." The information was acknowledged and one of the crewmembers commented, "Stuff is moving in."

Seconds later, the crew switched to the tower frequency and the captain told the controller, ". . . out here in the rain, feels good." They were cleared to land and advised that the winds were, "090 at five, gusts to 15." The first officer called for the "before-landing" checklist and the crew confirmed that the gear was down and the flaps were set at 33 degrees.

Almost immediately thereafter, the first officer said, "Lightning coming out of that one." The captain asked, "What," and the first officer repeated, "Lightning coming out of that one." The captain then asked, "Where," to which the first officer replied, "Right ahead of us." Less than a minute later, the captain called out "1000 feet" and he cautioned the first officer to watch his indicated airspeed. The sound of rain had already begun, when at 1805:21 the captain warned the first officer, "You're gonna lose it all of a sudden, there it is." The captain immediately added, "Push it up, push it way up." At 1805:29, the sound of engines at high RPM was heard on the CVR, and the captain said, "That's it."

About 20 seconds later, the ground proximity warning system's (GPWS), "Whoop, whoop, pull up," alert activated and the captain commanded, "TOGA." This is an acronym that stands for Takeoff/Go Around. When the airplane is flown manually the pilot can actuate a "TOGA" switch that provides flight director command bar guidance for an optimum climbout maneuver. At 1805:48, the GPWS again sounded, followed by other noises and the takeoff warning horn. When the controller saw the aircraft emerge from the rain at 1805:56, he told the crew to "go around." The CVR stopped recording at 1805:58.

Impact and wreckage path

The airplane initially touched down in a plowed field about 360 feet east of the extended centerline of runway 17L and 6336 feet north of the threshold. Witnesses on or near

State Highway 114, north of the airport, saw Flight 191 descend from the clouds about 1.25 miles from the approach end of runway 17L. The nose gear came down in the westbound lane, the airplane bounced back in the air, then knocked over a light pole and collided with an automobile. Surviving passengers then saw fire enter the left side of the mid-cabin area. Pieces of the number 1 engine inlet cowling were found in the car, and fragments of the car were discovered in the number 1 engine compressor inlet.

The airplane yawed significantly to the left when it crossed the highway. The jet began to break up as it skidded along the ground, and towards two water tanks located on the airport. A 45-foot by 12-foot crater was found about 700 feet beyond the highway. Sections of the nose gear, left horizontal stabilizer, engine components, and various pieces from the wing, were scattered along the wreckage path. The number 1 engine had separated from the airplane near the highway and had tumbled about 800 feet along the ground before coming to a stop.

The airplane grazed the north water tank and slammed into the second tank where the remaining fuselage broke apart. Amazingly, the jet had traveled 3194 feet beyond its initial touchdown point in the field. The forward portion of the fuselage was destroyed and both wing sections had separated and were burned extensively. Parts of the engines and other major components were strewn as far as 1125 feet past the second tank.

The aft fuselage section, containing the rear cabin and the empennage, came to rest in an upright position. Passengers and flight attendants recalled that this portion had originally stopped on its left side, but rolled upright by wind gusts. It was the only section that was found relatively intact.

Accident survivability. The flightcrew, 5 flight attendants, and 128 passengers sustained fatal injuries. The driver of the automobile that Flight 191 struck also died. After the aircraft hit the water tanks, the fuselage disintegrated from the nose section, aft to row 34. There were no survivors in the first 12 rows of seats. Some of the passengers in the mid-cabin area were ejected from the wreckage, many still strapped to their seats. Occupants in the rear of the cabin received mostly survivable injuries. Except for one flight attendant and three passengers, all were able to escape, unaided, through the large breaks and holes in the airframe.

The investigation

Based on the evidence, the Safety Board centered the investigation in several areas, including atmospheric conditions, airplane and flightcrew performance, air traffic control factors, and weather personnel considerations.

Weather analysis. Researchers at the National Oceanic and Atmospheric Administration (NOAA) conducted a multi-scale analysis of the weather conditions at the time of the accident. They concluded that the microburst-producing storm occurred almost in the center of a large-scale, high pressure area that extended through a deep layer of the troposphere. It was interrupted only near the surface by a thermal low-pressure area in combination with a low-pressure trough that was associated with a weak frontal boundary.

A computer-generated analysis of the pattern of deep tropospheric forcing revealed that a high lapse rate in the lower troposphere was generated by a pattern of vertical motion that tended to stretch the atmospheric column. At levels above 500 mb the forcing was upward; at 700 mb, it was downward. The lapse rate also would have been enhanced by strong solar heating. Weak subsidence above the surface boundary layer tended to cap it and preserve relatively high dew point temperatures near DFW, while heating southwest of the area tended to produce much drier surface conditions. Furthermore, a dry layer above 700 mb provided an elevated source of potentially cold air that would fuel strong downdrafts, which, penetrating into a deep, mixed subcloud layer, found a very favorable environment in which to become severe. In those two ways, the vertical thermodynamic structure of the DFW environment was a hybrid of a type that favored both dry and wet microbursts. The very weak front that lingered over DFW had resurged southward at about the time of the accident and lifted the shallow, moist surface layer. As a result, a line of discrete thunderstorm cells was triggered, one of which was located along the approach end of runway 17L. It was a microburst out of that storm that brought down Flight 191. See Fig. II-A.

Further evidence was provided by the FDR, which proved that the wind field within a microburst has a very complex structure of imbedded small-scale vortices. These vortices were so severe that once the jet penetrated the microburst and windshear environment, it caused serious control problems. During the final 38 seconds the airplane encountered a horizontal windshear of 72 knots. There were also six rapid reversals of vertical winds, in part causing the right wing to dip down 20 degrees. These events were indicative of a vortical wind flow.

Delta meteorology and dispatch. The Delta dispatcher told investigators that he had tried unsuccessfully to contact the Stephenville radar site between 1745 and 1750.

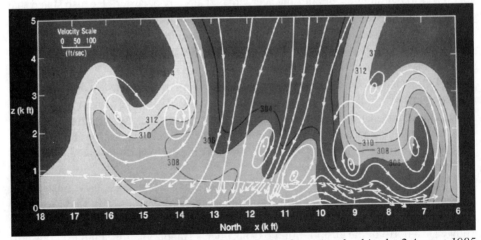

Fig. II-A. *A vertical cross section through the microburst involved in the 2 August 1985 airplane crash at Dallas-Fort Worth International Airport, constructed from the digital flight recorder data and based on the conceptual model.* Adapted from NOAA

Since he had no new or different weather information to provide to the crew of Flight 191, he did not attempt to call the crew as they approached DFW.

Fort Worth forecast office. The aviation forecaster on duty became aware of the storm cell northeast of DFW at about 1804. But this was only after he had overheard the radar specialist at Stephenville describe the cell to the public and state forecaster. He then observed the cell on his television monitor.

The forecaster testified that during the day, he had watched numerous cells build to Level 4 and then dissipate. The cell northeast of DFW did not, in his judgment, seem any different from those he had observed earlier. Therefore, he decided not to issue an Aviation Weather Warning to DFW. He further stated that he had considered the intensity of a radar weather echo to be "merely an indicator" of the severity of a storm and that in the absence of on-scene reports for verification, he would not label a Level 4 echo a thunderstorm. Based upon that testimony, the Board believed his actions to have been reasonable.

Center Weather Service Unit. The Fort Worth Center's CWSU was staffed by an NWS meteorologist and an assistant traffic manager serving as the weather coordinator. Since the ATC personnel assigned to that position are not trained or qualified to interpret the weather, no one was available to monitor the weather radar when the meteorologist took a 45-minute dinner break. According to the meteorologist, before he left his station he had made sure that there were no threatening thunderstorms to any of the area airports. Radar photographs later confirmed his evaluation of the situation.

However, during his absence, cells "A" and "B" developed and intensified. The meteorologist told investigators that he would have issued a warning if he had witnessed the rapid growth of the storms. Nevertheless, even if he had been on duty to notify the TRACON and tower personnel, routine notification procedures would have taken five to ten minutes, or after the accident had already occurred. He also stated that given the nature of the storm intensity he would have probably called the tower directly. The information would have reached ATC around 1802 or 1803. Although not optimum, the Safety Board did not believe that the NWS-to-ATC-to-pilot method of weather dissemination was a factor in this accident.

ATC decisions. The primary sources of weather information for approach controllers comes from the NWS, surface observations, PIREPs, eyewitness accounts by tower controllers, and precipitation returns on the radar. The radar returns are rather difficult, however, because precipitation degrades the returns, masking the actual intensity of the storm. In turn they must rely on other means, mostly from the tower controllers and pilots.

The Safety Board also noted that the approach controllers might have mislead the crew into thinking that the conditions weren't as severe as they actually were. At 1756, the controllers issued an "all aircraft listening" transmission describing, "a little rain shower just north of the airport." In the Controller's Handbook, the standard phraseology should have been light, moderate, or heavy. The controller later told investigators that he meant "a little area of precipitation," not the intensity of the rain.

Tower. At 1803, the crew contacted the tower controller and said, ". . . in the rain, feels good." The controller testified that he didn't advise the crew of the storm because they were obviously as aware of it as he was. Although a few controllers saw lightning, they did not pass the information along to the weather observer, TRACON, or to flightcrews.

PIREPs. Several crews saw lightning to the north of the airport. One pilot even thought he saw a tornado and still another described flying through a waterspout. Yet, there were no pilot reports to either the TRACON or tower personnel. Controllers are required by regulation to immediately report a PIREP over the appropriate frequencies.

Investigators believed that because the tower controllers had not received any PIREPs or concerns about the approach, they assumed the route was still acceptable. Based on that evidence the Safety Board concluded that the lack of PIREPs was causal to the accident.

Flightcrew performance. The Safety Board believed that the captain had sufficient information to adequately appraise the weather along the approach path, despite the lack of PIREPs and other ATC advisements. When the crew entered the area of lightning and heavy rain, they were within 4 nm, or two minutes, of the runway. Investigators suggested that no reports from preceding pilots might have influenced them to continue for the short distance that was remaining. The Safety Board also noted that the captain was known to encourage cockpit participation with any decision making. As a result, they believed that the first and second officers would not have been intimidated to speak up if either had thought the conditions were too severe. Therefore, the Safety Board determined that the entire flightcrew was responsible for the decision to penetrate the storm.

Airplane performance. Most major air carriers, including Delta, have taught their crews to trade airspeed for altitude if they inadvertently encounter low-level windshear. The techniques were practiced in the simulators, where pilots were instructed to increase pitch attitude and to add maximum thrust as necessary to control the airplane's flight path. In extreme cases the crews were also allowed to increase pitch to the point of activating the stickshaker. They were then to fly through the windshear area at a pitch angle that was just below that which would reactivate the stickshaker.

In the Safety Board's opinion, the first officer was apparently able to apply those techniques since the airplane remained on the glideslope as it entered the initial segment of the microburst. However, when the jet descended into the vortex, the combination of an airspeed loss of 20 KIAS and a strong updraft most likely caused the one-second activation of the stickshaker. Just for an instant the aircraft nosed over to a −8.5 pitch attitude with a 5000 fpm descent rate. At about 1805:19, the airspeed dropped 44 knots in ten seconds as the airplane traversed an area of increasing headwinds and downdrafts immediately followed by decreasing headwinds.

When the airplane was in the TOGA mode and the first officer was responding to a "fly-up" flight director command, the vertical wind changed from a 40 fps downdraft to a 10 fps updraft. The reversal in the wind component, combined with the substantial nose-up pitch rate, caused the AOA to rapidly increase. Since the aircraft's initial touchdown seemed to have been rather light, the Board speculated that the jet might have started to recover, albeit too late.

Lessons learned and practical applications

1. Don't underestimate thunderstorm activity. This can't be stressed enough. Seemingly benign cells can quickly develop into dangerous Level 4 or 5 thunderstorms. Unfortunately, you can't always rely on your fellow pilots to provide PIREPs, or that the NWS or ATC is going to have the latest weather information. So, you be the judge. Get out of the area if you see conditions conducive to microbursts, low-level windshear, lightning, and heavy rain.

2. Provide PIREPs. Reporting severe weather to ATC is imperative. Don't let your buds behind you get caught in something dangerous, or even fatal. Controllers also need this information to coordinate safe approaches and departures; and to determine if a runway change is in order.

3. Be prepared with correct flying techniques. If you can't practice in a simulator, study the AIM, talk with a CFI, and, if possible, pick the brains of several airline pilots. If you get in a predicament, it's vital that you're prepared with the proper procedures and techniques.

Case study references

Caracena, F., R. Ortiz, and J.A. Augustine. December 1986. *Crash of Delta Flight 191 at Dallas-Fort Worth International Airport on 2 August 1985: Multiscale Analysis of Weather Conditions*. National Oceanic and Atmospheric Administration. Boulder, Colorado.

National Transportation Safety Board. 15 August 1986. Aircraft Accident Report: Delta Air Lines, Inc., Lockheed L-1011-385-1, N726DA. Dallas/Fort Worth International Airport, Texas. August 2, 1985. Washington, D.C.

CASE STUDY II-3: Pan Am Flight 759

Safety issues: air mass thunderstorms, low-level windshear, effects of heavy rain on aircraft performance, role of NWS, role of ATC

On 9 July 1982, a Pan Am 727 crashed shortly after taking off from New Orleans International Airport, Louisiana.

Probable cause

The NTSB determined the probable cause of this accident was the airplane's encounter with a microburst-induced windshear, which imposed a downdraft and a decreasing headwind. The pilot would have had difficulty arresting the airplane's descent after recognizing and reacting to the windshear before impact.

Contributing to the accident was the limited capability of current ground-based, low level windshear detection technology to provide definitive guidance for controllers and pilots.

History of the flight

Flight 759 was a regularly scheduled passenger flight from Miami, Florida, to Las Vegas, Nevada, with an intermediate stop in New Orleans, Louisiana. The 727 departed at 1606 Central Daylight Time with 138 passengers and 7 crewmembers onboard.

Pilot experience

The captain had 11,727 total flight hours, 10,595 in the 727. He had been a qualified captain since 1972 and had completed his last line check seven months before the accident.

The first officer had 6127 total flight hours, 3914 in the 727. He had been with the airline since 1977 and had passed his last recurrency check two days before the accident.

The second officer had 19,904 total flight hours, 10,508 in the 727. He had been a qualified flight engineer since 1968 and had completed his last recurrency training about seven months prior to the accident.

Weather

The New Orleans area forecast included, "thunderstorms occasionally forming lines or clusters; thunderstorm tops to above 45,000 feet . . . possible severe or greater turbulence . . . severe icing and low level windshear." The terminal forecast called for, "scattered clouds, variable to broken clouds at 3000 feet; chance of overcast ceilings at 1000 feet; visibility two miles; thunderstorms, moderate rain showers." According to the NWS, there were no SIGMETs, severe weather warnings or watches in effect at the time and area of the accident.

However, at 1455 the National Severe Storms Forecast Center had issued a convective SIGMET for Alabama and Mississippi, including an area 60 miles southeast of New Orleans. About six minutes later the New Orleans clearance delivery transmitted the SIGMET to, "all aircraft" and advised them to, "monitor the VORTAC or check with flightwatch for further information." This message was also broadcast on the New Orleans' tower, approach, and departure control frequencies.

At 1531, an NWS station 30 nm northeast of New Orleans recorded the following radar weather observations: "An area of 3/10 covered [sky] by intense echoes containing thunderstorms with intense rain showers, no change in intensity over the last hour . . . cells were stationary. A maximum top of 50,000 feet was located at 060 degrees at 40 nm from the radar."

The surface weather observations for New Orleans International were updated at 1603, which stated: Special. Ceiling measured 4100 feet overcast. Visibility two miles, heavy rain showers, haze. Wind 070 degrees at 14 knots, gusting to 20 knots. Cumulonimbus overhead.

The accident

As the crew taxied towards runway 10, the first officer requested a wind check from ground control. The winds were reported out of 040 degrees at 8 knots. About four minutes later, at 1603, the first officer asked the controller for another wind check. He replied, "Wind now zero seven zero degrees at one seven . . . peak gusts, two three, and we have low level windshear alerts all quadrants. Appears to be a frontal passing overhead right now, we're right in the middle of everything." The captain then told the first officer to, ". . . let your airspeed build up on takeoff," and added that they would turn off the air conditioning packs so they could increase the EPRs on all three engines.

At 1606, the flight was cleared for takeoff as the crew completed the final items on the takeoff checklist. Seconds later, the local controller advised a landing Eastern flight that, "wind zero seven zero at one seven . . . heavy Boeing just landed said a 10-knot windshear at about a hundred feet on the final." According to the CVR, this advisory was received on Flight 759's radio.

As the crew began their takeoff roll, the captain called VR and V2. Based on eyewitness accounts, the aircraft lifted off about 7000 feet down runway 10, climbed to about 100 to 150 feet in a wings-level attitude and then began to descend. One of the witnesses was an airline pilot who reported seeing the airplane have a normal rotation, liftoff and initial climb segment. He did not, however, observe the windshear encounter.

Impact and wreckage path

Flight 759 initially hit three trees located about 2376 feet beyond the end of runway 10. According to the FDR, the aircraft struck the trees at 50 feet agl and in a 2- to 3-degree left-wing-down bank angle. The airplane then hit several houses and another group of trees as its bank angle increased to about 105 degrees. The fuselage exploded on impact and came to rest about 4600 feet from the departure end of the runway.

Except for the three engines, the landing gear, and parts of the horizontal stabilizer and flaps, fire damage was so extensive investigators were unable to obtain much useful information from the physical inspection of the wreckage.

Accident survivability. The accident was not survivable because impact forces exceeded human tolerances.

The investigation

The Safety Board determined that there were no powerplant or system malfunctions that caused the accident. Therefore, the investigation concentrated on the severity of the windshear encounter and flightcrew performance.

Meteorological observations. Between 1558 and 1627, four air carriers and one corporate jet departed New Orleans International. All of the flightcrews used their onboard weather radars to observe the conditions near the airport. A Delta Airlines crew reported a cell directly over the airport when they departed at 1558. They also observed other storm cells 25 nm away.

At 1601, a Republic Airlines DC-9 took off from runway 19. According to the captain, thunderstorms were all around the airport, including the east-northeast and south to west quadrants of the field. He also noted that the cell near the departure end of runway 10 had a "very steep" gradient. The captain told investigators that during their takeoff roll, they encountered heavy rain and windshear about half way down the runway. The visibility became very poor as the airplane began to drift to the right even after left rudder was applied. The captain testified that rather than abort the takeoff, and possibly run off the side of the runway, he elected to rotate about 11 knots prior to V1. After liftoff, the stall warning stickshaker activated for a brief moment as the landing gear was being retracted. The first officer recalled that the airspeed fluctuated between 100 KIAS and 110 KIAS during the takeoff roll. He also noticed that when the airplane passed over the end of the runway, the airspeed went through V1 (132 KIAS), V2 (140 KIAS), and 160 KIAS "almost simultaneously." He advised the departure controller that they, "had a windshear on the runway," but the report was not relayed to the tower controllers.

A few minutes later, a Texas International Airlines (TI) DC-9 lined up for a runway 19 departure. The crew stated seeing contouring storm cells 5 to 6 nm southwest of the airport. Compared with the Republic's encounter moments earlier, the TI crew's experience was amazingly different. They reported taking off in light rain with no turbulence or windshear during climbout. Studies have shown that the sudden onset and termination of heavy rain and strong, shifting winds are characteristic of microburst activity.

The crew of a Southwest Airlines 737 had stopped abeam the east end of runway 10 to await takeoff clearance. The captain told investigators that his radar showed a storm cell above his airplane that was, "between 5 to 6 miles wide, extending 2 miles east of the airport." He added that once he was cleared to runway 01, the same cell had hardly moved but showed a "heavy contour." As he observed Flight 759 pass over the departure end of runway 10, he thought the jet was about 200 feet agl, the landing gear was retracted, and that it was turning to the left. The captain remembered that the ceiling was around 3000 feet, light rain, and a 3 nm visibility. His attention was diverted back to his radar scope, and therefore, he did not witness the crash.

Around 1609, the crew of a Cessna Citation reached the apron of runway 19, and made a 360 degree turn to scan for local weather. The pilot noted two storm cells about 2 to 3 nm east of the airport that were about ¼ mile apart. Another cell was observed 7 nm southwest of the field. According to the pilot's testimony, each cell was about 3 to 4 nm in diameter and were depicted as sharp-edged red areas on his radar. He believed that they were either Level 4 or Level 5 radar echoes. Shortly thereafter, the crew was cleared to taxi and take off from runway 01. When investigators asked the pilot if he ever considered departing from runway 10, he stated that he did not, "primarily because of the weather east of the airport."

At 1608, a radarscope photograph from the NWS corroborated the observations made by numerous pilots. It showed a Level 2 echo located almost over the departure end of runway 10, and a similar echo about 4 nm east of the airport. The same photograph also showed Level 3 echoes at 4 nm north, 2 nm west, and 6 nm south of the departure end of runway 10.

Weather analysis. The National Oceanic and Atmospheric Administration (NOAA) conducted a multi-scale analyses of the meteorological conditions that affected Flight 759. Scientists noted that air mass thunderstorms had formed in the New Orleans area, caused, in part, by abundant low level moisture. A weak upper-air disturbance helped provide the lift that released the convective instability. At low levels, the interaction of several of these boundaries produced the local convection.

Satellite imagery clearly revealed three intersecting boundaries of air over New Orleans International. The first of these was a sea breeze front, formed by the difference in heating rates between land and the adjacent water, which moved north and was strengthened by downdrafts from thunderstorms that developed along it. The second was a lake breeze front that moved south off nearby Lake Pontchartrain and had initiated a line of showers that extended across the airport to the New Orleans' Lakefront airport, several miles away. The third was a weak thermal boundary that stretched from the International airport southwestward, which had caused early morning fog and stratus.

Those conditions were combined with a mass of mid-level dry air that had developed in the northern region of the city. This air, when cooled by precipitation evaporation, had the potential for generating strong downdrafts. It appeared that the air mass storms had sporadically ingested cold, mid-level air. In turn, this produced localized, intense microbursts within the broader areas of convective rain.

The researchers noted, that although these strong microbursts were not recorded, the available evidence "strongly supports these hypotheses." Infrared satellite photographs indicated that showers over the airport had "vigorously growing tops," as documented by the rapidly falling cloud top temperatures. The growth rate of these cumulonimbus clouds were much greater than that of any of the surrounding areas of showers. The researchers found that this growth rate began several minutes before Flight 759's departure and continued for a short time after the crash. During that period, two aircraft had encountered low level windshear.

The analysis also revealed that the radar imagery from the local NWS station confirmed the existence of a rain shower over the departure end of runway 10. As the sea breeze front approached, a Level 2 echo over the airport had intensified to Level 3 by 1558. This echo lingered over the area until 1614 when it suddenly dissipated. By this time the most vigorous echo was a rapidly growing Level 4 storm southwest of the airport. After the accident, this cell moved northeastward and produced widespread rain in the New Orleans area.

The scientists commented that despite this satellite analyses, the radar appearance of the shower over the departure end of runway 10 at the time of the accident was not noteworthy. When evaluated under the operational standards, the radar showed a small

Level 3 echo that did not display any classic severe turbulence signatures. However, from a research point of view, they believed it exhibited two potentially important characteristics.

First, the Level 3 portion of the echo was shaped like a spearhead at 1603. The distinctive shape might have been somewhat degraded relative to the actual cell shape because the two-degree-wide radar beam was just about equal to the width of the echo. Nevertheless, under similar conditions spearhead echoes have been associated with microbursts.

Second, at the time of the accident, an arc-shaped area of a rapidly-weakening echo was seen along the western boundary of the shower. Refer to Fig. II-B. A large Level 3 echo dissipated directly over the path of Flight 759 between 1603 and 1608. Renown experts on microburst activity have noted that rapid echo dissipation, which produces notches or arcs, might be associated with severe downdrafts.

The NOAA study also analyzed the probable winds that were in the vicinity of Flight 759's departure path. When the aircraft took off, a shaft of "extremely heavy" rain covered the intersection of runways 19 and 10. Eyewitnesses along the terminal concourses and on the airport access road that parallels runway 19 reported a sudden shift or increase in the wind. The pattern of this wind activity indicated that the heavy rain was the apparent source of its downward and outward flow near the surface. Based on this information there was a "potential" decrease in Flight 759's headwind component between 30 and 60 knots along its takeoff roll.

According to the data obtained from the FDR, shortly after the aircraft lifted off the wind shifted from a headwind of 16 knots to a tailwind of 23 knots within a dis-

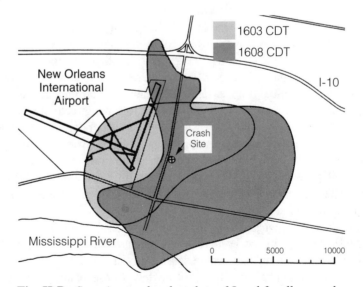

Fig. II-B. *Superimposed radar plots of Level 3 cells over departure end of runway 10 at 1603 CDT and at 1608 CDT.*
Adapted from NOAA

tance of 3400 feet. At the same time that the jet crossed this adverse headwind gradient it also ascended into the base of a downdraft. The maximum downflow was estimated at 7 fps at 100 feet agl. NOAA scientists applied linear extrapolation of the vertical wind gradient (assuming zero at the surface) to 300 feet agl and came up with a downdraft of 21 fps. Remember, that the definition for a downburst is a localized, severe downdraft with a vertical component that exceeds 12 fps at 300 feet agl. Furthermore, the width of the downburst was estimated at less than 4 km, which also met the accepted microburst criteria.

Additional computations that were derived from the FDR, indicated that Flight 759 was in a 12-degree, pitch-up attitude when it hit the first tree during climbout. The aircraft reached its highest altitude of 163 feet agl as it struggled to climb at 361 fpm. The performance analysis also determined that the aircraft's stall speed was around 122 KIAS and the stickshaker speed was at 138 KIAS.

From all of the data, analyses, and eyewitness accounts collected during the investigation, the researchers responsible for the NOAA study stated that it was "evident that Flight 759 experienced downdrafts and horizontal windshears exhibiting downburst characteristics for about 2000 feet of its flight path."

Aircraft performance in heavy rain. At the time of the accident scientists from the University of Dayton Research Center had obtained formidable data on the effects of heavy rain on aircraft performance. The theory stated that heavy rain can penalize performance in three ways: (1) some amount of rain adheres to the airplane and increases its weight, (2) the raindrops striking an airplane must take on the velocity of the aircraft, and the resulting exchange of momentum reduces the velocity of the airplane, and (3) the rain forms a water film on the wing, roughens its surface, and decreases the aerodynamic efficiency of the wing.

Calculations have shown that the landing weight of a large transport aircraft will increase by only one to two percent when flying through heavy rain. Therefore, researchers do not believe that the added weight is a significant factor.

The momentum penalty, however, is considered much more detrimental. When raindrops strike the surface of a fuselage, they cause the airplane to decelerate. The amount of velocity that is lost is dependent on the following factors: (1) airspeed, (2) rainfall rate, (3) raindrop size, (4) size distribution, (5) water content of the air, and (6) airplane configuration. Anytime the leading and trailing edge devices are extended, as in a landing or takeoff configuration, the penalty is the most severe. According to the study, the penalty becomes significant when rainfall rates approach 500 millimeters (mm)/hour. At those levels the rainfall could reduce airspeed at a maximum rate of about one-half knot per second.

The most dangerous penalty is the formation of water film on the wing. As a result, there might be a change in boundary layer flow, which is the fluid layer adjacent to the airfoil surface. The surface roughness would cause the boundary layer to transition prematurely from a smooth laminar flow to turbulent flow. Consequently, there would be an increase in skin friction drag. Due to this rough surface, scientists believe that drag can be increased "10 to 20 percent." They added, that depending on rainfall rate, a "lift penalty

[will] increase as the angle of attack increases, therefore the stall angle of attack could occur before the stall-warning system could activate." A rainfall rate between "150 mm/hour to 500 mm/hour" can produce enough surface roughness to cause these penalties.

In 1993, similar research was conducted at the University of Utah. The study confirmed that the roughening of an airfoil in heavy rain might cause an aircraft to encounter a decrease in lift, an increase in drag, and an earlier onset of a stall. This latest research, however, revealed an interesting phenomenon that causes the loss of boundary layer momentum through the acceleration of splashed-back droplets. It states, that when raindrops hit an airfoil, "some fraction of the incident mass is splashed back to form a *droplet ejecta fog* near the leading edge, while the remainder forms a thin water film on the airfoil surface."

Based on mathematical models, the study indicated that the vast majority of droplets near the airfoil are due to splashback. The formation of a highly concentrated area similar to the ejecta fog can significantly deform the shape of an airfoil. Scientists concluded that, "for a chord length of 1 meter, the thickness of this layer varies from approximately 2 centimeters (cm) at the leading edge of the airfoil, to approximately 10 cm further downstream along the upper surface."

As noted earlier, the angle of attack (AOA) is also affected by heavy rain. For example, an aircraft with an AOA of about 4 degrees, causes the stagnation point to be slightly below the leading edge. Thus, splashed-back droplets from raindrops impacting the leading edge tend to be carried over the upper surface of the airfoil more than the lower surface. The drag produced by these splashed-back droplets acts as a "sinking" momentum near the leading edge of the airfoil, which creates the potential to "de-energize" the boundary layer. Therefore, the combination of a momentum penalty and airfoil roughness is the most likely cause for the degradation of aircraft performance in heavy rain.

Flightcrew performance. Between 1600 and 1607, ATC transmitted nine windshear advisories. The Safety Board believed that the captain was well aware of the warnings and should have also noticed the cumulonimbus clouds directly overhead. Therefore, they concluded that the captain had received adequate weather information from Pan Am and ATC to have made a satisfactory assessment of the conditions at the airport.

It was further noted, that the captain had taken precautionary steps by instructing the first officer to "let your speed build up on takeoff," and deciding to turn off the air conditioning packs. He also planned to turn left after liftoff in order to avoid the cells at the departure end of the runway. Based on the captain's reputation for "superior judgment," and given his performance record in the previous ten years, the Safety Board believed that he would not have intentionally flown into dangerous thunderstorms. Based on the evidence, they concluded that the "captain's decision to take off was reasonable."

Lessons learned and practical applications

1. Recognize the clues: Dark cumulonimbus clouds overhead; pilot reports of windshear; ATC windshear advisories; onboard radar returns; NWS forecasts and observations. Each fills in a piece of a weather puzzle. Collectively, they indicate potentially dangerous conditions.

2. Don't become complacent. You might have successfully flown near thunderstorm activity hundreds of times before, but it takes only one misinterpreted encounter to bring down a perfectly good airplane. Take the clues seriously.

3. Update your personal weather knowledge. Make a point to seek out the latest information. In the area of weather research, a lot can change in just a few years.

Case study references

Caracena, F., R.A. Maddox, J.F. W. Purdom, J.F. Weaver, and R.N. Green. January 1983. "Multi-Scale Analyses of Meteorological Conditions Affecting Pan American World Airways Flight 759." National Oceanic and Atmospheric Administration. Washington, D.C.

National Transportation Safety Board. 21 March 1983. Aircraft Accident Report: Pan American World Airways, Inc., Clipper 759, Boeing 727–235, N3737. New Orleans International Airport, Kenner, Louisiana. July 9, 1982. Washington, D.C.

Valentine, James R. and Rand A. Decker. January-February 1995. "Tracking of Raindrops in Flow over an Airfoil." *Journal of Aircraft*: 100-5. American Institute of Aeronautics and Astronautics. Washington, D.C.

Related references

Bezos, G.M., R. E. Dunham, G.L. Gentry, and W.E. Melson, Jr. August 1992. "Wind Tunnel Aerodynamic Characteristics of a Transport-Type Airfoil in a Simulated Heavy Rain Environment." *NASA TP-3184*. Washington, D.C.

Haines, P.A., and J.K. Luers. 1983. "Aerodynamic Penalties of Heavy Rain on Landing Aircraft." *Journal of Aircraft*: Vol. 20, No. 2, 111–119. American Institute of Aeronautics and Astronautics. Washington, D.C.

Fig. II-C. *JAL B747 landing on runway 36 at Naha IAP, Okinawa, Japan.* Dan Simonsen/Check Six.

CASE STUDY II-4: Japan Airlines Flight 46E

Safety issues: mechanical turbulence, mountain wave turbulence, role of NWS, CRM

On 31 March 1993, a Japan Airlines (JAL) 747, encountered an in-flight engine separation shortly after taking off from Anchorage International Airport, Alaska.

Probable cause

The NTSB determined that the probable cause of this accident was the lateral separation of the number 2 engine pylon due to an encounter with severe or possibly extreme turbulence. As a result, the dynamic multi-axis lateral loadings exceeded the load-carrying capability of the pylon. The integrity of the pylon was already compromised by a fatigue crack near the forward end of the firewall web.

History of flight

Flight 46E was operating as a scheduled cargo flight from Tokyo-Narita Airport, Japan, to Chicago-O'Hare, Illinois, with an intermediate stop at Anchorage. The aircraft was leased under an agreement with Evergreen International Airlines. The 747 departed Anchorage at 1224 local time with three crewmembers and two nonrevenue company employees onboard.

Pilot experience

The captain had 10,000 total flight hours, 750 in the 747. His last simulator proficiency check was conducted about seven weeks before the accident.

The first officer had 10,500 total flight hours, 600 in the 747. He passed his last simulator proficiency check eight months prior to the accident.

The second officer had 2600 total flight hours, 1201 in the 747. Her last simulator proficiency check was completed six months prior to the accident.

Weather

The area forecast issued by the NWS stated: Scattered ceilings below 1000 feet and visibility below 3 miles. Light rain, light snow . . . 3500 to 5000 feet broken west side. Wind east-northeast 45 knots. Local strong gusts southern inlet and from passes and channels along east side.

At 1145, a SIGMET had been issued for moderate and frequent severe turbulence from the surface to 12,000 feet. It included a warning for moderate and severe mountain wave turbulence from 12,000 feet to 39,000 feet within a widespread area south of the airport. An AIRMET was also issued for low level windshear associated with the strong surface winds.

The accident

The scheduled departure time for Flight 46E was 1125, but as the crew taxied out of the ramp area, the start-valve light for number 2 engine illuminated, indicating "open."

As a precautionary measure the engine was shut down and the crew returned to the ramp for a maintenance inspection. Evergreen mechanics replaced the start valve. However, the light remained on. It was later determined that the indicator system was most likely the problem and the aircraft was released for the flight.

At 1221 the crew taxied toward runway 6R and tuned in the latest ATIS information. The ceiling was estimated 8000 feet overcast, visibility 60 miles, temperature/dewpoint was 49 degrees F/21 degrees F, and winds were 090 degrees at 7 knots. The crew also acknowledged receiving the current SIGMET. The ground controller advised them of, "Pilot reports severe turbulence leaving 2500 [feet] climbing on the KNIK [standard instrument departure (SID)] off runway 6R by company B-747."

As the captain climbed through 1000 feet the departure controller informed the crew to, "expect severe turbulence 2500 reported by 747 . . . continuous moderate 3000 to 10,000." The crew of Flight 46E reported encountering moderate "bumps" at 1500 feet, followed by "large wave action . . . with large vorticity."

At about 2000 feet the captain initiated a left 20-degree-bank turn to a heading of 330 degrees as directed by the SID. While in the turn the airplane entered an uncommanded left bank to approximately 50 degrees. At the same time the airspeed fluctuated about 75 knots, between 170 KIAS and 245 KIAS. The crew also reported a "huge" yaw, at which time the No. 2 throttle slammed to its aft stop, the No. 2 reverser indicator showed thrust reverser deployment and the No. 2 engine electrical bus failed. The crew immediately shut down the engine but were initially unaware that it had physically separated from the aircraft.

Several witnesses on the ground told investigators that the airplane went into several severe pitch and roll oscillations before the engine separated. The pilots of two U.S. Air Force F-15s were flying in the local area when they noticed something large had fallen from the 747. They contacted the nearby Elmendorf Air Force Base tower controllers who then notified Anchorage departure control. The crew was still going through their emergency procedures when they were told, ". . . something large just fell off your airplane." The first officer replied, ". . . we know that . . . we're . . . declaring an emergency." The controller then provided vectors for their return to Anchorage.

The captain instructed the second officer to manually lock down the leading edge devices and dump fuel. Momentarily, the captain was unable to stay at altitude as the aircraft descended 200 to 300 fpm. He told investigators that he used emergency/maximum power on the No. 1 engine, full rudder, and almost full-right aileron in order to maintain control. He also stated that the stickshaker and bank-angle warnings activated intermittently throughout the remainder of the flight.

The controller coordinated the two F-15 pilots to fly close to the 747 and inspect the damage. Since the fighter pilots were talking with the controller on a different frequency, their input was not included in the accident report's ATC transcripts. Therefore, according to the Safety Board, the pilots reported that the No. 2 engine was missing; and all of the leading edge devices and trailing edge flaps between the No. 1 and No. 2 engines were damaged. In referencing the report's transcripts, however, the full extent of that transmission was not relayed to the crew. Instead, the controller told the crew, ". . . you've lost . . . approximately 50% of the leading edge slats on the left wing, and structural damage to the trailing edge flaps."

As the captain flew a large radius turn back towards the airport, the No. 1 engine was maintained at emergency/maximum power for control purposes. While on downwind, the bank angles of the airplane momentarily exceeded 40 degrees, causing the captain to use manual steering, and at times, full right rudder.

The second officer stopped dumping fuel as they turned on final approach, and the landing gear was lowered shortly thereafter. At 300 feet the flaps were extended to 25 degrees, and the airplane landed seconds later. It was estimated that the jet weighed about 685,000 pounds when it landed. The normal maximum certificated landing weight is 585,000 pounds. As a result, the aircraft experienced hot brakes.

Engine impact area

The debris from the flap and engine pylon was scattered over a localized area about 7 miles from the airport. The No. 2 engine came to rest a short distance farther. Several houses and automobiles were damaged from the impact, but there were no reported injuries on the ground.

Damage to the aircraft

The number 2 engine pylon had separated into four pieces as a result of three principle fracture areas. Metallurgical examination determined that the splits occurred due to overstress. However, investigators also found the presence of preexisting 2-inch-long fatigue crack in the pylon structure. Inspection of the other three pylons on the airplane revealed no similar fatigue cracks.

The upper surface fixed leading edge panels from a portion of the left wing were missing. Sections of the leading edge "D" beam, variable camber flaps, and the thermal anti-ice duct were all torn from the wing. Damage was also discovered to the left wing's trailing edge flaps, a spoiler, and the lower left side of the rudder. The total repair cost was estimated at 12 million dollars.

The investigation

The pylon is designed to carry the thrust and torque loads of the engine, as well as lateral, longitudinal, and vertical loads from maneuvers and wind gusts. Lateral loads are ultimately absorbed by the midspar fuse pins and side brace. According to Boeing, the fuse pins can withstand at least 2.8 Gs of lateral load on the engine. But the most critical structure under those types of loads is the firewall just aft of the forward engine mount. In reference to this accident, Boeing calculated that given the size and location of the fatigue crack, the firewall would have fractured between 2.35 Gs and 2.88 Gs.

The information obtained from Flight 46E's FDR indicated that the number 2 engine nacelle's center of gravity ranged between –2.5 Gs vertical, 2.1 Gs to 3.0 Gs outboard lateral, and 0.1 G to 0.3 G longitudinal. It was unknown if these loads were acting on the pylon at the same time.

There was no specific requirement to perform inspections in the area of the forward firewall web where the crack was found. However, since the accident, inspections of other 747s have found no additional evidence to suggest that this type of cracking was an inherent problem. Furthermore, Boeing proposed to the FAA several structural modifications to the 747 pylon to increase its load-carrying capability. The plans were to significantly strengthen the midspar fuse pins, which would enhance the pylon's vertical and longitudinal integrity.

Weather analysis. On the day of the accident, the interaction of strong easterly winds with the mountains east of Anchorage, was responsible for the production of moderate to severe mountain wave and mechanical turbulence. It was noted that these winds flowed around the mountains and through valleys before reaching Anchorage. This channeling effect caused the winds to rapidly accelerate in the lower layers of the atmosphere and created the severe turbulence at an altitude of a few thousand feet.

Several witnesses remembered seeing strong, gusty winds of around 62 knots just 10 miles southeast of the airport and at an elevation of 2500 to 3000 feet. A "funnel of rotating debris" was also observed 7 miles northeast of the airport, at a height of 500 to 1000 feet.

The pilot reports were also quite dramatic on the day of the accident. Around noon, the pilot of a U.S. Marshall Service Cessna 310 took off from a nearby airport and encountered a downdraft at 300 feet agl. He stated that the airspeed dropped from 120 knots to 90 knots, and the airplane lost about 200 feet in altitude. He managed to get out of the downdraft and climb to 900 feet, but quickly entered an updraft of 4000 fpm. Although his throttles were at idle, he could not keep the airspeed below 160 knots. The pilot was able to return to the airport and later stated in a written report, ". . . in 20 years of flying up here, this was the worst turbulence I have encountered, and it was the first time I have ever wondered if I would make it back because, at times, I was not really flying this aircraft."

Just 10 minutes later at Anchorage International, a 747 encountered severe turbulence at 2500 feet and moderate turbulence between 3000 and 10,000 feet during the climbout to the north. About 30 minutes after Flight 46E landed, the crew of a DC-8 also reported severe turbulence and low level windshear on their departure from Anchorage.

Flight operations. According to the NWS at Anchorage, the strong winds that produce significant turbulence occurs about 15 times a year. Interviews with local meteorologists and pilots revealed that the weather and turbulence on the day of the accident were fairly typical, and that airplane operations were routinely carried out on similar days. The Safety Board believed that since the captain had experienced turbulent conditions at Anchorage before and that other aircraft were safely flying out of the airport, they found no reason for the captain to have suspected that his airplane would have been damaged during climbout.

The Safety Board also believed that it was unnecessary to suspend flight operations at Anchorage when similar turbulence has been reported. However, they noted that since most of the intense turbulence tends to occur near the mountains at low altitude, they recommended that the FAA consider modifying the departure routes at Anchorage.

Lessons learned and practical applications

1. Recognize the importance of CRM. All three crewmembers had to immediately work together as a team. If each person's role had not already been clearly defined, the flow of duties and responsibilities might have been disrupted by unnecessary confusion.

2. Communicate clearly and directly by using standard phraseology. Be specific with your directions and intent. When asked a question, avoid answering with a, "yeah," or even a "yes." Respond with an, "affirmative," or "negative." This eliminates misinterpretations to very important inquiries.

3. Be organized. Data cards, checklists, and any other quick reference guides should be at your fingertips. During an emergency you don't want to find yourself fumbling through papers and junk.

4. Maintain conservative control inputs. Understanding your airplane's flight characteristics and limitations is the first step in maintaining control in unusual attitudes. Be careful to not overcorrect.

Case study reference

National Transportation Safety Board. 13 October 1993. Aircraft Accident Report: In-Flight Engine Separation. Japan Airlines, Inc., Flight 46E Boeing 747–121, N473EV. Anchorage, Alaska. March 31, 1993. Washington, D.C.

CASE STUDY II-5: EASTERN AIRLINES FLIGHT 66

Safety issues: severe thunderstorm activity, microburst, downburst, role of ATC, PIREPs

On 24 June 1975, an Eastern 727 crashed while attempting to land in a thunderstorm at John F. Kennedy International Airport, New York.

Probable cause

The NTSB determined that the probable cause of this accident was the aircraft's encounter with adverse winds associated with a very strong thunderstorm located astride the ILS localizer course, which resulted in a high-descent rate, and the flightcrew's delayed recognition and correction of the high-descent rate.

A contributing factor was the continued use of Runway 22L when it should have become evident to air traffic control personnel and the flightcrew that a severe weather hazard existed along the approach path.

History of flight

Flight 66 was a regularly scheduled passenger flight from New Orleans, Louisiana, to JFK International Airport. The 727 took off at 1319 Eastern Daylight Time with 116 passengers and 8 crewmembers onboard.

Pilot experience

The captain had 17,381 total flight hours, 2813 in the 727. He was promoted to that position in 1968.

The first officer had 5063 total flight hours, 4327 in the 727. He had been a second officer with Eastern since 1966 and became a fully qualified first officer three months before the accident.

The second officer (primary crewmember) had 3910 hours total flight hours, 3123 in the 727. He had held that position since 1968. Another second officer (a check flight engineer) was onboard giving the primary second officer his annual line check. He had 5369 total flight hours, 676 in the 727, and had been with Eastern since 1970. He was also a current pilot with the Air Force Reserves. Of his total flight hours, 1379 were logged in heavy military aircraft.

Weather

At the time of the accident there was scattered thunderstorm activity in the New York City area. The thunderstorm near JFK was reported to be very strong with heavy precipitation. Surface observations were as follows:

1550: 3000 feet scattered, estimated 6000 feet broken. Visibility was five miles in light rain showers and haze. Winds were 300 degrees at 6 knots. North of the field, visibility was 2 miles with towering cumulus.

1602: Special. 3000 feet scattered, estimated 5000 feet broken. Visibility was 2 miles in a thunderstorm and light rain showers. Winds were 210 degrees at 7 knots. The thunderstorm was overhead and moving northeast, with occasional lightning from cloud to cloud. Visibility to the south was 5 miles.

The accident

At 1535:11: Kennedy approach control (Southgate arrival controller) provided radar vectors to Flight 66 to sequence it with other traffic and to position it for an ILS approach to runway 22L. The ATIS broadcast, "Kennedy weather, VFR, sky partially obscured, estimated ceiling 4000 broken, 5 miles with haze. . . wind 210 degrees at 10, altimeter 30.15. Expect vectors to an ILS runway 22L, landing 22L, departures are off 22R . . ."

The Southgate-arrival controller revised the ATIS at 1551:54, and issued the updated weather to all aircraft on his frequency, ". . . we're VFR with a 5 mile . . . very light rain shower with haze, . . . ILS 22L . . ." Only 50 seconds later the same controller transmitted, "All aircraft this frequency, we just went IFR with 2 miles, very light rain showers and haze. The runway visual range is not available, and Eastern 66 descend and maintain four thousand, Kennedy radar one three two four." The crew of Flight 66 acknowledged the transmission.

At 1553:22, Flight 66 contacted the Kennedy final-vector controller who continued to provide radar vectors around thunderstorms in the area and position the flight on the localizer course. A few minutes later, the crew discussed among themselves the problems associated with carrying minimum fuel loads when confronted with delays in terminal areas. One crewmember stated that he was going to check the weather at their alternate, which was La Guardia Airport, New York. Less than a minute later, one of the pilots remarked, ". . . one more hour and we'd come down whether we wanted to or not."

The final-vector controller, transmitted to all aircraft on his frequency at 1559:19 that a "severe wind shift" had been reported on the final approach and that he would report more information shortly.

White-knuckled flying. Flying Tiger Flight 161, a DC-8, had landed on runway 22L at 1556:15. After the jet cleared the runway a little more than a minute later, the captain reported to the local controller: "I highly recommend that you change the runways and . . . land northwest, you have a tremendous windshear down near . . . the ground on final." The local controller responded, "Okay, we're indicating wind right down the runway at 15 knots when you landed." The captain replied, "I don't care what you're indicating, I'm just telling you that there's such a windshear on the final on that runway you should change it to the northwest." The controller did not respond.

During the accident investigation of Eastern 66, the Flying Tiger captain testified that during his approach to runway 22L, he entered precipitation at about 1000 feet msl. He experienced severe changes of wind direction, turbulence, and downdrafts between the outer marker and the airport. He observed airspeed fluctuations of 15 to 30 knots and at 300 feet he had to apply almost maximum thrust to stop his descent and attempt to maintain 140 knots. At that point, the captain believed the conditions had become so severe the airplane would not respond to the thrust needed for a missed approach, so he landed.

Eastern Flight 902, an L-1011, had abandoned its approach to runway 22L at 1557:30. A little more than two minutes later, Eastern 902 reestablished radio communication with the Kennedy final-vector controller. The flightcrew reported, ". . . we had . . . a pretty good shear pulling us to the right and . . . down and visibility was nil, nil out over the marker . . . at 200 feet it was . . . nothing." The final-vector controller responded, "Okay, the shear you say pulled you right and down?" The captain replied, "Yeah, we were on course and down to about 250 feet. The airspeed dropped to about ten knots below the bug [airspeed indicator] and our rate of descent was up to 1500 feet a minute, so we put takeoff power on and we went around at a hundred feet."

Testifying during the accident investigation, the captain of Eastern 902 stated that on his approach to runway 22L, he flew into heavy rain near 400 feet. The indicated airspeed dropped from about 150 knots to 120 knots in seconds and his rate of descent increased significantly. The aircraft moved to the right of the localizer course, so he abandoned the approach. He was unable to stop the aircraft's descent until he had established a high noseup attitude and applied near maximum thrust. He thought the aircraft had descended to about 100 feet before it began to climb.

Two other aircraft, Finnair Flight 105, a DC-8 and a Beechcraft Baron, followed Eastern 902 on the approach. All of the pilots involved testified that they also experienced significant airspeed losses and increased rates of descent. However, they believed that they were able to cope with the problem because they had been warned of the windshear condition and had increased their airspeeds substantially to compensate for the expected shear. Neither pilot reported the windshear conditions. One pilot stated that he did not report the encounter because it had already been reported and he believed that the controllers were aware of the situation.

Final approach for Flight 66. As the captain of Eastern 902 was making the windshear report, the captain of Flight 66 said, "You know this is asinine." Another crewmember responded, "I wonder if they're [the crew of Eastern 902] covering for themselves."

The controller asked Flight 66 if they had heard Eastern 902's report. The captain replied, "Affirmative." The controller then established the flight's position as being 5 miles from the outer marker, and cleared the flight for an ILS approach to runway 22L. The crew acknowledged the clearance at 1600:54, and added, "Okay, we'll let you know about the conditions."

At 1602:50 the first officer said, "Gonna keep a pretty healthy margin on this one." Another crewmember replied, "I . . . would suggest that you do." The first officer responded, "In case he's [Eastern 902] right."

Eight seconds later, the crew reported over the outer marker and the controller told the flight to contact the Kennedy tower. At 1603:44, the local controller cleared the flight to land. The captain acknowledged the clearance and asked, "Got any reports on braking action?" The controller eventually responded, "No . . . approach end of runway . . . about the first half is wet. We've had no adverse reports."

Following Flight 66 was National Flight 1004. When the National jet reached the outer marker, the crew asked the local controller if there were any subsequent pilot reports concerning the weather conditions on final approach. At 1604:58 the controller responded, "Eastern 66 and National 1004, the only adverse reports we've had about the approach is a windshear on short final . . ." National 1004 acknowledged that transmission, but Flight 66 did not.

Impact and wreckage path

Both surviving flight attendants, who were seated in the aft portion of the passenger cabin, described the approach as normal with little or no turbulence. According to one, the aircraft rolled to the left and she heard a significant increase in engine power. The aircraft then rolled upright and rocked back and forth. She was suddenly thrown forward and then upright.

Witnesses located along the localizer course, from about 1.6 miles from the threshold of runway 22L to near the middle marker, described the weather conditions as severe when Flight 66 passed overhead. Heavy rain was falling with lightning and thunder. The wind was blowing hard from directions ranging from north through east.

At 1605:11, the aircraft first struck an approach light tower located 2400 feet from the threshold of runway 22L. A fire erupted when the outboard section of the left wing was severed, igniting fuel as the airplane continued to slice through additional towers. The jet then rolled in excess of a 90 degree left bank, struck several more towers and hit the ground. The impact made a 340-foot long gouge as the aircraft skidded to nearby Rockaway Boulevard and came to a fiery stop.

Three large pieces of the left wing were found near the point of impact. The number 1 engine was severely damaged, and parts of it were scattered near one of the remaining approach towers. The number 2 engine had separated and was located next to the empennage, and the number 3 engine remained attached to the tail section.

Accident survivability. The flightcrew and 2 flight attendants sustained fatal injuries, along with 107 passengers. The Safety Board classified this accident as nonsurvivable due to the near total destruction of the fuselage. However, the aft section of the aircraft remained relatively intact. The investigators noted that all of the survivors were seated in the inverted rear portion of the passenger cabin.

The seats in the cockpit, forward flight attendant positions, and most of the passenger cabin were torn from their supporting structures. They were found mangled and twisted, strewn over the last 500 to 600 feet of the wreckage path. Although nearly every passenger seatbelt remained attached and fastened, when the fuselage began to disintegrate and the seat anchors snapped, those occupants were hurled against objects with a tremendous force. Many died of multiple extreme impact injuries.

The only seats that remained attached to the floor structure were the two aft flight-attendant positions. They were both able to escape without assistance.

Airport firefighters arrived at the scene about three minutes after the crash, followed shortly thereafter by the New York City fire department. The main fire was brought under control and extinguished just five minutes after the first units arrived.

Personnel from a local medical clinic quickly arrived at the site to administer first aid to the survivors. Only one ambulance was available and it was used to transport six survivors to an area hospital. The remaining survivors were transported to the hospital in a fire truck.

The investigation

The Safety Board concentrated their investigation in three main areas: Weather, aircraft performance, and flightcrew performance.

Weather forecasts and observations. It is important to note that the following weather reports, issued by various agencies and radar facilities, were not disseminated by ATC to flightcrews operating in the New York area.

While Flight 66 was en route to New York, the NWS issued a terminal forecast for Kennedy Airport. It called for thunderstorms and moderate rain showers after 1515. The forecast was updated with a strong wind warning that was valid from 1600 to 2000. The warning called for gusty surface winds to 50 knots from the west and thunderstorms in the New York City terminal area. At 1545, the forecast was amended to include thunderstorms, heavy rain showers with visibility down to ½ mile. After 1615 the winds were to come from 270 degrees at 30 knots with gusts to 50 knots.

About eight minutes before the accident, the NWS weather radar showed that an area of thunderstorm activity was centered along the northern edge of Kennedy Airport. The area was 30 to 35 miles long and about 15 miles wide. A few minutes later, the surface weather observation at Kennedy Airport included, "thunderstorm overhead moving northeast, occasional lightning cloud to cloud." At 1604, the NWS weather radar tracked a large group of cells that merged over the approach course to runway 22L.

The only forecast that the flightcrew received was prior to departure in New Orleans at 1208. The company-generated forecast predicted widely scattered thunderstorms from 1215 to 2000. After 2000, thunderstorms were possible with light rain showers.

Satellite data. Researchers from the University of Chicago analyzed weather satellite imagery to determine the life cycle and severity of the JFK thunderstorm. Detailed examination of the meteorological conditions revealed that the growth rate of the storm was at its peak when the accident occurred.

At 1353, a very complicated thermal structure of a weak cold front extended from central Pennsylvania to Rhode Island. North of the front, the temperatures were in the 90s, but to the south they were in the 70s and 80s.

Due to solar heating, a Long Island, New York, sea breeze was blowing inland across the Atlantic beaches. This was producing cumulus clouds to form along the island's north coast. The early stage of the storm was moving toward the east-southeast at 16 knots. By 1451, it was located on the cold front in north-central New Jersey.

As the storm moved away from the front, it split into two cells. One hovered over lower Manhattan, New York, and the other was located northeast of La Guardia Airport. The storms quickly intensified, and between 1500 and 1600 there were dramatic changes in the echo pattern.

The JKF thunderstorm moved very rapidly toward the western tip of Long Island. A line of arc clouds developed along the leading edge of the storm's outflow—the south edge was being held back by the cold sea breeze from the Atlantic. Interestingly, the sea breeze temperature was cooler than that of the thunderstorm. JFK was reporting 77 degrees F, and La Guardia, while in the outflow, was reading 86 degrees F.

The squall line activity in eastern Pennsylvania and northern New Jersey was also intensifying. As a result, a surge of northwesterly winds generated in advance of a line of echoes. At 1505, the echoes near JFK were moving between only 15 and 17 knots. However, a severe change took place in the following 11 minutes which accelerated the storm towards the airport.

Between 1505 and 1516, an appendage formed near the east end of the major echo. The first appendage was 3 miles long with a sharp point. This point developed into a spearhead and extended very rapidly. By 1540, the spearhead had become so large that the "parent" echo was quickly drawn into it. The spearhead echo had dominated the area and within one minute had suddenly merged with a small echo located to the north of JFK. At 1602, the spearhead echo was just north of the airport, and about 15 miles long and 5 miles wide. Although the life cycle of a spearhead is relatively short, about 50 to 60 minutes, the severity of the JFK thunderstorm is proof of the dangerous conditions it produces throughout each of its stages.

Downburst encounters. It is important to know what conditions were encountered by other aircraft that were landing on 22L before Flight 66. Downburst and windshear activity can be quite insidious. Note how rapidly the conditions changed in a matter of seconds and over a short distance. You will find a similarity between these findings and those discussed in case study II-3 (Pan Am Flight 759). The following are excerpts of pilot reports made to the NTSB during the investigation.

1544: 747— Moderate rain at outer marker. No turbulence. Some windshear on final approach, but not significant enough to report it to tower.

1546: 707—Smooth air all the way down to the final approach. Downdraft after passing 500 feet. Added power to maintain ILS. Possibly an increased headwind.

1548: DC-9—Downdraft around one mile from the threshold. Landed in light rain.

1549: 707—Moderate rain between outer and middle markers. Landing normal. Dry runway.

1551: 747—Heavy rain at 1000 feet. Some windshear. Little rain on touchdown.

1552: 747—Slight increase in headwind and left drift near the outer marker. Airspeed dropped at 300 feet, requiring power. First half of runway was wet, other half dry.

1554: 707—Extremely heavy rain at 800 feet. At 200 feet, right drift and 18-knot crosswind. No drift correction needed at touchdown. Rolled for about 1000 feet, and broke out on dry runway in sunlight. While on the taxiway, pilot observed the Flying Tiger DC-8 struggling on final approach.

1556: DC-8—Strong, sustained downdraft from about 700 feet to 200 feet. Pilot used abnormal amount of power for an unusually long period of time. Strong right crosswind. Wind was blowing at 50 to 55 knots just above the ground, and then stopped at the surface.

1558: L-1011—Smooth air and no rain. At 400 feet, entered into extremely heavy rain and visibility dropped to zero. Aircraft started to sink and drift to the right. Twenty knot drop in airspeed. Added full power, but aircraft continued to descend to 60 feet. Initiated a missed approach by using considerable power and abnormally high AOA.

1559: DC-8—Heaviest rain between three and six miles on final approach. Wind at 1500 feet was 230 degrees at 30 knots. Slight left drift. At two mile final, airspeed dropped 25 knots.

1602: Beech—Light turbulence, and moderate to heavy rain from just outside the outer marker to halfway to middle marker. Heavy sink rate at 200 or 300 feet. Airspeed dropped 20 knots. Applied power to recover from sink. Remainder of approach was normal.

Aircraft and flightcrew performance. The correlation of the CVR, FDR, and radar data showed that Flight 66 intercepted the glideslope at an altitude of 3000 feet at 1601:20. The captain commented, "Just fly the localizer and glideslope," and the first officer replied, "Yeah, you save noise that way and get a little more stability." The flaps were extended to 15 degrees and the landing gear was lowered. The crew engaged in final checklist duties for the next 30 seconds, and the aircraft was bracketing the glideslope. The airspeed varied between 160 and 170 knots. At 1603:05, the first officer requested 30 degrees of flaps. The aircraft continued to bracket the glideslope and the airspeed oscillated between 140 and 145 knots.

At 1603:57, the second officer called, "One thousand feet," and at 1604:25, the sound of rain was recorded. Thirteen seconds later, Flight 66 was nearly centered on the glideslope when the second officer called, "Five hundred feet." The airspeed was oscillating between 140 and 148 knots. The sound of heavy rain could be heard as the aircraft descended below 500 feet, and the windshield wipers were switched to high speed.

At 1604:40, the captain again said, "Stay on the gauges." The first officer responded, "Oh, yes. I'm right with it." Twelve seconds later, the captain called, "I have approach lights," and the first officer replied, "Okay." As the aircraft passed through 400 feet, its rate of descent increased from an average of about 675 fpm to 1500 fpm. Flight 66 rapidly began to deviate below the glideslope, and four seconds later, the airspeed decreased from 138 knots to 123 knots in a period of 2.5 seconds. The aircraft continued to deviate further below the glideslope, and at 1605:06, when the jet was at 150 feet, the captain called, "runway in sight." The first officer immediately replied, "I got it." At 1605:10 the first officer commanded, "Takeoff thrust." The sound of impact was recorded at 1605:11.4.

Aircraft performance analyses. This study was conducted to determine the extent to which Flight 66, Flying Tiger 161, and Eastern 902 were affected by the winds they encountered during their approaches to runway 22L. Boeing, Lockheed, Douglas Aircraft, and the NASA Ames Research Center participated in the analyses.

The airplane's theoretical performance capability for a given set of conditions (weight, configuration, thrust, airspeed, and altitude), was established by a specific plot of vertical speeds versus longitudinal accelerations. When the values for the airplane's rate-of-altitude change and rate-of-airspeed change at a given instant were not compatible with the calculated theoretical performance capability, the differences were attributed to external forces on the airplane which were produced by changes in the vertical and horizontal components of the wind.

For Flight 66 and Flying Tiger 161, certain thrust settings and airplane configurations were assumed as a function of time. The analysis of Flight 66's data was based on cockpit conversations, other sounds recorded on the CVR, and standard operating procedures for Eastern's pilots.

Because Eastern 902 had the most sophisticated FDR of the three aircraft involved, researchers used those components to design a wind model. The total performance degradation caused by this wind model was nearly identical to the *calculated* performance degradation attributed to wind in the analysis of Flight 66's flight path. It appeared that Flight 66 and Eastern 902 encountered similar wind environments, except that Flight 66 experienced the conditions closer to the runway threshold.

The results of these analyses showed that Flight 66 probably encountered an increasing headwind as it descended on the ILS glideslope. The wind changed from about a 10-knot headwind at 600 feet to an approximate 25-knot headwind at 500 feet. The aircraft was approximately 8000 feet from the runway threshold when it descended through 500 feet. It then encountered a downdraft with peak speeds of about 16 fps, and the headwind diminished to about 20 knots as the aircraft descended to 400 feet. The speed of the downdraft abruptly increased to about 21 fps, and the headwind suddenly decreased from 20 knots to 5 knots over a four second period.

After close examination of this data, it was determined that the combination of downdraft speed and the rate-of-airspeed change *might* have exceeded the aircraft's static performance capability. Therefore, the aircraft could have still lost airspeed and/or altitude regardless to maximum thrust or compensatory flight control inputs.

Simulator tests. A Boeing 727 flight simulator, programmed with the dynamic winds and the flight characteristics of Flight 66, was used to assess the influence of pilot responses on aircraft control and performance. The approach environment, including the approach light system and ILS glideslope/localizer geometry, was modeled from runway 22L. The simulator was also modified to accept various combinations of vertical and horizontal wind changes, similar to those deduced from the foregoing performance analyses.

The objectives of the simulator tests were: (1) To examine the flight conditions that probably confronted the crew, and (2) to observe the difficulties that a pilot has in recognizing the development of an unsafe condition and in responding with appropriate corrective action.

Fourteen pilots participated in the tests; nine pilots were either currently or formerly qualified in 727 aircraft. Each pilot flew several approaches, beginning at the outer marker through one or more of the wind models. The pilots were told to maintain an airspeed of 140 to 145 knots, which was 10 to 15 knots above reference speed. They were given the option of attempting to land or executing a missed approach.

Of the 54 approaches that were flown, 18 reached an altitude which corresponded to an impact with the approach lights. Thirty-one missed approaches were flown successfully. Five approaches were completed, but only when the simulator coordinates were placed (unrealistically) over the runway threshold.

None of the pilots had problems bracketing the glideslope while the simulator descended to 500 feet. At 400 feet the simulator deviated rapidly below the glideslope and produced a 20 knot decrease in airspeed. Although the pilots were prepared for these cues, and most responded immediately with thrust increases and noseup control movement, there was still hesitation on the part of many of the participants.

The pilots who flew approaches which ended in a crash, were reluctant to add thrust because they were uncomfortable interrupting their instrument scan. Consequently, most of the pilots actually added less thrust than they thought they had added. Also, on several approaches the pilots did not rotate the aircraft to the nine-degree-noseup attitude commanded by the flight director, in part, because the back pressure needed to maintain the control column was more than they had anticipated.

Following the simulator tests, comments were solicited from the pilots. Seven of the 10 pilots who responded believed that their recognition of the effects produced by the wind would have been delayed had they disrupted their instrument flying to "go visual" during the descent through 400 feet. Eight of the 10 pilots believed that they might have crashed during the actual flight.

Eastern Air Lines' procedures. Eastern's altitude awareness procedures required the pilot not flying the aircraft call out the following information during an instrument approach: (1) 1000 feet above field elevation, airspeed, and rate of descent, (2) 500 feet above field elevation, airspeed, and rate of descent, (3) 100 feet above decision height or minimum descent altitude, and (4) decision height or minimum descent altitude.

Eastern Air Lines' bulletins. During the year preceding the accident, Eastern issued a number of bulletins on low-level windshear associated with thunderstorms and frontal-zone weather. Although the bulletins were informative and contained suggestions on how to anticipate low-level windshear, they did not provide specific flying techniques on how to overcome these effects. The bulletins implied that higher approach speeds should be used, but cautioned that when runways are wet excessive landing speeds should be avoided because of hydroplaning.

ATC decisions. The tower supervisors, aided with computer-generated analyses, were responsible for runway selection at Kennedy Airport. Criteria included: (1) safety, (2) noise abatement, and (3) operational advantages.

The assistant chief of the Kennedy tower testified that the 1500- to 2300-duty period generally was very busy. Shortly after 1500, he observed thunderstorms to the northwest of Kennedy. Thereafter he was busy coordinating various activities and did not notice the rain and lightning northeast of the airport. He was aware that Eastern 902 had abandoned its approach to runway 22L but did not know why. The assistant chief also did not know that Flying Tiger 161 had reported windshear and recommended that the runway be changed. He stated, however, that had he known of Flying Tiger 161's report, he would not have changed the runway because the surface wind was nearly aligned with 22L.

The local controller testified that he was aware of thunderstorms to the north of Kennedy about 15 minutes before the accident, but he considered them to be weak. He was very busy with his duties and did not have time to pass the reports from Flying Tiger 161 or Eastern 902 to the assistant chief. He stated that he did not consider a change of runway in response to Flying Tiger 161's recommendation because the official wind instrument was indicating the surface wind was nearly aligned with runway 22L. He further stated that it would take up to 30 minutes to change the runway.

The local-control coordinator testified that at approximately 1551 he observed a thunderstorm with considerable lightning north of the airport, and that there was heavy rain just off the approach end of 22L. He further described the rain as forming a solid wall beyond which he could not see. Although he witnessed severe weather at the approach end of 22L, he said that he and the local controller were very busy, and continued with their duties.

The final-vector controller testified that he saw on his radar screen the image of a small thunderstorm cell centered on the localizer course about the time he cleared Flight 66 for the ILS approach. The cell was located midway between the outer marker and the airport. He said that he was very busy with his duties and that he had received no report that windshear had affected Flying Tiger 161. The only report he had received was from Eastern 902.

Since the thunderstorm astride the localizer course to runway 22L was obvious, and since there was a relatively clear approach path to at least one of the northwest runways (31L), the Safety Board sought to determine why approach operations to runway 22L were continued.

In reviewing the testimonies of the controllers involved, the Safety Board was concerned that the runway-use program at Kennedy Airport was not being properly implemented. As noted earlier, the criteria for runway selection did not include wind direction. It did, however, include a "noise abatement" section. Under that criterion, a runway could not be used for more than six hours at a time. Since runways 31L/R had already been used in excess of the six-hour rule, the tower supervisor elected to continue 22L operations.

The Safety Board concluded it should have been apparent that the approach to runway 22L was unsafe had the thunderstorm activity been evaluated properly. Investigators further believed that ATC did not consider a runway change either before or after the Flying Tiger captain's recommendation because a change of runways would have increased traffic delays and ATC workload.

Flightcrew decisions. Although the crew of Flight 66 was made aware of the windshear through Eastern 902's report, the Safety Board believed that they remained skeptical ("I wonder if they're covering for themselves.") as to the severity of the conditions. Had the captain known that Flying Tiger 161 had also reported adverse weather, he might not have been so quick to question Eastern 902's encounter. However, the crew of Flight 66 did decide to increase their approach speed 10 to 15 knots ("In case he's right."). According to the Board, the other concern the captain might have had was his landing speed on a wet runway. He had asked the local controller about, ". . . reports on braking action . . .", and he might have been aware of the bulletin Eastern had issued on the hazards of "excessive landing speeds . . . wet runways . . . hydroplaning . . . ."

There were no comments from the cockpit until shortly before impact, which indicated that the crew was neither aware of nor concerned about the increased rate of descent. The Board believed that throughout this time period the captain probably was looking outside, because 6 seconds before they departed the glideslope, he called, "I have the approach lights." Seven seconds after they deviated from the glideslope, he called, "runway in sight." By this time, the aircraft was descending rapidly through 150 feet and was about 80 feet below the glideslope. Investigators further believed that the first officer's immediate response, "I got it," to the captain's identification of the runway, indicated that he too had probably been looking outside, or at best was alternating his scan between the instrument panel and the approach lights. Although Flight 66 was in heavy rain, the absence of significant turbulence might have caused them to underestimate the severity of the winds' effects.

The Safety Board recognizes the tendency of a pilot who is flying the aircraft to transfer from instruments to visual references at the earliest opportunity. This tendency is noted to be even greater when operating on approaches to runways like 22L because the ILS glideslope is designated unusable below 200 feet.

The final analysis

According to the Safety Board, the loss of Flight 66 and the near-accidents involving Flying Tiger 161 and Eastern 902 were the result of an underestimation of the significance of severe and dynamic weather conditions. In their conclusion, they stressed the need for ATC personnel, especially supervisors, to fully recognize the serious hazards associated with thunderstorms in terminal areas. When these conditions appear likely, ATC must be capable of adjusting the flow of traffic into terminal areas so that timely actions and rational judgments, in the interest of air safety, are primary to moving the traffic. In addition, the Safety Board said air carrier and NWS forecasters must emphasize the accurate and timely forecasting and reporting of severe weather conditions. Furthermore, the NWS must provide this information and other weather radar information to the ATC system expeditiously. The Safety Board also highlighted the continuing need for air carrier operations managers and dispatchers, in conjunction with captains of flights destined for high density terminal areas, to plan their operations around possible extensive delays.

Lastly, the Safety Board reiterated that the pilot-in-command is the final authority, including whether to conduct an instrument approach through a thunderstorm or other adverse conditions. They emphasized that pilots must exercise more independent judgment when confronted with severe weather conditions.

Lessons learned and practical applications

1. Keep your authority of pilot-in-command. With or without pilot reports, you are still your own best judge as to the situation at hand. If it looks bad, or you get that weird gnawing in the pit of your stomach, take a safer course of action.

   As the Safety Board noted in this case, when it comes to rapidly developing and maturing thunderstorms, the conditions can change significantly within a brief period of time and/or a short distance. Therefore, pilots must be aggressive in their decision-making process, and practice more independent judgment.

2. Do not underestimate the significance of severe and dynamic weather conditions. According to the Safety Board, ATC personnel, the crews of Eastern Flight 66, Flying Tiger Flight 161, and Eastern Flight 902 failed to recognize the seriousness of the thunderstorm activity that was along the approach path to 22L. As the investigation indicated, decades of flying experience and a thick logbook are no guarantees that you won't find yourself in a hairy predicament. A false sense of security ("If these guys ahead of me made it, so can I.") and complacency ("I've flown through horrific storms before.") are common attitudes. However, they serve no purpose in the cockpit, especially when confronted with potentially dangerous weather activity.

3. Become a student of meteorology. Understanding weather phenomena and how it relates to flight operations is probably one of the weakest areas of study for many pilots, regardless of experience levels. Yet, pilots' decisions often are based on that understanding. Therefore, a strong comprehension of meteorology is paramount to a safe flight. Consider doing some independent research on the latest discoveries pertaining to windshear, microbursts, thunderstorm activity, turbulence, and icing. Think about the benefits in attending a weather workshop or seminar.

4. Always make a pilot report (PIREP) after you've encountered windshear or any other potentially dangerous weather condition. PIREPs are a valuable resource in developing a keen sense of situational awareness. Pilots of the Finnair jet and the Beech Baron that landed just prior to Flight 66 testified that from the information obtained from the windshear alert they were able to plan their approaches accordingly, and land safely. Conversely, neither the crew of Finnair Flight 105 nor the pilot of the Baron reported their windshear encounters with ATC. The reason, according to one of the pilots, was that he knew the weather conditions had already been reported and that he believed the controllers were aware of the situation. The Safety Board acknowledged that pilots commonly rely on the rate of successes achieved by fellow pilots of preceding flights when confronted with common hazards. Therefore, the Board believes it was likely the captain of Flight 66 continued the approach dependent on additional PIREPs. When no reports were provided from the Finnair flight or the Baron, the captain apparently committed himself to the approach. You can never sound like a "broken record" when advising ATC and your fellow pilots of potentially dangerous weather activity. Pilot and controller workloads, and communication frequency congestion, can lead to omissions, assumptions, and confusion about who is aware of what. Therefore, consider being the eyes and ears for the airplane directly behind you. Just think of the better decisions you could make if each aircraft ahead of you reported an updated PIREP as it passed through a dynamic weather environment. Overall, everyone's situational awareness would be greatly enhanced. Additionally, providing crucial weather information to ATC enables them to get a clearer picture of a rapidly-changing flight environment. In turn, they will be able to make more timely and appropriate decisions pertaining to runway changes, traffic separation, and flow control.

5. Take PIREPs seriously. The crew of Flight 66 questioned the validity of Eastern 902's PIREP concerning the near catastrophic windshear they had encountered. Although Flight 66 eventually kept "a healthy margin" on the airspeed, it was still viewed with skepticism. Believe your fellow airman.

6. Ask for specifics when given a vague severe-weather advisory. When National 1004 requested pilot reports concerning weather conditions on final approach, the reply was only that there was, ". . . a windshear on short final." Since Eastern 902 and Flying Tiger 161 had provided ATC with graphic details concerning their windshear encounters, it was a good possibility that Flight 1004 could have received additional information had they pursued it.

7. Anticipate a busted forecast. The only forecast the crew of Flight 66 had received was just prior to takeoff. It predicted widely scattered thunderstorms turning into light rain showers. However, the Safety Board believed that by the time Flight 66 arrived in the New York terminal area, it would have been quite evident to the crew that they had a busted forecast. Consequently, the crew discussed their minimum fuel situation associated with landing delays, and checked the weather at their alternate. From the remark of one pilot, ". . . one more hour . . . we'd come down whether we wanted to or not . . ." provides a clear understanding of their time constraint. A runway change, at that point, would have taken about 30 minutes leaving only another 30 minutes of fuel remaining. Hence, the Safety Board addressed the need for air carrier dispatchers and captains to anticipate extensive delays in high density terminal areas and plan their flight operations accordingly.

8. Compensate for the greater of two evils. A portion of an Eastern Air Lines' bulletin that had been distributed the year preceding this accident implied that higher approach speeds should be used when windshear is anticipated, but warned of the dangers associated with hydroplaning due to the excessive landing speeds. The Safety Board believed that since the captain asked ATC about "braking action," he might have been concerned about those hazards. Consequently, the crew might have been in a more conservative mindset when setting their approach speeds. In this case, the "greater of two evils" was the *known* severe weather and *reported* windshear at the approach end of the runway. The *possibility* of hydroplaning should have taken a lessor priority.

9. Do not disregard your instrument scan during the final minutes of *any* approach. In their investigation the Safety Board recognized the temptation for a pilot flying the aircraft to glance away from the instruments to outside the cockpit, especially while on an instrument approach. However, they emphasized that visual references that are hindered by rain and poor visibility will only increase the chances of a pilot making inadequate and untimely decisions. From the CVR and FDR readouts, it appeared likely that the captain and first officer had momentarily shifted their attention from the instrument panel to visual references, shortly before impact. Seven seconds after the rate of descent

began to increase, the captain called, "runway in sight." The first officer immediately replied, "I got it," indicating to the Safety Board that he had probably been looking outside or was alternating his scan between the flight instruments and approach lights. Nevertheless, by that time the aircraft was rapidly descending through 150 feet and was about 80 feet below the glideslope. Provided the glideslope signal was working properly, there should have been a full-scale "fly up" indication from the flight director. All of which went undetected, most likely, because the captain and first officer had directed their attention toward the runway.

10. Ensure that the appropriate altitudes, airspeeds, and descent rates are called out. According to Eastern Air Lines' altitude awareness procedures, the non-flying pilot should make 1000 feet and 500 feet call outs, including airspeeds and descent rates. The second officer of Flight 66 partly complied with the 1000 feet and 500 feet calls but without the airspeed and descent rate information. As previously discussed, the captain and first officer had, at least momentarily, diverted their attention to outside the cockpit. Had the second officer added the airspeeds, and most importantly in this case the descent rates, the first officer might have had enough time to correct the situation.

11. Safety first. The criteria for runway selection at JFK listed safety as the first priority. According to the investigation, the controllers decided to disregard pilot reports and their own radar for the sake of noise abatement procedures and workload considerations. The Board concluded that had the overall situation been evaluated properly, runway 22L would have been closed to landing traffic. Furthermore, they considered the actions of the controllers were based on their own desire to not create additional work for themselves, thereby eliminating safety as their primary responsibility.

Case study references

Frost, Walter. 1983. *Flight in Low-Level Wind Shear*. NASA Contractor Report 3678. Washington, D.C.

Fujita, T. Theodore. 1976. *Spearhead Echo and Downburst Near the Approach End of a John F. Kennedy Airport Runway, New York City*. The University of Chicago.

National Transportation Safety Board. 12 March 1976. Aircraft Accident Report: Eastern Airlines, Inc., Boeing 727–225. JFK International Airport, Jamaica, New York. June 24, 1975. Washington, D.C.

CASE STUDY II-6: USAir Flight 1016

Safety issues: microburst, low-level windshear, weather dissemination, role of ATC, CRM

On 2 July 1994, a USAir DC-9 crashed after it encountered a windshear near Charlotte, North Carolina.

NOTE: At the time this book went into production the final NTSB report had not been released to the public. The information obtained in the following case study is a compilation of the NTSB's Factual Report of the accident, and the Air Line Pilots Association's analysis of the events.

Probable cause

The NTSB determined that the causes of this accident were, (1) the flightcrew's decision to continue an approach into severe convective activity that was conducive to a microburst, (2) the flightcrew's failure to recognize a windshear situation in a timely manner, (3) the flightcrew's failure to establish and maintain the proper airplane attitude and thrust setting necessary to escape the windshear, and (4) the lack of real-time adverse weather and windshear hazard information dissemination from air traffic control. These factors led to an encounter with a microburst-induced windshear that was produced by a rapidly developing thunderstorm located at the approach end of runway 18R.

Contributing to the accident were, (1) the lack of air traffic control procedures that would have required the controllers to display and issue ASR-9 radar weather information to the pilots of Flight 1016, (2) the Charlotte tower supervisor's failure to properly advise and ensure that all controllers were aware of and reporting the reduction in visibility, the runway visual range information, and the low-level windshear alerts that had occurred in multiple quadrants, (3) the inadequate remedial actions by USAir to ensure adherence to standard operating procedures, and (4) the inadequate software logic in the airplane's windshear warning system that did not provide an alert upon entry into the windshear.

ALPA submission

The Air Line Pilots Association's (ALPA) team of accident experts participated in this investigation under full-party status with the Safety Board. Although they concurred with a few of the points in the official report, the Association took strong opposition to the allegations against the flightcrew. In their 91-page document of findings, they provided the Board with a formal submission that detailed their extensive analyses with regards to the events that led to the accident.

ALPA's position is quite clear: Based on the limited weather information presented to them [the flightcrew] and the absence of specific windshear warnings, the crew initiated a normal go-around, not a windshear escape maneuver. The misleading and incomplete weather information [that] was relayed to the aircraft seriously biased the flightcrew's decision-making process, and thereby, encouraged them to proceed unknowingly into a rapidly deteriorating situation."

History of flight

Flight 1016 was a regularly scheduled passenger flight from Columbia, South Carolina, to Charlotte. The DC-9 departed Columbia at 1823 Eastern Daylight Time with 52 passengers and 5 crewmembers onboard.

Pilot experience

The captain had more than 9000 total flight hours. He had been with USAir since 1985 and was upgraded to DC-9 captain in 1990.

The first officer had approximately 13,000 total flight hours. He was hired by Piedmont Airlines in 1987, and had been with USAir since the two airlines merged.

Weather

Weather observations for the Charlotte area around the time of the accident were as follows:

1836: Special. Measured ceiling 4500 feet broken. Visibility 6 miles. Thunderstorms, light rain showers, haze. Winds from 170 degrees at 9 knots. Thunderstorms overhead, occasional lightning cloud to cloud.

1840: Special. Measured ceiling 4500 feet overcast. Visibility 1 mile. Thunderstorm, heavy rain showers, haze. Winds from 220 degrees at 11 knots. Thunderstorm overhead, occasional lightning cloud to ground.

The accident

NOTE: The following are excerpts from the NTSB Factual Report.

During the flight from Columbia to Charlotte, the crew told investigators that they saw no significant weather, although they did avoid some buildups. About 30 minutes from Charlotte, the crew performed their preliminary checklist, briefed for a visual approach, and obtained the ATIS information. According to the captain, ATIS was calling for a broken ceiling of 4500 and 5500 feet, that is was hazy and hot, and visual approaches were in progress.

The crew began their descent profile and were vectored on the west side of the airport for the downwind leg of an approach pattern to runway 18R. While south-southwest of the airport they noticed two cells. One south of the field and a very small one east of the airport, that they considered not to be a factor to the flight. The airborne weather radar showed the cell to the south as red in the center and surrounded by yellow. The crew said it did not look threatening and that there were no other cells on either side of it.

As they joined the localizer, the captain and first officer discussed the cell south of the airport. They decided that if they had to execute a go-around, they would turn right rather than fly straight ahead, as was called for in the published missed approach procedure. That way, they would avoid the cell. Shortly thereafter, the crew turned onto the final approach course and could see the airport and the runway environment. The first officer was manually flying the airplane and maintained a speed of Vref + 10 knots. He continued to reference the instruments while the captain looked outside. The captain was also checking the radar and monitoring the cell south of the airport.

As the flight passed the outer marker, the crew completed the final checklist and set the flaps to 40 degrees. They could see a rain shower trailing from the clouds between their aircraft and the runway but they could still see through it. The captain asked for a wind check and learned the wind had changed direction. There was no turbulence as the jet entered the rain. However, the captain asked ATC for any pilot reports (PIREPs) from aircraft that were ahead them. The response was, "smooth rides."

As the crew continued the approach there was still no turbulence, despite very heavy rain. The captain stated that he had not previously experienced rainfall as heavy. Seconds later, at 1200 feet msl, the captain commanded a go-around. He said that his decision was based on the sudden deterioration of visibility, the intense rain, the wind shift, and the potential hazards in landing on a wet runway with a cross wind.

The captain saw the first officer advance the throttles. As a procedural habit, the captain voiced the go-around setting of maximum power and flaps 15 degrees. The first officer was still manually flying the aircraft. Neither the captain nor the first officer were using the flight director.

The first officer went to maximum power and climbed and turned to the right, bringing the nose up to 15 degrees as he made the turn. He called for flaps to be positioned to 15 degrees. The aircraft was in heavy rain when he noticed a rapid decrease in airspeed. The captain called for firewall power as he placed his hand over the first officer's hand, which was already on the throttles. The first officer could not recall if the engine spooled up because the entire event happened so quickly.

Both pilots described feeling as though the aircraft had dropped out from under them. The captain said that he took control of the aircraft from the first officer, without announcing that he was doing so. He said this was not a conscious decision but that he perceived the situation was going badly. When asked if the first officer could also have been on the controls, he said that he did not believe so, because he did not feel any contrary inputs. The first officer, however, believed that he retained control of the aircraft.

Neither pilot remembered seeing a positive rate of climb on the vertical speed indicator and did not recall raising the landing gear. The captain could not give a rate of descent, but said there was a rapid decrease in airspeed.

The crew heard the stickshaker activate and the captain said he checked the yoke to recover. From the airspeed indication, the captain believed that they were experiencing a windshear. He looked out and saw they were below the tree tops. The ground proximity warning system's (GPWS) aural warning alerted, "TERRAIN." The captain said he knew that they would hit the trees but he tried to keep the wings level and control the aircraft.

The captain described the initial impact as not too hard. Then he saw the ground and the road. He noticed the nose dip and tried to pull back the yoke to keep from hitting in a nose-down attitude. The aircraft then struck the ground "extremely hard," followed by another impact, and then stopped. The crash was recorded at 1842:55.

Accident survivability

Thirty-seven passengers sustained fatal injuries. The captain and one flight attendant received minor injuries. The first officer, two flight attendants, and 15 passengers suffered serious injuries.

The investigation

NOTE: The following are paraphrased excerpts taken from ALPA's official submission, as summarized in the June/July 1995 issue of *Air Line Pilot* magazine.

ALPA believes that several ATC factors proved causal to the accident. A little more than five minutes before Flight 1016 was to land, ATC changed the approach from visual to an ILS. Information the crew received was, ". . . they got some rain just south of the field. Might be a little bit coming off north, just expect the ILS." However, just two minutes earlier, at 1835, the tower supervisor remarked to other controllers that it was, "raining like hell," on the south end of the airport. About the same time, the final-west controller noted a Level 3 cell "pop up" over or very close to the field. More state-of-the-art weather radars showed that a Level 5 storm was over the airport.

At 1838, the presence of lightning was discussed in the tower but not relayed to Flight 1016. One minute later, the crew of a USAir jet reported the storm and delayed their departure from runway 18R. That transmission was not on the same frequency as Flight 1016. By 1840, visibility had decreased to 1 mile, and a special weather observation was issued for thunderstorms and deteriorating visibility. This too, was not given to the crew. The runway visual range (RVR) of runway 18R had also decreased to about 500 feet. The ATIS was then updated to reflect the rapidly-changing weather.

At 1841, another USAir flight delayed its departure due to the weather. Seconds later, Flight 1016 entered heavy rain and a wind shift. About 17 seconds before the crew initiated the go-around, the tower supervisor noted windshear alerts in multiple quadrants but did not relay the information to Flight 1016.

Although the pilots were aware of the potential for windshear, ALPA believed the limited information they were provided suggested nonthreatning conditions. Their onboard radar did not display significant precipitation between the airplane and the field. When the captain asked for PIREPs, he was given two "smooth ride" reports.

Approximately 40 seconds before landing the aircraft entered light rain. Within seconds, the intensity of the rain increased dramatically. The captain lost sight of the runway. Crosswind reports were 100 degrees at 19 knots and 110 degrees at 21 knots. Based on the overall poor conditions and landing environment, he told the first officer to go-around. The jet was about 200 feet agl and over the middle marker when the first officer advanced the power, pitched up to 15 degrees, retracted the flaps from 40 degrees to the 15-degree go-around setting, and began a climbing turn to the right. This was in accordance with the crew's plan to avoid the cells at the departure end of the runway. The airplane climbed approximately 200 feet before the airspeed and vertical acceleration began to decrease rapidly. It was then that the crew firewalled the throttles. Twenty-two seconds after starting the go-around the airplane hit the ground.

Weather analysis. The NASA Langley Research Center later analyzed the microburst and windshear activity that Flight 1016 had encountered. The study found that a rainshaft, approximately 1 to 3 nm in diameter, preceded the downburst. The microburst produced a peak windshear of about 70 knots over a distance of one-half-mile. The downward vertical winds were around 14 knots, or 1400 fpm. The microburst reached peak intensity within two minutes of starting, just as Flight 1016 entered it. Within five minutes the microburst had essentially dissipated.

Researchers also calculated the Charlotte microburst by using an F-factor. This is a nondimensional value used to quantify the effect of a microburst on aircraft performance and is a function of horizontal shear, vertical velocity, and airplane velocity. F-factors are given as one-kilometer (km) averages. The Charlotte microburst had an F-factor of 0.3. The FAA considers an F-factor of 0.1 to be hazardous. By comparison, the microburst that caused the crash of Delta Flight 191 (refer to case study II-2) was calculated to have an F-factor of 0.25.

Windshear alerts. According to ALPA, the pilots did not recognize that they were in a windshear until a few seconds before impact. Their onboard Honeywell windshear-detection system never produced an alert and the northwest low-level windshear alert system (LLWAS)—located directly below where the pilots began their go-around—did not sound an alarm until the airplane had already crashed.

ALPA commissioned a special study of the Charlotte Phase II LLWAS and discovered two important points. One was that the northwest sensor was positioned in a wind "shadow," shielded from southerly winds by nearby trees that had grown taller since the sensor was originally installed. And two, internal FAA documents revealed that the agency had known about problems with the Charlotte LLWAS for more than a year but had failed to correct them.

The NTSB investigation also exposed the delay in the installation of the new, state-of-the-art terminal Doppler weather radar (TDWR) system at Charlotte. Designed to detect windshear and microburst activity, it was originally scheduled to be installed in early 1993. The field had been number 5 on an FAA list of 41 U.S. airports slated to have a TDWR installed, but was slipped to number 38 after several problems. The primary difficulty was due to private land acquisition for the TDWR site.

Flightcrew performance. Based on evidence obtained from the FDR, investigators questioned why the crew had increased power to only 1.83 EPR when the go-around procedure called for 1.93 EPR. In response, ALPA noted that: "This crew was conducting a normal go-around and did not initially perceive a significant threat. Standard procedures on aircraft without autothrottles is to bring the power levers to their approximate required position, allow the engines to spool up, and then 'fine tune' the power settings." They added, "In accordance with this procedure, this crew had accomplished the first two steps. Time and events prevented the orderly accomplishment of the third step; the throttles were firewalled when the crew recognized that the aircraft was not responding as required."

Another issue pertained to the actual control inputs made by the crew. During the go-around, the aircraft's pitch attitude decreased from 15 degrees nose-up to nearly 5 degrees nose-down. The Board believed that the captain suffered a somatogravic illusion at the beginning of the go-around, causing him to think that the airplane was pitching up too high. He then told the first officer to push the nose down, but, in their opinion, the first officer failed to follow proper CRM principles and complied without asking the captain why he should deviate from the prescribed go-around attitude.

ALPA believed the Safety Board did not adequately explore other possible reasons for the pitch change. In their opinion, there were two valid points overlooked. One, the pitch stability of the trimmed airplane reacted to the rapidly decreasing airspeed. Two, the first officer might have reflexively lowered the nose to slow or stop the decay in airspeed.

ALPA's final analysis

In ALPA's view the most important of the causal factors associated with the accident are:

1. Failure of the ATC system to relay numerous pieces of readily available, vital weather information to the crews.

2. Lack of systematic flow of weather information from the weather radar to aviation users.

3. Lack of terminal Doppler weather radar to detect microburst and windshear activity at the airport.

4. Lack of onboard predictive windshear-alerting system.

5. A failure of the onboard reactive windshear-alerting device that should have alerted the crew when they entered a windshear.

6. Deficiencies in the ground-based, low-level windshear-alerting system at the airport.

Lessons learned and practical applications

Until the NTSB final report is made available to the public, lessons derived from the specifics of this accident are unobtainable. However, there are a few general lessons that can be addressed.

1. Conduct crew briefings. The captain and first officer anticipated a go-around long before the situation arose. Therefore, they had the luxury of discussing modified procedures that they believed would meet their requirements to maintain a safe flight. It's very important to plan ahead to minimize miscommunication and miscoordination.

2. Never underestimate thunderstorm cells. Convective activity can spawn dangerous conditions in a matter of seconds. Level 2 or 3 cells can quickly turn into Level 4s or 5s.

3. Microbursts are sneaky. As we've discussed in other case studies, aircraft flying ahead of the accident airplane often have uneventful rides.

4. Typical "clues" might be absent. Even though an airplane is about to enter a microburst, and is already in rain, there might be little turbulence. This catches many pilots off guard into thinking that the storm is not that severe.

5. Issue PIREPs. Although in this case ATC did not relay this information to Flight 1016, it's still a good reminder as to the importance of pilot reports.

Case study references

National Transportation Safety Board. Preliminary and Factual Reports. USAir Flight 1016, DC-9-30. Charlotte, North Carolina. July 2, 1994. Washington, D.C.

Steenblik, Jan W. June/July 1995. "Mugged in Charlotte." *Air Line Pilot*: 16–19.

PART III

Collision avoidance

THE FOLLOWING CHAPTERS DETAIL THE BASIC FUNDAMENTALS AND technological advancements that comprise a total midair collision-avoidance concept. They include see and avoid: the myth; see and avoid: the reality; the role of air traffic control in a collision-avoidance environment; and the traffic alert and collision-avoidance systems (TCAS).

12
See and avoid: The myth

Aᴛᴛᴇʀ READING THIS CHAPTER, YOU SHOULD BE ABLE TO:

1. Understand the physical limitations in a collision-avoidance environment.
2. Apply effective scanning techniques.
3. Discuss optical illusions and other visual phenomena that affect the see and avoid concept.
4. Discuss the recognition and reaction factors in a collision-avoidance environment.
5. Explain how flying against a complex background affects see and avoid.
6. Explain the significance of the Minimum Visual Detection Angle theory as it pertains to collision avoidance.

See and avoid is an integral part of a total collision-avoidance concept and should never be taken lightly or disregarded in any flight regime. Most pilots are aware that the FAA has mandated the enforcement of see and avoid in FAR 91:67, which states:

> When weather conditions permit, regardless of whether an operation is conducted under instrument flight rules or visual flight rules, vigilance shall be maintained by each person operating an aircraft so as to see and avoid other aircraft.

Pilots might also be familiar with the FAA Aviation Circular (AC) 90-48C that describes recommended see-and-avoid techniques:

Effective scanning is accomplished with a series of short, regularly-spaced eye movements that bring successive areas of the sky into the central visual field. Each movement should not exceed 10 degrees and each area should be observed for at least one second to enable detection. Although horizontal back-and-forth eye movements seem preferred by most pilots, each pilot should develop a scanning pattern that is most comfortable and then adhere to it to assure optimum scanning.

There are, however, physical limitations that are intrinsic to this concept and its associated techniques. Although we are unable to change these inherent flaws, it is helpful to understand what the body can and cannot do in a collision-avoidance environment.

PHYSICAL LIMITATIONS

The myth behind see and avoid is based solely on physical limitations. Therefore, it's important to learn how each problem area affects the collision-avoidance environment, and how it relates to our scanning ability.

Let's begin with the myth behind the recommended scanning techniques mentioned in AC 90-48C. Remember, we aren't discarding these techniques altogether, but rather looking at the limitations that are associated with these methods. You'll see a couple of modified versions at the end of this chapter and in the next.

The side-by-side (Fig. 12-1A) and front-to-side (Fig. 12-1B) scanning methods are both based on the theory that ". . . traffic detection can only be made through a series of eye fixations at different points in space." For each technique, pilots should break the field of vision in blocks of 10 to 15 degrees, totaling 9 to 12 blocks in the scan area.

Notice any similarities between these methods and your own? Most likely, yes. But what are you really seeing with these scanning techniques? Not as much as you think or have been led to believe.

THE HUMAN EYE

Since these limitations center around our vision, it's important to start with the basics on how the human eye works. Figure 12-2 illustrates the primary parts of the eye.

It takes both the cornea and the lens to create an image on the retina. The iris forms the pupil, which protects the eye itself from changes in light by widening when it's dark and contracting in a brighter light. The retina is a delicate, multilayer membrane that provides the vital link between the eye and the brain. It consists of a network of minute receptors, otherwise known as *rods* and *cones*. The rods are used for night or low-intensity-light vision. They inhibit depth perception and supply no detailed acuity or color (only shades of gray) capability. The cones, however, are used in day or high-intensity-light vision. They give us the greatest level of visual acuity, the ability to detect small objects and distinguish fine detail, and depth perception, as well as the ability to see color. A combination of rod and cone vision is used at light intensities equivalent to dusk and dawn.

232

Fig. 12-1A. *Side-by-side scanning method. The pilot is to start at the far left of the visual area and make a methodical sweep to the right, pausing in each block one or two seconds. At the end of the scan, the pilot is to refocus to the instrument panel.* Adapted from FAA

Fig. 12-1B. *Front-to-side scanning method. The pilot is to start in the center block of the visual field [center of windshield] and move to the left while focusing in each block one or two seconds. Pilot is to then quickly shift back to the center block after reaching the last block on the left. The pilot should scan instrument panel before repeating the process in the right blocks.* Adapted from FAA

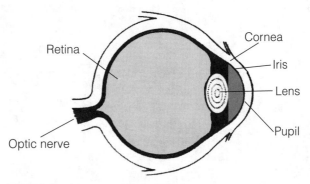

Fig. 12-2. *A cross-section of the human eye.*

These specialized nerve endings—rods and cones—convert light from the image that is on the retina to neural impulses. These impulses are then transmitted to the brain through the optic nerve. In essence, your brain, not your eye, processes and interprets the images that are displayed on your retina.

The center of the optic field, in the retina, is called the *fovea*. This is an extremely small region that covers slightly less than two degrees of visual field in which only cones are present. This is where visual acuity is at its peak. When we look directly at an object we are actually moving the image of that object onto the central fovea. Because that area provides the most lighted (cones) imagery, our best vision, theoretically, should be what appears directly in front of us. This however, does not always hold true for aircraft detection, which I will discuss later in the chapter.

The natural blind spot

The eye also has a natural blind spot that exists in the visual field of the retina through which the nerve fibers leave the retina to join the optic nerve. It is usually about 5 to 10 degrees in width. Refer to Fig. 12-3 and follow the directions. You might be surprised to see how wide your blind spot is.

Because the blind spot is on opposite sides of the visual field in the two eyes, each eye compensates for the other. This compensation is why we are generally unaware of its existence, although it does create a certain degree of visual deterioration.

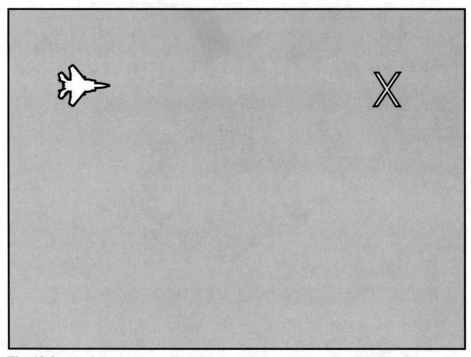

Fig. 12-3. *Blind-spot exercise. Cover your right eye and focus your left on the X. Move the diagram toward you until the airplane disappears.*

Clear vision

Now that we have a general idea of how the different parts of the eye function, it's important to understand how the eye actually focuses. The muscles that control the curvature of the lens tighten when there is a visual stimulus, such as an airplane. The lens adjusts, or accommodates, for the retina to focus on these objects. When an image is created on the retina the muscles tighten, which results in a focused vision of the target. Without this stimulus these muscles tend to relax, eliminating the ability to adequately focus on the object.

In 1966, Lockheed Aircraft Company conducted a study to determine the physical limitations of pilots in a dynamic environment. Although the research is not recent, it is classic and still provides valuable information concerning our visual scanning abilities. In part, the report stated that viewing targets at infinity creates an empty-visual field myopic [nearsightedness] effect. According to ophthalmology, infinity is any unlimited area of space greater than 20 feet. Because the eye does not have a sufficient visual stimulus, it tends to remain in constant movement, eliminating the ability to focus. This is caused by the image on the retina viewing nothing more than blank space. The brain, therefore, is unable to process a clearly defined image, which ultimately results in blurred vision.

Flying in total darkness, fog, a uniformly overcast sky, or a cloudless sky, all create a potential for this situation to occur. It can also happen when a pilot focuses on near objects, such as the instrument panel or dirt on the windshield. Most of the time we aren't even aware of this effect occurring, so we continue our scan of what seems to be several miles forward of the airplane when actually our search is effective only a short distance off the nose.

How does this relate to our ability to scan for aircraft? Very simply: Our eyes cannot focus from a close range object like the instrument panel to infinity of a featureless sky, then back inside the cockpit. That is a conclusion from the Lockheed report and a more recently conducted U.S. Air Force study on visual acuity.

Let's discuss the connection between the findings of the air force's study and the FAA-recommended scanning methods. Here's the scenario: You've just checked your altitude and heading while sitting in a shaded cockpit. You then look up and out into a clear-blue, cloudless sky, or a gray-white, solid overcast sky, and begin a side-by-side scan. What are you actually looking at? Infinity. Nothing. With no background or line of reference in which an airplane, especially one on a collision course, could stand out, you are literally staring into space. After the recommended 9 to 12 seconds, you complete your scan and then lower your eyes back inside the cockpit for an instrument and navigation check. A couple of minutes later, you repeat the process.

That probably sounds like a fairly routine flight, one that all of us have experienced countless times. With regards to see and avoid, however, it's quite misleading. Clearly, we are unable to rapidly, at one to two second intervals focus from an intricate object that's two feet in front of us to a blank sky several miles away.

Far-sighted fixation

Often pilots go to the other extreme and stare too long at objects or blocks of sky, in hopes of enhancing their ability to spot traffic. According to the same U.S. Air Force study, approximately 60 seconds after looking at a distant object, the eyes' focal points slips to less than ten feet. Take the previous scenario, and add a radio tower at your 12 o'clock position. You know the tower is about 10 miles away, and is near an uncontrolled airport. You realize that there's potential traffic in the vicinity, but you're also concerned about avoiding the tower. As you try to remain focused on the airspace above the tower, your sight becomes blurred and your field of vision quickly shifts from 10 miles to just off the nose of the airplane. Not good.

Saccadic eye movement

The Lockheed report examined the fact that human eyes cannot physically scan in a smooth rhythm. Rather, when searching for a target, they jump from one focal point to another in a number of fixations or *saccades.* Saccadic eye movement, as illustrated in Fig. 12-4, is actually a combination of four different types of eye motion ranging from very slow (2 to 5 times/second) to extremely fast (30 to 70 times/second). The distance between each saccade is as varied, extending from only 1 minute of arc to as much as 26 minutes of arc. While the eyes are in saccadic motion visual acuity is sharply decreased, leaving large gaps in the distant field of vision. From this evidence, the study concluded that during saccadic eye movement there is only a 35 percent probability of detecting another aircraft, even when the location of the target is known.

Not until the eyes have focused on an object and begun to track it does the saccadic eye movement suppress and the eyes automatically shift to a smoother scan. To demonstrate this, a U.S. Navy study on visual search used the following example: Hold up your index finger and have someone focus on it while you move it horizontally. You'll see that their eyes move smoothly as they follow your finger. Now, have the same person guide their eyes over the identical path, this time without focusing on any one object. Although there is no way for your friend to physically feel the saccadic movement in the eyes, you can see that person's eyes make a series of discrete jumps. Apply this example to scanning for other aircraft and you can begin to understand how these gaps in visual acuity occur.

This visual phenomenon obviously creates a real problem for the pilot while searching for traffic. A one- or two-second scan is not sufficient time for proper focusing or saccadic suppression. This short scan, therefore, further deteriorates our ability to see objects at distances greater than two miles.

Peripheral vision

Visual acuity is at its highest level when viewing objects directly in front of us. Since the fovea is only about two degrees in width, this leaves up to 178 degrees of detection range available through peripheral vision. This is one of the reasons why we tend to spot traffic or obstacles out of the corner of our eyes. Think back to those

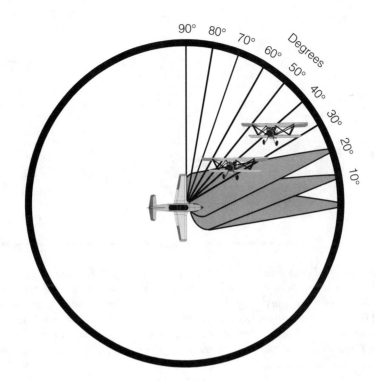

Fig. 12-4. *Saccadic eye movement. The shaded areas are the field of vision while scanning. The spaces between the tips of the cones are the visual gaps during saccadic motion. The gaps can be as much as 10 degrees, and at a distance of only one to two seconds.* Adapted from Naval Aerospace Medical Research Laboratory

times you have spotted other aircraft out of the corner of your eye. How close was that other airplane?

A person's acuity steadily drops the farther a target is from the central visual axis. However, as Fig. 12-5 illustrates, peripheral vision is the primary means of traffic detection. The U.S. Navy study included research on the relationship between acuity and central and peripheral vision. The visual detection lobe depicts this relationship, and also represents the range where visual perception of a target can occur. It's important to remember that the lobe represents the probability of detection, not a guarantee of detection.

Although the study concluded that peripheral vision is the key to rapid detection, it emphasized that peripheral vision does not have the detection range of central vision. Remember, that the primary reason we are able to see the other airplane peripherally is because it is closer, and obviously a larger target. But the probability that we would be able to avoid it decreases sharply.

Just how close to a target do we need to be before we have a chance of seeing it? First, you must know that peripheral vision is measured in degrees off the fovea. According to the Lockheed findings, at only six degrees off the fovea, an object must be

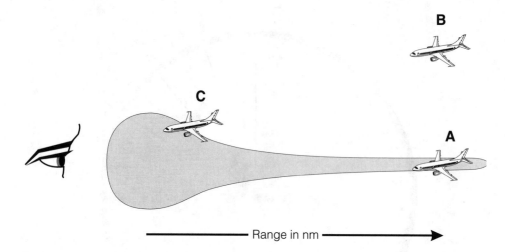

Fig. 12-5. *The visual detection lobe. The three positions—A, B, C— are of the same type of aircraft. Target A theoretically can be seen because it lies within the lobe, in the central field of vision. Target B is outside of the detection lobe but at the same range as Target A. Because it is not being directly viewed and acuity decreases rapidly once off the visual axis, the target cannot be detected at that angle and distance. Target C is at the same degree off the axis as Target B, but because it is closer it can theoretically be seen in your peripheral vision.* Adapted from Naval Aerospace Medical Research Laboratory

about twice as large as that seen in the central field of vision. At 20 degrees, the target must fill the width of a 10 minute arc, and at 70 degrees from the fovea the aircraft must fill the width of a 100-minute arc in order to be seen. In all three examples the threat aircraft must be uncomfortably close.

Eyesight and traffic detection

The condition of your eyesight is another factor concerning your overall ability to detect other aircraft. Throughout the years, there have been several studies conducted, primarily by the U.S. military, that have found pilots with normal 20/20 vision are twice as likely to spot targets as those with 20/30 or 20/40 vision. Additional evidence suggests that pilots with 20/10 or 20/15 vision further improves aircraft detection. There's no question that the better eyesight you have, the better chance you have to see other aircraft. Let's put this in perspective. If you were going to take an eye test with the chart placed one nautical mile away, the letters on the 20/20 line would have to be about nine feet tall. That's just 3 inches shorter than the tail of a Cessna 182. Now imagine trying to read those letters on the 20/20 line in two seconds, after first focusing on a chart on your knee. Having 20/20 vision does not guarantee collision avoidance.

Eyesight deteriorates with age. According to the U.S. Air Force study, peripheral vision can be up to 180 degrees when at age 20, but can drop to about 140 degrees by the time we're around 50.

Let's do a quick experiment. Look up and straight ahead. Is your peripheral vision close to 180 degrees? If so, place your hands at chin level and above your shoulders. Now slowly bring them forward and stop at approximately the 140 degree point. At a certain age, and it's different for everyone, your peripheral vision from where your hands are currently positioned to back where they started at the beginning of the experiment would be lost. It's easy to see what a big chunk of sky you would be missing if, or when, your peripheral vision drops 40 degrees. Our ability to detect targets peripherally is obviously reduced substantially.

Optical illusions and other visual phenomena

The Lockheed study also examined how the brain interprets the images displayed on the retina. Without the brain to process this information, we simply can't see. Sometimes, however, the brain gets fooled by the projected image, or might take a long time to evaluate the data. In these cases, valuable seconds that should be spent scanning, can be wasted.

Just like a mirage in the desert, optical illusions and other forms of visual phenomena are very common in a flight environment. In the air, these illusory images can easily create erroneous aircraft attitude orientation. By not being aware of these potential hazards, you can quickly find yourself in a turn when you think you're straight and level, or following traffic that doesn't exist. Therefore, your attempts at see and avoid can become totally ineffective.

Most illusions are day or night specific, so it's important to know what kind of phenomena can appear when.

Daylight illusions. *False reference.* According to the Lockheed study, the most frequently observed daytime illusion is slanting cloud layers. This illusion includes cloud formations, ceilings, or decks with sloping lines that are mistaken for horizontal. Here is what can happen: You have a widely spread cloud layer directly in front of you that looks level with the horizon. You check your ADI and notice you're in a steady, slow turn. You again look out ahead and the cloud reference still appears horizontal. But then you look out the side for a more panoramic view to discover the true horizon is not in the same position as the one straight ahead. The cloud formation that you've been using as a reference is actually slanted, but the illusion makes it appear as a level horizon. Before you caught this error and corrected for it, what were you unknowingly doing? Probably you were turning off course and possibly off altitude. Not a good situation.

Oculogyral. As you experience relatively fast acceleration and at the same time rotate your head, objects can appear to shift slightly away from their actual position. This can occur while you're scanning and rapidly increasing your speed at the identical moment. The airplane that you have spotted might seem to move to a different location. Once you stop rotating your head, even if you're still accelerating, your eyes will refocus and the airplane once again will be seen at its true position.

Oculogravic. This illusion is observed mostly while flying in a high performance (jet) aircraft. Viewed objects can appear to rise from their exact position during moderately high acceleration speeds. This is similar to an oculogyral illusory motion, but the acceleration speeds must be much greater. Instead of objects shifting laterally, they appear to slightly gain altitude. Although this illusion is not that common, even among fighter pilots, it is a visual phenomena that can skew an otherwise effective scan.

Night illusions. *Autokinesis.* This illusory effect occurs when a single stationary light appears to be moving. Here's the scenario: You are on a routine night flight. Earlier, as you were nearing the airport, ATC told you to "follow traffic" and you replied "traffic in sight." Suddenly, you realize that the aircraft you've been pursuing for the past several miles is a blinking red light on top of a radio tower. Shaking your head in disbelief, you keep asking yourself, "How could I have mistaken a stationary light for that of a moving, anti-collision light?" You just experienced an autokinetic illusion. Now for the really important question: "If that light wasn't my traffic, then where is it?" Above all else, notify ATC immediately when you have lost sight of your traffic.

False reference. It's not clouds this time that creates the illusion at night but rather stars or lights near the horizon. Because of atmospheric interference, stars tend to twinkle or appear in motion when they are near the horizon and can easily be mistaken for aircraft lights. This is a common occurrence, especially in the winter, when Sirius, the brightest star in the northern hemisphere, appears over the entire continental U.S. at one to three fist-widths above the horizon. On a night with scattered clouds, all you might see is a shaky bright light peeking in and out of the clouds, looking just like an aircraft, when it's actually Sirius.

Another similar event can occur at dusk. Venus is quite prominent at that hour, and pilots have mistakenly reported seeing everything from airplanes to satellites to even UFOs.

Night visual illusions have also led fighter pilots, who believed that they were rejoining their leader in a fingertip formation, to attempt a rejoin on a train or an isolated house with a porch light burning. Another hair-raising example is the general aviation pilot who landed on a road with streetlights. He thought it was a runway.

Oculogyral. The effects of this illusion at night are identical to those that occur during the day. However, the acceleration needed to induce the illusory motion is much lower.

Night myopia. You can experience this visual phenomena even with normal eyesight. According to the Lockheed report, the cause of night myopia is involuntary accommodation and a distortion of the retinal image due to the largely dilated pupil. Although the effects are not as severe as those experienced with empty-field myopia, night myopia still inhibits your ability to properly focus on objects.

Flash blindness. We have all experienced this phenomenon at one time or another. Ever walk into a dark room and look at the ceiling light as you turn it on? Or have you stared at a camera flash when your picture is being taken? The blurred vision, the white haze and even the jumping blue lights, are all symptoms of flash blindness. The same effects apply while night flying, prompting Lockheed to study the causes. Research found that it takes up to 30 minutes for eyes to adapt to darkness, but that ability can be totally lost after only 5 seconds of exposure to a bright light. Just by glancing at the moon or city lights—especially stadium lights—can be enough of a flash to blur your vision and temporarily blind you. This means you might experience inferior vision for the next 30 minutes. Not a good situation for a pilot to be in, especially at night.

If you must turn on the lights while flying, the general rule of thumb is to close or cover one eye so you can still retain some night vision capability. This presents a dilemma. Omitting about 90 degrees of visual field—central and peripheral—when you close one eye, or running the risk of seeing nothing but blurry bright spots if you keep both eyes open does not create an optimum see-and-avoid environment. Take a small flashlight, preferably with a red lens, with you while night flying. You eliminate the need to turn on the cabin lights to read a map, change a frequency, or cross check a darkly lit instrument panel. Your night vision can stay intact.

RECOGNITION AND REACTION

The see-and-avoid concept is comprised of four elements: eye perception, three neuron-reflex arc, muscle reflex, and the limitations of the aircraft.

For the eye to see—but not recognize—an object takes 0.1 seconds. The three neuron-reflex arc, a scholarly way to say how fast the brain is able to determine the object is truly a threat, is an amazing 10 seconds. That time includes recognizing the object you see is actually an aircraft, realizing you are on a collision course, and making a decision on how to avoid it. Take a look at the significance of these numbers. It takes slightly more than 10 seconds to visually and mentally process the situation—and here's the kicker—you haven't even begun to move your airplane out of his way. Add 0.4 seconds for muscle reflex time—and that's on a good day—for a grand total of 10.5 seconds.

OTHER VARIABLES

Let's take into consideration the numerous physical variables that might extend those 10.5 seconds. The list could easily go on at tremendous length, but for now, let's just highlight the more obvious ones.

1. *Task saturation.* High workload is a very big factor that prevents pilots from looking outside the cockpit. This is true anytime checklists or frequency changes are required in a critical phase of flight, especially on climbout and prelanding. Getting behind the airplane, through poor judgment and planning, distraction, or lack of knowledge and skill, can also create unnecessary amounts of work that take your attention away from your visual scan.

2. *Poor meteorological conditions.* Your ability to effectively see and avoid is obviously diminished when there is a cloud or layer of haze outside your window.

3. *Stress.* Research conducted at the Aviation Psychological Laboratory at Ohio State University has proven that pilots under a lot of stress, either personal or flight-related, often change their focus from the normal to a level of panic.

4. *Fatigue.* Tedious flights or long duty days are definitely fatigue-inducing. By the time you were setting up for the approach, you most likely were no longer in a vigilant state for recognizing a potential threat. Making good decisions were probably not high on your list, either. Inactivity produces fatigue. Studies have found that the visual effects of fatigue include reduced night vision and an increased susceptibility to vertigo.

Mental stimulation, in moderation, usually keeps the blood pumped, however, student pilots or pilots who fly infrequently often get overtaxed very quickly. Becoming acclimated to a flight environment requires a certain level of stamina, and before you achieve that, you can easily become fatigued.

5. *Smoking.* The most important reaction to smoking is the loss of sensitivity to light and a restriction of the visual field. According to a Lockheed study, carbon monoxide that is inhaled saturates the blood 210 times more than oxygen. It also takes between 6 and 24 hours for the carbon monoxide to leave the bloodstream. Because there is a decreased supply of available oxygen, the effects of smoking closely parallels those of hypoxia.

Even a mild case of hypoxia is detrimental to vision, especially at night. There is a 5 percent loss of night vision as low as 4000 feet, at 6000 feet the loss increases to 10 percent, and at 10,000 feet the visual loss jumps to 20 percent.

6. *Alcohol.* The same study also concluded that alcohol reduces depth perception, the ability to distinguish different brightnesses, and produces effects similar to hypoxia. Often, the only chance to detect other aircraft is through observing a target in peripheral vision—a time when depth perception becomes important—or in a luminescent environment where it would be best if we could still differentiate levels of contrast. Remember, we have only two degrees of foveal area to begin with, and drinking or smoking further shrinks our overall visual field.

7. *Legal drug and medications.*The Lockheed report also found that certain over-the-counter drugs and prescription medications can affect vision. Antihistamines and sulfonamides, a sulfa-based antibiotic used for broad range illnesses such as upper respiratory infections and bladder infections, can alter depth perception and acuity. These drugs can also enlarge our natural blind spot, which can limit our peripheral vision and create a loss of muscular coordination of the eyes. Losing muscular coordination can hinder our ability to properly focus.

Additionally, aspirin used in excessive quantities can have these same negative effects on your vision. For example, persons with painful arthritis might take three times the amount of aspirin as one would normally take for a bad headache.

Aircraft lag time

The study allotted a minimum of two seconds for aircraft lag time, measured under ideal conditions. Numerous factors could easily prolong that lag time. The size and wingspan (the bigger the airplane, the greater the chance of it not completely clearing the threat aircraft), the turn radius (don't expect every airplane to be able to turn on a dime), and the condition of the aircraft (an older and ill-maintained airplane is not going to maneuver like it did when it was new).

A final thought

The study concluded that it takes a minimum of 12.5 seconds to recognize a threat (see) and react to it (avoid). That's a sobering thought when you think you have spotted all of your traffic only to realize that the bug-splatter on your windshield has turned into another airplane coming straight at you. When you're in the middle of the see-and-avoid process, 12.5 seconds can truly go by with a blink of an eye.

BACKGROUND CLUTTER

You might take for granted that, at the very least, you would be able to recognize an object as that of an airplane. That depends on whether the view behind the target has a distinct and clear contrast, or one that is complex and not uniform.

Flying against a clear contrast

It's important to understand how and why contrast affects ability to recognize a target. The scientific answer is that contrast is a ratio between the difference in luminance of an object and the background it is viewed against; and the luminance of the background itself. The Lockheed study proved that the threshold contrast, the least contrast required for an object to be detected against its background, decreases as luminance increases. This simply means that the ability of the eye to observe changes in the brightness of objects increases as illumination increases. That finding—it's easier to see in the light than the dark—was, of course, expected. But, when a color contrast was added to the experiment, notable data were collected. When there is a high brightness contrast composed of different colors between the target and the background, visual acuity improves only slightly. However, when the brightness contrast is low, color contrast can improve visual acuity appreciably. Brightly colored aircraft are easier to notice in an overcast sky, and white aircraft are best detected in a darkened sky. A white- or tan-colored airplane can be extremely difficult to detect against a bright, sunny sky.

Flying against a complex background

Visual search of an aircraft over a heavily populated area or downtown sections of a city is extremely difficult. At times the target can disappear altogether. These regions have the greatest diversity in background and contrast. They are filled with varying building heights, houses, streets, shadows, reflections, and dozens of other city landscapes. Unlike the previously discussed conditions where a pilot has the chance of noticing a discontinuity against a more uniform background, flying amidst buildings and in and out of shadows, makes recognizing an aircraft on a collision course nearly impossible.

Helicopters frequently fly in a complex background in and around cities. They have unique shapes and flight characteristics and not necessarily a form with which you are familiar. For a different, yet realistic, look at how difficult, and sometimes impossible, it is to recognize aircraft against complex backgrounds, look at the following illustrations.

Figure 12-6A depicts the baseline image of how a helicopter should appear at ½ mile against a bluish-gray sky with a bright contrast level (100 mL) of luminosity. Although we can't identify any specific details unless there is bold lettering on the side, we can, at least, recognize the object has an outline of a helicopter. Vertical-flight buffs could even venture to guess the make and model of the target.

Figure 12-6B shows the same helicopter flying in the same conditions at one mile away. All detail has been lost, but the rotor wing characteristics are still marginally evident and the general shape is intact. The image can be perceived as that of an helicopter.

Figure 12-6C illustrates the same helicopter viewed at two miles. This image, at best, would be observed as a discontinuity against the background sky. With that in mind, you could take measures to monitor the path of this potential traffic. Once the target got closer, it would be likely that you could recognize the shape as a helicopter, and most importantly, its course of flight.

Figure 12-7A illustrates that same helicopter at ½ mile, but this time it's set against a complex-patterned background of two buildings with shadows casting near the helicopter's tail rotor, and behind the fuselage. You might be able to identify this target if you catch a glimpse of movement, or if the sun just happens to reflect off the aircraft at just the right angle. What is most likely to occur, however, is that the helicopter will blend in with the buildings. Don't even hope to see a white aircraft against a white building. A detection of relative motion will be lost, and the sun will not only be reflecting off the aircraft but every other piece of glass and metal structure.

Look at Figures 12-7B and 12-7C. They depict the aircraft at one and two miles, respectively. Either image only remotely resembles a helicopter. Remember what we need in order for our eyes to focus: a visual stimulus. Do we have one? No. Instead, we have a blob looking like an ink spot that wouldn't trigger any eye-brain response, and therefore, would go undetected during our visual search. Likewise, there are no distinct discontinuities against a complex background because everything is viewed as a discontinuity.

THE FINAL MYTH

The Lockheed study also presented analyses concerning the two primary visual detection principles. The Minimum Visual Detection *Angle* theory (myth) and the Minimum Visual Detection *Area* theory (reality) have been debated over the years as to which one is the most accurate in a see-and-avoid environment. The evidence that evolved from this research can logically support only one theory. We'll discuss the relevance of those principles in a moment.

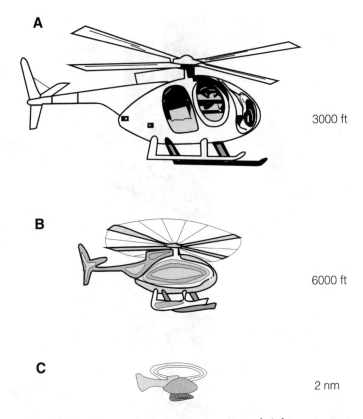

Fig. 12-6A. *Image of a helicopter against a bright contrast at a distance of 3000 feet.* **Fig. 12-6B.** *Image of a helicopter against a bright contrast at a distance of 6000 feet.* **Fig. 12-6C.** *Image of a helicopter against a bright contrast at a distance of 2 nm.*

A practical look at visual detection

Because these theories are founded on rudimentary principles of geometry, it might be best to have a quick review. Figure 12-8 illustrates degrees, minutes, and seconds of arc and how they relate to seeing close or distant objects.

Let's apply this concept to real life. Without getting too cosmic, visualize the above diagrams in relation to flying. We already know that there are 360 degrees in a full circle of arc. As you sit in your airplane, you have a 360 degree view of the sky. The wedges in Figs. 12-8 A, B, C, and D are progressively smaller amounts of sky you can see within in that 360 degree circle around your aircraft. As you can see, 1 minute of arc is a very small piece of sky; one second of arc is even smaller. With this information, let's discuss the visual detection theories.

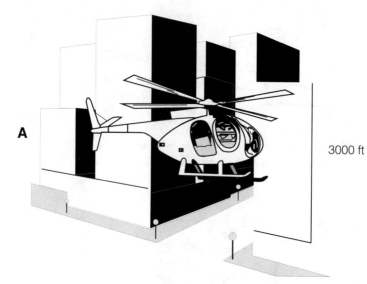

3000 ft

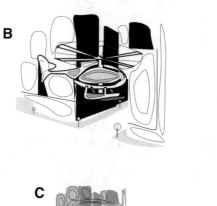

6000 ft

2 nm

Fig. 12-7A. *Image of a helicopter against a complex background at a distance of 3000 feet.* **Fig. 12-7B.** *Image of a helicopter against a complex background at a distance of 6000 feet.* **Fig. 12-7C.** *Image of a helicopter against a complex background at a distance of 2 nm.*

The Minimum Visual Detection Angle theory (refer to Fig. 12-9) is measured as a single dimension. This simply means that an object is viewed as a straight line, without any depth or height. Obviously, a real airplane is not two-dimensional and cannot be referred to as simply a drawing, therefore, this theory loses credibility.

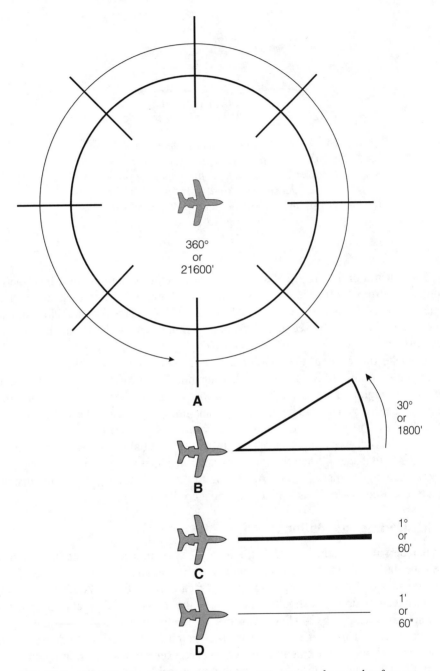

Fig. 12-8. *The relationship between degrees, minutes, and seconds of an arc. A depicts a circle of 360 degrees. B illustrates an arc (or "wedge") of 30 degrees from the 360-degree circle. C shows 1 degree of arc. D represents 1 minute of arc.*

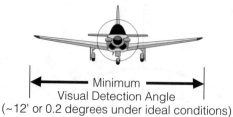

Minimum
Visual Detection Angle
(~12' or 0.2 degrees under ideal conditions)

Fig. 12-9. *Minimum Visual Detection Angle theory. According to the theory, in a head-on or overtaking encounter, the opposing pilot should detect the threat aircraft when its wingspan fills 12 minutes (0.2 degree) of arc. To scale with Fig. 13-1.*
Adapted from Lockheed Corporation

The illustration in Fig. 12-9 shows a general aviation aircraft in a head-on encounter. According to the theory, we would have a high probability of seeing this airplane as soon as it fills a visual angle of 12 minutes (0.2 degrees) of arc. Now, quickly refer back to Fig. 12-8C to see just how small 12 minutes of arc is. Then, visualize that area of sky in relation to your 360 degree field of view.

To make this concept even easier to understand, let's do a little exercise right where you're sitting. Make a circle with your thumb and forefinger (an "OK" sign), and look through it at arm's length. That's about a 3 degree circle. Now according to the Angle theory, you should be able to spot an entire airplane in a head-on encounter at 0.2 degrees. That equates to ¼₄ the size of your "OK" sign. Here's a good example of what ¼₄ of 3 degrees might look like. Take a typical ball-point pen, turn it upside down, and hold it at arm's length. Spot a distant object that is the same size as the pen's clicker. Amazingly small, isn't it? According to the Angle theory, that is the distance we are supposed to see a head-on airplane on a collision course.

Misleading calculations

The numbers used to support the Angle theory can be misleading. For this example, I've used the dimensions of a Grumman Cheetah. The aircraft's wingspan is 32 feet and the width of the fuselage, including prop clearance, is 6 feet. After doing a little long-hand calculation, you should expect to see the threat at 9169 feet (1.5 nm), or at 120 knots for each plane (240 knots closure), a time to impact of 22.6 seconds.

But, beware. This is one of those sneaky "gotchas" that can fool you unless you know on which theory the calculations were based. Remember how small your object was in the ball-point exercise, even though you probably chose something quite substantial like a house on a hill or a gigantic oak tree? Just think of how much more difficult it would have been to spot your object if you had chosen a high-tension wire. That's exactly what an airplane looks like as it appears motionless off your nose. The thickness of wings are so narrow, it makes them nearly invisible to detect at a safe distance. And the width of the fuselage, still using the Cheetah as an example, is approximately one-fifth the size of the wingspan.

CHAPTER REVIEW

See and avoid

- Evolves from FAR 91.67
- AC 90-48C recommendations

Scanning methods

- Side-by-side method. Pilot is to start at far left of visual area, and make a methodical sweep to the right, pausing in each block one or two seconds. Pilot ends each round of scan by refocusing on instrument panel.
- Front-to-side method. Pilot is to start at center of windshield, move left of visual field, focusing in each block one or two seconds. Pilot ends each round of scan by refocusing on instrument panel.

Physiology of the human eye

- Cornea and lens create image on retina.
- Iris forms pupil.
- Pupil protects eye from changes in light, by widening when it's dark and contracting when it's bright.
- Retina provides eye-brain connection.
- Retina consists of receptors (specialized nerve endings), known as rods and cones.
- Rods are used for night or low intensity light vision.
- Visual acuity is the ability to detect small objects and distinguish fine detail.
- Rods inhibit depth perception, supply no detailed visual acuity, or color capability.
- Cones are used in day or high intensity light vision.
- Cones provide the greatest level of visual acuity, depth perception, and the ability to see color.
- The combination of rods and cones are used at light intensities equivalent to dusk and dawn.
- Rods and cones convert the light from the retinal image to neural impulses.
- The brain processes and interprets the retinal images.
- Fovea is the center of the optic field, in retina.
- Fovea is about 2 degrees in width.

- Only cones are present in fovea, which produces the best level of vision and acuity.
- Natural blind spot exists in the retina where the nerve fibers leave the retina to join the optic nerve.
- Natural blind spot is usually 5 to 10 degrees in width.
- Eye focuses when the muscles that control the curvature of the lens tightens due to a visual stimulus.
- Accommodation is when the lens adjusts to pet retinal focus of images of objects at different distances.
- Without a visual stimulus, the eye cannot properly focus.
- Infinity is any unlimited area of space greater than 20 feet.
- Empty-visual field myopia is caused by viewing objects at infinity.
- Flying in total darkness, fog, a uniformly overcast sky, or a cloudless sky, creates the potential for empty-visual field myopia to occur.
- Frequent focusing of the instrument panel or dirt on the windshield can also produce empty-visual field myopia.
- Eyes cannot focus from a close range object to infinity and back to a close range object instantly.
- Staring at a distant object for one minute, allows the eyes' focal points to drop to under 10 feet.
- Saccadic eye movement occurs when the eye jumps from one focal point to another, while searching for a target to focus on.
- Saccadic eye movement sharply decreases visual acuity and leaves large gaps in the distant field of vision.
- During saccadic eye movement, there is only a 35 percent chance of target detection, even when the location of the threat aircraft is known.
- Saccadic eye movement is suppressed when the eye is focused on an object.
- Peripheral vision is measured in degrees off the fovea.
- Visual acuity drops the farther away from the fovea.
- Peripheral vision aids in rapid target detection because the threat aircraft is closer [appears bigger], and is flying at an angle.
- Visual detection lobe represents the relationship between acuity and central and peripheral vision.
- 20/20 eyesight enhances visual search by twice as much as 20/30 or 20/40.

Optical illusions and other visual phenomena

- False reference (day) includes cloud formations, ceilings, and decks with slanting cloud layers.
- False reference (day) creates the illusion that a sloped horizon is actually level.

- Oculogyral (day) occurs when pilot is accelerating and rotating head simultaneously. Objects appear to shift laterally.

- Oculogravic (day) occurs when pilot is rapidly accelerating and rotating head simultaneously. Objects appear to rise vertically.

- Autokinesis (night) occurs when a single stationary light appears to be moving.

- False reference (night) occurs when stars or lights near the horizon are mistaken for aircraft.

- Oculogyral (night) occurs when pilot is accelerating at a moderate speed, and rotating head simultaneously. Objects appear to shift laterally.

- Flash blindness (night) occurs when the dark adaptation of the eye is interrupted by bright lights.

- The eye can take up to 30 minutes to adapt to darkness, and lose it after only 5 seconds of exposure to bright lights.

- When flying at night, use a flashlight (red if possible) for cockpit tasks.

Recognition and reaction factors

- Seeing an object is not the same as recognizing what the object is.

- Three physical elements: Eye perception, three neuron-reflex arc, and muscle reflex.

- Fourth element: Aircraft limitations.

- Minimum recognition and reaction time for physical elements is 10.5 seconds.

- Other factors that might increase the 10.5 second baseline are:
  ~Task saturation
  ~Poor meteorological conditions
  ~Stress
  ~Fatigue
  ~Smoking
  ~Alcohol
  ~Medications

- Minimum aircraft lag time

Background and contrast

- Contrast is a ratio between the difference in luminance of an object and the background it's viewed against; and the luminance of the background itself.

- Brightly colored aircraft are easier to detect in an overcast sky.

- White colored aircraft are easier to detect in a darkened sky, but more difficult to observe against a bright background.

- Distant aircraft might be detected as discontinuities against a clear or uniform background.

- Aircraft viewed against complex backgrounds are extremely difficult to detect, and might entirely disappear from view.

Geometry

- 1 minute of arc equals 60 seconds of arc.
- 1 degree of arc equals 60 minutes of arc.
- 0.2 degrees of arc equals 12 minutes of arc.
- 360 degrees is equal to a full circle of arc.

Minimum visual detection angle theory

- Theory is measured in a single dimension, therefore, an object is viewed as a straight line.

- Theory states that entire threat airplane can be seen when it fills a visual angle of 12 minutes of arc.

- Calculations can be misleading.

CHAPTER REFERENCES

Edwards, Gerald, James Harris, Sr. 1972. *Visual Aspects of Air Collision Avoidance: Computer Studies on Pilot Warning Indicator Specifications*. NASA.

George, Fred. April 1991. "Can You See in Time to Avoid?" *Flying*: 81-4.

Hanff, G.E. 1966. *Collision Avoidance Visibility*. Lockheed Corporation.

Howell, Wayne. 1957. *Detenation of Daytime Conspicuity of Transport Aircraft*. Civil Aeronautics Administration.

Krause, Shari Stamford, Ph.D. 1995. *Avoiding Mid-Air Collisions*. Blue Ridge Summit, Pa.: McGraw-Hill.

Lorenz, Fred H. May 1992. "Visual Approaches." *Air Line Pilot*:17-21.

Rosdahl, C. *Textbook of Basic Nursing*, 4th edition. Philadelphia: J.B. Lippincott, 1985: 118.

Steenblik, Jan W. October 1988. "The Eyes Don't Have It." *Air Line Pilot*: 10-16, 57.

Selected Statistics Concerning Pilot-Reported Near Midair Collisions (1984-1987). 1989. U.S. Department of Transportation. Federal Aviation Administration.

"Vision System Puts Eyesight in Blind Spots." 27 April 1991. *Science News*: 262.

13
See and avoid: The reality

After completing this chapter, you should be able to:

1. Discuss the Minimum Detection Area theory as it pertains to collision avoidance.
2. Apply effective scanning techniques.
3. Prepare a personalized collision-avoidance checklist.

Now that the myths of see and avoid have been exposed, learning how to realistically compensate for those potential hazards will further enhance your ability to avoid mid-air collisions.

The Minimum Visual Detection Area theory (Fig. 13-1) is based on measuring a circle with the diameter of the Minimum Visual Detection Angle. Very simply, it states that the fuselage, not the entire airplane, of the threat aircraft must fill 12 minutes of arc before there is a high probability of seeing it. By using a Cheetah as an example, we can determine just how terrifyingly close an airplane has to be before you recognize the seriousness of the situation. As discussed earlier, the wingspan is not a practical variable when calculating the range of visual detection. Therefore, using the six-foot diameter of the fuselage and prop clearance area, the distance between both airplanes equates to approximately 1720 feet away.

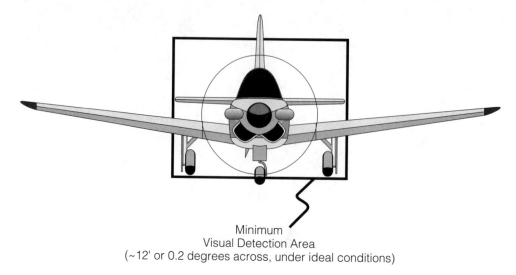

Minimum
Visual Detection Area
(~12' or 0.2 degrees across, under ideal conditions)

Fig. 13-1. *Minimum Visual Detection Area theory. According to the theory, in a head-on or overtaking encounter, the opposing pilot should detect the threat aircraft when its fuselage fills 12 minutes (0.2 degree) of arc. To scale with Fig. 12-9.* Adapted from Lockheed Corporation

Now let's put that 1720 feet range into perspective. If each general aviation aircraft is traveling at 120 kts head-on, at the same altitude, their closure rate is 240 kts or 405 fps. An aircraft that size, with that closure, is 4.2 seconds away from impact. If it takes slightly more than 10 seconds to recognize and react to an object, you can see that the impact would occur six seconds before you could respond.

THE SCANNING TECHNIQUES: A revised version

A clear sky, without features upon which to focus the eye, frequently is a challenging environment for maintaining a disciplined visual lookout. When there is no outstanding visual reference at a distance, our eyes refocus to a position about 10 to 20 feet away. When we glance back inside the cockpit for an instrument or map check, our eyes once again must refocus, this time at only one to two feet. In addition to our eyes having to focus and refocus between our scanning and flying the airplane, there are many common annoyances that redirect our attention. That bug smashed on the windscreen directly in front of our faces can even cause us to refocus our eyes. It also holds true that when we bring the aircraft's wings, or cowling, into focus, distant objects lose their detail. It is no wonder that our eyes can be in an almost constant focus-refocus mode throughout an entire flight. Therefore, a balance must be met with seeing both distant and very near objects.

The constant refocusing is not bad; if we didn't refocus, an instrument crosscheck would be impossible. But to ensure safe separation from other aircraft, that all important visual scan must also be tuned to the appropriate range to be most effective. Refer once again to Fig. 12-1A. The FAA says that these patterns can be accomplished in about 10 seconds. Conversely, a navy study on visual acuity says that five seconds are needed for each block of the scan. Obviously, there is a discrepancy. This difference of opinion becomes quite apparent when scanning in a featureless sky.

It is possible to do complete and disciplined scan patterns in every 10 to 20 seconds as the FAA recommends, but since the eye isn't focused beyond the length of the plane, a long range visual contact on a relatively small object, without the assisting cue of motion, is unlikely. Therefore, that head-on Bonanza, not moving on the canopy because it is on a collision course, keeps closing in range, unnoticed.

So how can we conduct a focused, but complete *and* rapid scan pattern in a disciplined manner? How can we follow FAA guidelines and respect the research done on scanning and focusing? One possible solution is a technique taught to military fighter pilots for detecting fast-moving, enemy aircraft.

Figure 13-2 demonstrates the technique of using clouds, cultural features, or constructed objects to focus the eye during a scan pattern. In each scan block, note an object, maybe a cloud, building, or pier. Pick an object several miles away, about the distance you want to search for other planes. The faster your plane travels, the farther away in front of your plane you want to look—about 5 to 10 miles is a compromise distance for light, general aircraft. According to the navy study, this should take three to five seconds, after which, you immediately scan your block. Move to the next block, find an object, and repeat the procedure. Vary distances to ensure a thorough scan and to reduce fatigue. Remember to also look above and beneath your altitude.

A common mistake pilots make is to concentrate too long on only the horizon. The reason we fly with a stick or yoke instead of a steering wheel is because planes operate in three dimensions. This is very important because the big-sky theory breaks down without warning. Even fighter planes collide, and they have multi-million dollar radars to avoid each other. There is no substitute for looking out the windows and scanning correctly and intently.

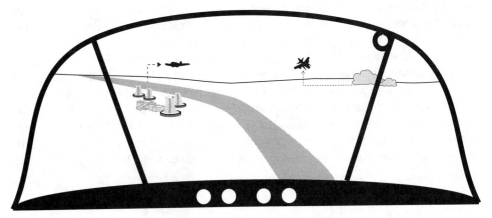

Fig. 13-2. *Example of a scan pattern in a clear, featureless sky.*

Another problem with a featureless sky is that it is usually associated with a nice day. On nice days the air is calm, the sun is bright, and everything is wonderful and trouble free at 4500 feet. Then . . . BAM! It's called complacency and must be guarded against. Performing a disciplined visual search pattern will provide ample opportunity to look around, but it will also prevent midairs.

Entering the pattern of an airfield, particularly an uncontrolled airport on a clear day, is notoriously hazardous. There are more aircraft airborne in fine weather than on a dreary day, and more complacent, unfocused pilots flying carelessly. Use proper focusing and scanning techniques to enter the pattern. Focus on the runway or tower from a distance, then look up to pattern altitude for your initial entry. Also look for high-to-low and low-to-high planes on takeoff or landing. Expect the occasional but usually unexpected blundering pilot who stumbles through the pattern without ever realizing other planes are aloft. Proper scanning, particularly in clear weather, will avoid unpleasant surprises caused by unfocused eyes and complacency.

More on scanning techniques

A 1990 Naval Aerospace Research Laboratory study noted that the fixations that occur during a scan require about ⅓ of a second. As we saw in Fig. 12-5, the closer a target is to your plane, the less fixations you need to scan due to the larger visual detection lobe. Besides the fact that closer planes are bigger than distant ones, fewer scans are required, and a visual pick-up is much more likely than during a long-range acquisition process.

The naval study goes further to recommend that visual search should be spaced widely apart to help cover a large area of the sky in a minimum of time. This also agrees with the FAA's basic premise of rapid scanning. But the navy study tells us to not expect long-range visual contacts; inside of 2 nm is more likely. To help make the most of the eyes we have, a *systematic* visual search is recommended.

A *sector scan* is one method of breaking-up the sky. A scan pattern recommended by the FAA appears in Figs. 12-1A and 12-1B, but the navy study divides the sky into horizontal and vertical sectors (see Fig. 13-3). Although all aircraft do not have two sets of eyes, a bubble cockpit, and the ability to roll upside down to scan 360 degrees in all directions, it is possible to see how we might divide the sky into more manageable sections to concentrate our scan patterns.

The FAA Aviation News reported that: "You can generally avoid the threat of an inflight collision by scanning 60 degrees to the left and right, and 10 degrees up and down." Figure 13-4 graphically depicts the areas recommended by the FAA to concentrate your visual search. By combining the naval sector idea (Fig. 13-3) with the side-to-side or front-to-side scanning patterns (Figs. 12-1A and 12-1B) with a concentration of scanning in the direction of your flight path (±60 degrees azimuth and ±10 degrees elevation), you might develop a personalized scanning technique. Be sure to

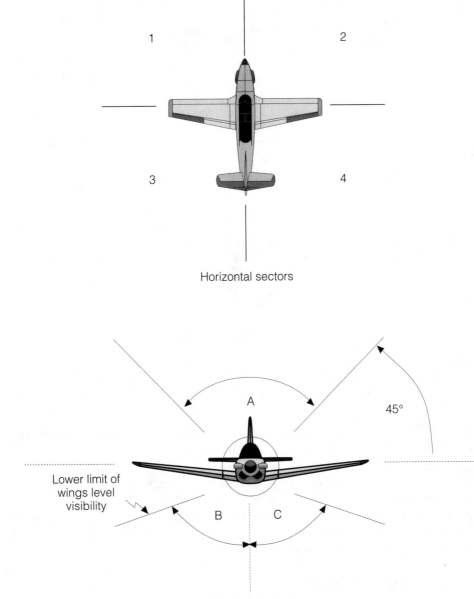

Fig. 13-3. *U.S. Navy-recommended visual-search sectors. Divide the sky into horizontal and vertical sectors.* Adapted from Naval Aerospace Medical Research Laboratory

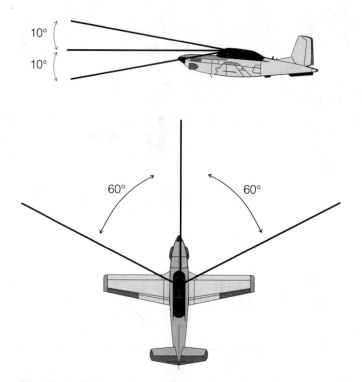

Fig. 13-4. *FAA-recommended visual-search sectors. Scan 60 degrees to the left and right, and 10 degrees up and down.*
Adapted from FAA

include or try techniques presented by CFIs, and build habit patterns for yourself. The best technique in the world does no good if you forget to use it.

Clearing turns, S-turns on final or after takeoff to look under your nose, and belly checks, rolling out of bank part way through a turn to see if someone is going to hit your plane's belly, are all techniques to add to your usual scan. Check your six-o'clock position as well as up and down, at least once per critical phase of flight. This might prevent that fast twin from descending on top of you during a localizer approach, or let you see that jerk trying to cut you off on final.

After acquiring traffic

Scan techniques are great, but seeing the other guy is only half of the equation. Illusions might cause you to hesitate taking evasive maneuvers. For instance: an aircraft more than a mile distant might appear above you in altitude, but it might actually be level with your airplane. Also, a plane on a collision course might appear to be remaining the same size and not closing—until those final few seconds when it gets big, really fast. An early evasive maneuver to cause the traffic to begin moving on your canopy can help ensure a greater miss distance. If an aircraft remains motionless for several seconds in the same position on your windscreen, you and it are probably on a collision or near collision course. Remember: An early maneuver to avoid traffic is better than a late one.

PUTTING IT ALL TOGETHER:
A collision avoidance checklist

The FAA has some good ideas to prepare and execute safe collision avoidance techniques. After some adaptation, below is a simple guide to think about using before your next flight. In fact, I would incorporate many of these questions and points into all flight planning. These are techniques only, not a be-all, end-all checklist.

1. Self-Check.
    How is your physical and emotional condition? Get enough sleep? Are you in the mood to fly? Mental preparation? Under a lot of personal stress?

2. Preflight Planning—accomplish.
    Did you get weather and Notices to Airmen (NOTAMS)? Do you have all the correct maps and publications? Have you highlighted your route on the maps and folded them neatly? Have you calculated gas? Do you know information about unfamiliar airfields? Have you checked frequencies? Did you call ATC if planning multiple approaches to avoid bad timing? Have you checked for student flying areas? Will you be crossing military Low Level flying routes or active Military Operating Areas (MOAs)? Doing these events on the ground, before the flight, saves heads-down time in the cockpit while airborne.

3. Aircraft Preflight—accomplish.
    Obviously you should follow all published aircraft checklists, and make sure you have a clear windshield. Remember, those "bug smashes" might turn out to be a 120-knot "bug smasher." Also, avoid solid sun visors and shades because they block areas of the sky from your now practiced scan pattern.

4. FARs and Standard Operating Procedures (SOPs)—know and follow.
    In your planning, research the SOPs for your airfield and any you might fly to. Use ATC to the maximum extent available, and file a flight plan when possible. Know how to enter unusual field patterns, and consider routes where conflict aircraft might be coming from.

5. Congested areas—avoid.
    When planning, expect navigational aids, major landmarks, and local attractions to have other traffic in the area. Avoid military airfields by as much as 25 miles because many fighters have waivers to operate faster than 250 knots below 10,000 feet, and during climbout they could be going as fast as 400 knots. Avoid the extended approach areas for fields with instrument approaches. Don't even try to do multiple patterns at very busy commercial fields.

6. Aircraft limitations—review.
    Every plane, besides having 20 airspeeds to memorize, has certain characteristics that aid or limit a visual search. Know these limits for your plane and cre-

ate a strategy for compensation. For instance, when flying a Cessna 152, you might have trouble seeing at your altitude during turns, but looking down is no problem. For a 90 degree turn, you might roll out of bank every 30 degrees or so to clear the flight path. Conversely, when flying a low-wing, such as a Grumman Tiger, it is easy to scan into a turn. But, you might do more belly-checks when entering a pattern to look under that low wing for unannounced traffic.

7. Lights—on.
The FAA reports that high intensity strobes could increase your plane's contrast by up to 10 times, day or night.

8. Transponder—on.
Use Mode-3 and -C with both the four-digit code and the altitude function activated. When possible, file a flight plan or at least pick up flight following for traffic advisories. Ask ATC to give you a discrete transponder code, to highlight yourself, particularly in a busy area. ATC wants to help you, so you help them.

9. Radio—use.
Learn how to talk and listen to a VHF radio. If you can afford a headset, it could be worth the price by allowing you to keep scanning while you talk, instead of hunting around with your head inside the cockpit looking for a mike to respond to a traffic advisories. Call early to airports that have radar service. If flying to a unicom airport, tune the frequency at least ten nm from the area and listen to the traffic. Put together a mental picture of where the traffic is, and where it is going to be. Use that situational awareness to plan your scan for entry into the pattern. Tell people your exact location.

10. Scan—accomplish.
Learn and practice scan techniques. Try new ones. Talk with experienced pilots about their close calls, and chair fly how you would handle the situation. Build a scanning habit pattern so you concentrate on scanning in all phases of flight. It must be a conscious effort, before you strap on that aircraft. Just when you think you can take a breather from your cockpit duties, you're not done yet—SCAN!

CASE STUDY REFERENCES

III-1, III-2, III-3, III-4, III-5, III-6, III-7

CHAPTER REVIEW

Minimum Visual Detection Area theory states that the fuselage of the threat aircraft must fill 12 minutes of arc.

Featureless Sky: Aircraft are very difficult to detect in a clear, featureless sky. Our eyes are almost in a constant state of focus and refocus. Scan patterns in a featureless sky include:

- Focus on distant object, then immediately scan block in that area.
- Faster your plane flies, the further in front you must scan.
- Vary distances to ensure thorough scan and to reduce fatigue.
- Always scan above and below your altitude.
- Don't stay focused on one object for too long.
- Maintain vigilance and don't get complacent.
- Maintain proper focusing and scanning techniques when entering the pattern.

Scan techniques. Sector scanning

- Degrees off nose
- Degrees above/below

After acquiring traffic: Illusions

- S-turns
- Belly checks

Putting it all together.

- Checklist

CHAPTER REFERENCES

Harris, Randall, Bobby Glover, Amos Space. 1986. *Analytical Techniques of Pilot Scanning Behavior and Their Application.* NASA.

Krause, Shari Stamford, Ph.D. 1995. *Avoiding Mid-Air Collisions.* Blue Ridge Summit, Pa.: McGraw-Hill.

Proctor, Paul. 27 April 1987. "FAA Investigates Seek Pattern in Near Midair Collisions." *Aviation Week and Space Technology*: 63-4.

Schallhorn, S., K. Daill, W.B. Cushman, R. Unterreiner. 1990. *Visual Search in Air Combat.* Naval Aerospace Medical Research Laboratory.

Steenblik, Jan W. October 1988. "The Eyes Don't Have It." *Air Line Pilot*: 10-16, 57.

"Four Eyes are Better Than Two—Or Are They?" September-October 1987. *FAA Aviation News*: 7-9.

"How to Avoid a Midair Collision." September-October 1987. *FAA Aviation News*: 7-9.

14
The role of
air traffic control

AFTER COMPLETING THIS CHAPTER, YOU SHOULD BE ABLE TO:

1. Discuss the latest FAA airspace reclassification.
2. Understand the full meaning of "radar contact."
3. Understand a pilot's responsibility in controlled airspace.
4. Explain the role of ATC in VFR flight operations.
5. Discuss the significance of "diffusion of responsibility" in an ATC environment.
6. Explain the ramifications of a reduced state of vigilance in an ATC environment.
7. Implement good radio communication skills, including standard phraseology.

The role of air traffic control (ATC) is easily one of the most misunderstood and misinterpreted concepts in flying. Pilots are often confused as to the scope of ATC coverage and services, proper communication techniques, and correct phraseology. The

ATC system as a whole is one of the most valuable resources we have to ensure a safe flight, yet it is sometimes disregarded or misused simply because pilots have never really learned how it all works. This text is full of accident case studies that will attest to that.

An ATC environment, however, is only as safe and efficient as the controllers and pilots operating in it. Each has a role, and if one fails to perform properly in that role, the system can quickly break down to dangerous levels. Midair collisions, near-misses, and many other accidents and close-calls have occurred during these breakdowns.

A major principle often overlooked is that we pilots must live up to our end of the bargain, which means controllers can't always help us if we are violating established rules. Instances of flying in the wrong airspace, at the wrong altitudes, at the wrong headings, and without clearance are not rare. So, let's begin with the basics of airspace management.

AIRSPACE CLASSIFICATION

A new definition of airspace classification went into effect 16 September 1993 in the United States. The familiar acronyms, PCA, TCA, ARSA, ATA, and the terms of controlled and uncontrolled airspace are no longer used. They are now letter-designated, A through E and G. The letter F is used internationally but not in the United States. Refer to Fig. 14-1 for this part of the discussion.

A, B, C, D, E, AND G

The new classification of airspace conforms to the international definitions and contain certain changes from the previous classification.

Class A airspace

Class A is now the name for what was called a Positive Control Area (PCA). The airspace is still controlled from 18,000 msl to FL 600 and includes the jet routes. Just like before, airspace entry is only through ATC clearance, the pilot must be instrument-rated, file an IFR flight plan, and maintain two-way radio communication.

Class A airspace quick reference

Class A airspace	IFR	VFR
Operations permitted	Yes	No
Entry requirements	ATC clearance	n/a
Minimum pilot qualifications	Instrument rating	n/a
2-way radio communications	Yes	n/a
Aircraft separation	All	n/a
Changes from existing rules	No	n/a

Derived from FAA Aviation News—Reprint

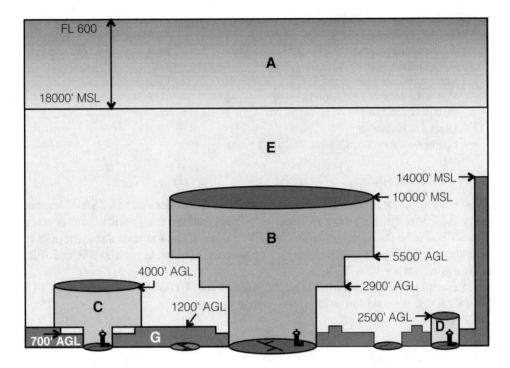

Fig. 14-1. *Airspace reclassification.* Adapted from FAA

Class B airspace

Class B is the new name for a Terminal Control Area (TCA). ATC clearance is required for airspace entry, and it's still available to IFR and VFR traffic, provided the pilot is at least an instrument-rated pilot when flying in IMC, and a student pilot when operating under VMC.

Traffic advisories and safety alerts are, of course, available from ATC. The minimum flight visibility for VFR is still 3 statute miles, but the big difference is in the minimum distance from the clouds. The old rule, under TCA airspace, required VFR traffic to be at least 500 feet below, 1000 feet above, and 2000 feet horizontal from clouds. Under the new classification, when VFR in Class B airspace, you just need to be clear of clouds.

Class B airspace quick reference

Class B airspace	IFR	VFR
Operations permitted	Yes	Yes
Entry requirements	ATC clearance	ATC clearance
Minimum pilot qualifications	Instrument rating	Student

2-way radio communications	Yes	Yes
Aircraft separation	All	All
Traffic advisories	Yes	Yes*
Safety alerts	Yes	Yes
Minimum flight visibility	n/a	3 statute miles
Minimum distance from clouds	n/a	Clear of clouds
Changes from existing rules	No	Yes

*Workload Permitting

Derived from FAA Aviation News—Reprint

Class C airspace

Class C airspace formerly was known as Airport Radar Service Area (ARSA). Airspace entry for IFR traffic requires ATC clearance; VFR traffic must provide radio contact. You still must be instrument-rated to fly IFR in Class C, and at least a student pilot to operate under VFR. Aircraft separation services are only between IFR-IFR and IFR-VFR traffic, not for VFR-VFR.

Remember the speed restrictions: 250 KIAS below 10,000 feet msl, and 200 KIAS below 2500 feet agl within 4 nm of the primary airport.

Class C airspace quick reference

Class C airspace	IFR	VFR
Operations permitted	Yes	Yes
Entry requirements	ATC clearance	Radio contact
Minimum pilot qualifications	Instrument rating	Student
2-way radio communications	Yes	Yes
Aircraft separation	All	IFR-VFR
Traffic advisories	Yes	Yes*
Safety alerts	Yes	Yes
Minimum flight visibilities	n/a	3 statute miles
Minimum distance from clouds	n/a	500' below 1000' above 2000' horizontal

*Workload Permitting
Derived from FAA Aviation News—Reprint

Class D airspace

The old classifications of Class D airspace were Airport Traffic Areas (ATA) and Control Zones (CZ). Entry prerequisites, minimum flight visibility and minimum distance from the clouds have remained the same. The big changes pertain to the new ceilings in the Class D airspace. The upper limit of the Control Zone has been lowered from 14,500 feet msl to 2500 feet agl. The upper limit of the Airport Traffic Area has been lowered from 2999 feet agl to 2500 agl.

Class D airspace quick review

Class D airspace	IFR	VFR
Operations permitted	Yes	Yes
Entry requirements	ATC clearance	Radio contact
Minimum pilot qualifications	Instrument rating	Student
Minimum flight visibility	n/a	3 statute miles
Minimum distance from clouds	n/a	500' below
		1000' above
		2000' horizontal
Changes from existing rules	No	No

Derived from FAA Aviation News—Reprint

Class E airspace

Class E airspace was previously known as Controlled Airspace—General. Primary requirements have remained the same but the specifics of flight operations can still be a little confusing, so it's worth taking the time to go over them.

- Minimum flight visibility: 3 statute miles below 10,000 feet msl. 5 miles at or above 10,000 feet msl.

- Minimum distance from clouds: 500 feet below, 1000 feet above and 2000 feet horizontal below 10,000 feet msl. 1000 feet below, 1000 feet above and 1 statute mile horizontal at or above 10,000 feet msl.

Class E airspace quick reference

Class E airspace	IFR	VFR
Operations permitted	Yes	Yes
Entry requirements	ATC clearance	None
Minimum pilot qualifications	Instrument rating	Student
2-way radio communications	Yes	No
Aircraft separation	IFR/SVFR**	n/a
Traffic advisories	*	*
Safety alerts	Yes	Yes
Minimum flight visibility	n/a	see previous text
Minimum distance from clouds	n/a	see previous text
Changes from existing rules	No	No

*Workload Permitting
**Special VFR
Derived from FAA Aviation News—Reprint

Class G airspace

Class G airspace was once known as Uncontrolled Airspace. There are changes. However, it's important to note that there will continue to be airports in Class G airspace. If there are instrument-approach procedures at those airports, the floor of controlled airspace will generally be a Class E area extending upward from 700 feet agl. As with Class E airspace, there are very definite requirements pertaining to flight operations.

- Minimum Flight Visibility. Night operations below 10,000 feet msl—3 statute miles. Day or night operations at or above 10,000 feet msl—5 statute miles.

- Minimum Distance from Clouds. Operations above 1200 agl but below 10,000 msl—500 feet below, 1000 feet above, 2000 feet horizontal. Operations at or above 10,000 feet msl—1000 feet below, 1000 feet above, 1 statue mile horizontal.

Class G airspace quick reference

Class G airspace	IFR	VFR
Operations permitted	Yes	Yes
Entry requirements	None	None
Minimum pilot qualifications	Instrument rating	Student
2-way radio communications	No	No
Aircraft separation	None	None
Traffic advisories*	*	
Safety alerts	Yes	Yes
Minimum flight visibility	n/a	1 statute mile
Minimum distance from clouds	n/a	see previous text
Changes from existing rules	No	No

*Workload Permitting
Derived from FAA Aviation News—Reprint

Entering Class B airspace

As a pilot it is important to know what is expected from you before entering Class B airspace. Say you've been talking with approach control, and you've told them of your intentions. Most likely you either want to land at an airport within that airspace, or you just want to fly through it en route to your destination. After receiving a new transponder code and squawking ident, approach says, "Radar contact."

What does "radar contact" mean to you?

1. I'm not cleared into Class B airspace, ATC just has me on their radar.
2. Because approach knows of my intentions, "radar contact" automatically means that I am cleared into Class B airspace.
   The correct answer is "1".

Many pilots have long misunderstood the phrase "radar contact" to mean that they have received ATC clearance into that airspace. It does not. According to the Airman's Information Manual (AIM), there are actually two parts to the definition of "radar contact." Most pilots don't get past the first section, but it's the second statement that gets pilots in some pretty hairy, and illegal, predicaments.

1. Controller-to-pilot. "Used by ATC to inform an aircraft that it is identified on the radar display and radar flight following will be provided until radar identification is terminated." When informed of "radar contact," a pilot automatically discontinues reporting over compulsory reporting points. Nowhere in the definition does it say ". . . proceed into controlled airspace."

2. Controller-to-Controller. ". . . to inform the controller that the aircraft is identified and approval is granted for the flight to enter the receiving controller's airspace.

All this means is that your aircraft is identified on the radarscope and that a handoff into the next controller's airspace is approved. Again, no additional sentences give the pilot permission to enter Class B airspace.

The only way that you're allowed to enter B airspace is when you hear the words, "Cleared to enter (or "into") B airspace." Until that happens, you must remain outside of the controlled airspace.

Radar following services

Before we move on, another term needs to be addressed. In the first part of the "radar contact" definition, there is a statement ". . . radar following will be provided . . . ." Does this mean, relax and let ATC do the flying? Absolutely not. Controllers weren't put on this earth to hold the hands of lazy pilots.

According to the AIM, the definition of "radar flight following" is: "The observation of the progress of radar-identified aircraft, whose primary navigation is being provided by the pilot . . . ." Therefore, it's still up to the pilot to maintain the correct heading and altitude and always know the exact position.

PILOT RESPONSIBILITIES

The safety of a flight is a pilot's responsibility. The following are a few more points to remember when operating under VFR.

1. If flying through Class B or Class C airspace, you might be advised to remain VFR. What if there are clouds around? That's why it is so important to understand the legal requirements for the minimum distance from clouds (refer back to the "Quick Reference" guides). ATC is not going to do it for you.

2. Likewise, a controller could vector you too close to an area of clouds. It's up to you to advise the controller that you will violate the VFR cloud-separation minimums if you accept the vector.

3. Another common occurrence is ATC assigning a non-VFR altitude. That's fine, the controller put you there for a reason. But when you are told to "resume own navigation," or "maintain VFR altitude," that means to climb or descend to the appropriate VFR altitude and stay there.

VFR ALTITUDES

While on the subject of VFR altitudes, let's refresh your memory on what they are. When flying VFR, in level flight above 3000 feet agl and below 18,000 feet, the rule is as follows:

Magnetic Course: 0—179 degrees, odd msl + 500 feet.
For example: 3500 feet, 5500 feet.

Magnetic Course: 180—359 degrees, even msl + 500 feet.
For example: 4500 feet, 6500 feet.

CLEARING UP MISCONCEPTIONS

The intended purpose of ATC is to promote the safe and orderly separation of traffic. Sounds good. But there's a catch. According to FAA Order 7110.65E: "Controllers shall give priority to separating IFR aircraft . . . ." It further states that, ". . . traffic advisories are provided as an additional service . . . workload permitting and contingent upon higher priority duties."

In the wake of a 1984 midair collision between a Wings West commuter and a Rockwell Commander, the FAA reemphasized, through a prepared statement, that:

. . . assigning of discrete transponder codes in this situation [under ATC supervision] could lead pilots into believing they are receiving a service that the controller might not be able to provide.

The FAA added: ". . . advisory services provided to VFR traffic are predicated on workload, frequency congestion, radar limitations, and traffic volume.

In a nutshell, VFR traffic will receive advisories *only* if ATC is not busy or there is a potential safety conflict. Unfortunately, this regulation is sometimes misunderstood or misused by the pilot community.

AIR TRAFFIC CONTROL AND THE VFR PILOT

Let's put this in perspective by using a few examples.

1. "Well, I'm squawking a discrete transponder code, and I've been periodically talking with ATC. So it goes without saying that I'll receive traffic advisories."

No, you won't. Even though you're "squawking and talking," remember, you're still VFR. Don't expect to automatically get traffic advisories. It's neither a regulation nor obligation for ATC to provide that service to you. However, you have a regulation to follow. It is that you: ". . . shall maintain vigilance to see and avoid other aircraft."

A suggestion: After you've made your initial radio call and hear "radar contact," request traffic advisories. There's still no guarantee that you will receive any, due to ATC workload at the time, but at least if the controller has a few extra minutes, you might be at the top of the list to get one or two. On the other hand, the controller might not be busy at all, so you could be lucky and end up receiving more advisories than you know what to do with. In the world of collision avoidance, you can never have too many traffic advisories.

    2. "I'm not really sure where Class B airspace begins, but I'm sure approach will tell me if I accidentally fly into it."

Don't count on it. First, an attitude like that is irresponsible. Second, it goes back to the reason controllers were put on this earth, and it wasn't to watch out for pilots who don't bother to figure out their position. If you do get a response from ATC, you might get a call for a violation notification or to tell you that you've just been involved in a near-miss with a 727.

Another suggestion: Know your exact position and where you are in relation to other airspace. There are at least two reasons why this is so important. One, you must be aware of the surrounding airspace boundaries. It's just too dangerous to be aimlessly buzzing around those areas. And two, when you supply ATC with a position report, they are probably vectoring traffic dependent on that information. If you're only guessing, or you think a good ballpark estimate is good enough, it's not. Midairs and near-misses have directly resulted from erroneous position reports from pilots.

Diffusion of responsibility

The previous two scenarios are actual examples of a concept known as *diffusion of responsibility*. This is simply a tendency for pilots to relax attentiveness and vigilance when under radar control. It's safe to say that all of us have been guilty of this from time to time. Remember, the blinking, amber light on the transponder in the cockpit does not mean sit back and relax.

Years ago, NASA conducted a study on near-midair collisions and found a definite connection between those incidents and the state of vigilance, or lack of vigilance, of the pilot in a controlled environment. The researchers viewed this condition as a subconscious idea of shared responsibility between pilot and controller. The study states that many pilots, when under ATC supervision, believe that they will be advised of traffic that represents a potential conflict. Therefore, they tend to relax their visual scan for other aircraft until warned of their presence.

Remember, when you're looking out the window more to sightsee than to watch for traffic, you have just demoted yourself from pilot-in-command to baggage.

A reduced state of vigilance

While on the subject of responsibility, there's yet another condition that needs to be mentioned. A reduced state of vigilance can affect both pilots and controllers.

A pilot's perspective. Anytime a pilot shares responsibilities with a controller, in effect he is also reducing his state of vigilance. No longer is the pilot being attentive to his visual search, navigation, systems checks, fuel checks, and just about anything else. "ATC will tell me when there's traffic or when I'm off course. And besides, I'm only 10 minutes from landing, I'm home free!"

A controller's perspective. The NTSB conducted a study on a reduced state of vigilance after noticing a distinct pattern in the circumstances surrounding many controller mistakes. This situation, especially during light to moderate traffic workloads, had contributed to five midair collisions between April 1988 and April 1989.

Results from this study show that controllers have a tendency to relax their level of alertness in a low-workload environment which makes them vulnerable to operational errors and omissions. In addition, because of their FAA training and experience levels, controllers tend to focus an inordinate amount of attention to targets that have been identified (tracked/IFR) to those that are unidentified (untracked/VFR).

In yet another study, the Safety Board conducted a special investigation on the causes of ATC-related runway incursions, ground collisions between either two aircraft or an airplane and an airport vehicle. The report concluded that heavy traffic and reduced visibility rarely were factors. On the contrary, traffic was considered light or moderate at the time of most incursions where controller actions were involved. In some of those instances, the controllers were working as few as two aircraft.

RADIO COMMUNICATION AND PHRASEOLOGY

In most collision-avoidance environments, effective radio communication combined with good scan techniques might be the only two available resources that you have. We already know about the physical limitations of see and avoid. Take away effective radio communication and you have just increased your chances of being involved in a midair collision.

Radio voice

If you mumble in person, you most likely mumble over the radio. If you have a high-pitched voice in person, you probably sound like a "screaming Mimi" over the radio. If you're a fast talker in person, you might sound like a fast-talking, mumbling, "screaming Mimi" over the radio. A speech pattern that is understandable in a face-to-face conversation doesn't always come across that way over the radio.

Here are a few tips on effective radio communication:

1. Think before you talk. If you know what you're going to say before you depress the mike button, it will eliminate a lot of the "ahs" and "ums."

2. Speak in a clear and slow tone of voice. Not only will you be more understandable, but you probably won't have to repeat yourself as often.

3. Slightly lower your voice. Sometimes when you speak slowly, your voice will automatically sound lower, but not always. It's especially true if you naturally have a high-pitched voice. If that's the case—and this goes for women, in particular—make sure your "radio voice" sounds a little deeper than your normal conversational level.

4. Be direct and concise; short and to the point.

5. Avoid using unfamiliar acronyms and nonstandard phraseology. Effective radio communication means that you must first be understandable. If no one knows what you're talking about, you're not understandable.

What you say is not always what you mean

Effective oral communication is not just what you say, but also how you say it. Ambiguous and indirect phraseology is quite common, especially when a pilot is in a critical phase of flight and is being inundated with other tasks. But a departure or arrival stage of flight is exactly the wrong time to make poor radio calls. Remember, there's a much greater possibility for a midair to occur at those busy times.

Avoid ambiguous phraseology

The following is an actual example of ambiguous phraseology. Refer to case study III-1. Midair collision between PSA Flight 182 and a Cessna 172.

ATC: PSA . . . traffic twelve o'clock, one mile, northbound.
PSA: We're looking.
ATC: PSA . . . additional traffic's . . . twelve o'clock, three miles just north of the field northeast bound, a Cessna one seventy-two climbing VFR, out of one thousand four hundred.
PSA: Okay, we've got that other twelve.

Within 20 seconds, Flight 182 had received two traffic advisories pertaining to aircraft at their twelve o'clock position. Therefore, when Flight 182 said they had "that other twelve," which traffic advisory were they referring to? The controller assumed that they had the 172 in sight, and although that was the most likely conclusion, there was no way to tell for sure since there were other aircraft in the vicinity.

Be clear and direct. Anytime you are given more than one traffic advisory in a short period of time, or when there is confusion over which aircraft you have in sight, you should always clarify your response. Instead of ". . . we've got that other twelve," what would've been a clearer and more direct reply?

PSA: PSA one eighty-two has the northeast bound Cessna in sight. Still looking for northbound traffic.

Listen, understand, and remember. Often we don't hear all the information that is given in a traffic advisory. Heading, altitude, attitude, or aircraft type can easily be missed when you're busy with checklists and there's a lot of activity on the radio. This is why it's so important to pay attention and listen for your traffic advisories.

Just as common—and this doesn't happen exclusively to new pilots—is misunderstanding the heading of your traffic. Say you were given the same two traffic advisories that Flight 182 had received. The first was northbound at one mile and the second was northeast bound at three miles. You detect only one aircraft, which seems to being flying in a northerly direction, so you report, "Northbound traffic in sight."

Since you weren't really listening to either traffic advisory but remember hearing something about northbound traffic, you assume the aircraft you're looking at is the northbound traffic. Actually it was the northeast bound traffic.

There are two easy steps you can take that will help avoid this mistake.

1. Learn and memorize your compass headings and the reciprocals. That way, when you hear a traffic advisory that includes the aircraft's heading, you can immediately determine if it is flying away from you or towards you. Say you were flying in a northeasterly direction and your traffic is flying southwest bound. You are then aware that an aircraft will be coming in your direction. Likewise, if you're still northeast bound, and your traffic is heading west at your 2 o'clock position, you can be on the lookout for an aircraft that will be crossing your flight path from the right.

2. Know what view of the threat aircraft will be visible from your position. Visualizing your traffic's position in relation to your own enables you to understand what view of the aircraft you'll be looking for. Front view, belly view, rear-quarter view, etc.

A simple rule to help with figuring reciprocals is to:

1. Add two to the first number, then subtract two from the second, or,
2. Subtract two from the first number, then add two to the second.

Remember, in either case you might have a negative number or you might have a first digit larger than a 3 if you do not use common-sense mathematics.

For example: What is the recip of 3-6-0 degrees?
Answer: 3 – 2 = 1, 6 + 2 = 8, and 0 on the end; 1- 8-0

Another example: What is the recip for 1-1-0 degrees?
Answer: 1 + 2 = 3, 1 – 2 = 9 and a 0 on the end; 2-9-0 (The middle number 1 must become an 11 from which 2 is subtracted to become 9, leaving the first number to be 2.)

Another easy method is to add 200 if the first digit is a 1, then subtract 20. If the first digit is a 2 or 3, subtract 200 and then add 20.

Example: What is the recip for 180 degrees?
Answer: 180 + 200 = 380 – 20 = 360

Another example: What is the recip of 210 degrees?
Answer: 210 – 200 = 10 + 20 = 030.

Nondirective phraseology

The following is an actual example of nondirective phraseology.
Midair collision of PSA Flight 182 and a Cessna 172.

ATC: PSA . . . traffic's at twelve o'clock, three miles out of one thousand seven hundred.

PSA: Traffic in sight.
ATC: . . . maintain visual separation, contact Lindbergh tower . . . .
PSA: Okay.
PSA: Lindbergh PSA . . . downwind.
ATC: PSA . . . traffic twelve o'clock, one mile, a Cessna.
PSA: Okay, we had it there a minute ago.
ATC: . . . roger.
PSA: I think he's pass(ed) off to our right.
ATC: Yeah.
Cockpit: He was right over here a minute ago. Yeah.

Notify ATC as soon as you've lost sight of your traffic. "We had it there a minute ago," and "I think he's passed off to our right," are not examples of directive communication. Be clear, direct, and concise. "PSA 182 has lost the traffic, request advisories," is a more directive, no-nonsense message.

Controllers: do not be part of the problem. The tower controller made only two comments after the PSA's transmissions. "Roger," and "Yeah." Noncommittal, nondirective, and no help. Don't wait to hear the magic words, "traffic no longer in sight," before you speak up and take charge of the situation. Remember, this is a team effort. When the pilot fumbles the ball, pick it up.

A final thought

Work on your radio voice. Be clear, direct, and concise. Avoid any words or phraseology that might not be understandable to others.

RADIO DISCIPLINE

Pilots aren't the only ones guilty of miscommunicating on the radio. Controllers can also create or exacerbate a situation to the point of causing a potentially serious threat to safety. Some of these problems stem from a disregard for radio discipline, and shouldn't be tolerated. Others are the result of controllers not being able to communicate in a clear and direct manner. In any event, when pilots are subjected to this type of ATC breakdown, one of their most important resources for collision avoidance is eliminated.

The following are a few examples of actual transmissions made by Chicago-O'Hare approach controllers:

1. "Listen up gentlemen, or something's gonna happen that none of us wants to see. Besides that, you're [expletive] me off."

2. "For radar identification, throw your jump-seat rider out of the window."

3. "I am way too busy for anybody to cancel on me."

4. "Don't anybody maintain anything."

5. "Climb like your life depends on it, because it does."

6. "Leave five on the glide, have a nice ride, tower inside, twenty-six nine, see ya."

7. "Turn in and take over . . . you know the rest."

FINAL THOUGHTS

Maintain radio discipline. Communicate clearly and directly. Always be professional. And, be a help, not a hindrance.

CASE STUDY REFERENCES

I-1, I-2, I-3, I-7, II-1, II-2, II-5, II-6, II-8, III-1, III-2, III-4, III-5, III-6, III-7, IV-1, IV-3

CHAPTER REVIEW

Airspace reclassification

- A airspace (Positive Control Area).
- B airspace (Terminal Control Area).
- C airspace (Airport Radar Service Area).
- D airspace (Airport Traffic Areas and Control Zones (CZ).
- E airspace (Controlled Airspace—General).
- G airspace (Uncontrolled Airspace).

ATC services

- "Radar contact" does not mean a pilot is automatically cleared into controlled airspace.
- Radar following services can be beneficial to VFR pilots.

VFR altitudes

- Level flight, above 3000 feet:
  ~0–179 degrees, odd msl + 500 feet.
  ~180–359 degrees, even msl + 500 feet.

Purpose of ATC

- Promote safe and orderly separation of traffic.
  ~Priority to IFR traffic.
  ~Additional services to VFR traffic on a workload permitting basis.

ATC and the VFR pilot

- Pilots still have responsibilities and obligations even when under ATC supervision.
- Ask for additional services.

Diffusion of responsibility

- A belief of shared responsibility when under ATC supervision.

A reduced state of vigilance

- Lack of attention and situational awareness.
  ~Shared responsibility.
  ~Boredom.
  ~Complacency.

Radio communication

- Speak with a radio voice.
- Be clear, direct, and concise.

Phraseology

- Avoid ambiguous phraseology.
- Avoid indirect phraseology.
- Avoid unfamiliar acronyms.
- Be understandable.

Radio discipline

- Communicate clearly and directly
- Be professional

CHAPTER REFERENCES

Andrews, J.W. 1991. *Air-to-Air Visual Acquisition Handbook*. Lincoln Laboratory. Massachusetts Institute of Technology.

Cardosi, Kim, Pamela Boole. 1991. *Analysis of Pilot Response Time to Time-Critical Air Traffic Control Calls*. U.S. Department of Transportation. Federal Aviation Administration.

Hopkin, V. David. "Air Traffic Control." *Human Factors in Aviation*. Ed. Wiener, Earl L. and David C. Nagel. San Diego: Academic Press, 1988: 639–663.

Keesling, Otto. Federal Aviation Administration Air Traffic Representative. Personal interview. 7 August 1991.

Krause, Shari Stamford, Ph.D. 1995. *Avoiding Mid-Air Collisions*. Blue Ridge Summit, Pa.: McGraw-Hill.

"Say Again." July/August 1995. *Airways*.

Stein, Earl. 1992. *Air Traffic Control Visual Scanning*. U.S. Department of Transportation. Federal Aviation Administration.

15
Traffic alert and collision avoidance systems

AFTER COMPLETING THIS CHAPTER, YOU SHOULD BE ABLE TO:

1. Explain the principles of an airborne collision avoidance system.
2. Explain the relevance of the three types of transponders in a collision avoidance environment.
3. Describe TCAS I and II.
4. Discuss the events that drove the operational TCAS program.
5. Explain how TCAS affects the entire flying community.
6. Discuss the future of TCAS.

A Traffic Alert and Collision Avoidance System, or TCAS, is a system in the cockpit that is independent of, but compatible with, the ground based ATC network. With the use of an onboard computer, a TCAS processes transponder signals from other aircraft to determine their positions, altitudes, and rates of closure. These signals can be received and analyzed up to 40 miles away. That information is presented to the flightcrew on either the cockpit weather-radar scope, or a specially designed TCAS color display.

Now that you have a basic understanding of how TCAS operates, let's briefly discuss the beginning of airborne collision-avoidance development. As you might have already guessed, the concept relies on a little mathematics and physics.

TAU PRINCIPLE

Besides transponder interrogation, virtually all airborne collision avoidance systems designed in the last 30 years have used a mathematical equation known as the *TAU Principle*. In 1955, Dr. J.S. Morrell of the Bendix Communication Division developed the TAU Principle and applied his findings to learn more about the physics of collisions.

According to Dr. Morrell's concept, each aircraft flies inside an invisible envelope which varies in shape and size depending on its speed and the speed of other airplanes. Figure 15-1 illustrates how the ratio of range to range rate, represents the time period in which two aircraft on a collision course would intercept. A pilot's TAU is the minimum time needed to recognize the threat and take evasive action, typically about 25 seconds.

TRANSPONDERS

Because transponder interrogation is integral to a TCAS, let's take a quick look at the three types of transponders and see how each works, or doesn't work, with the TCAS system. A transponder is an onboard device that transmits a coded signal back to ATC. That numbered code appears as a distinct pattern on the controller's radar scope. The same transponder signal, however, provides slightly different information to TCAS.

Mode-A

The signal shows only position, not altitude. Therefore, it reports only range and bearing information. Mode A is sometimes referred to as Mode 3.

Mode-C

Transmits range, bearing, and altitude information.

Mode-S

Mode-S is the most sophisticated of the transponders and is an integral part of TCAS II. We'll get to TCAS II in a minute. In addition to the Mode-C capabilities, Mode-S provides two critical functions; discrete addressing and data linking. From ATC's viewpoint, Mode S-equipped aircraft can be interrogated separately, eliminating garbled radar returns. The datalink function is actually two-fold. ATC is able to relay messages faster, via cathode ray tube (CRT) computer screen presentations, with less frequency congestion and most importantly, TCAS II-equipped aircraft can detect each other's presence. Mode-S transmits signals every second, known as a *squitter pulse,* making the interpretation of a traffic conflict nearly instantaneous. If both threat aircraft are equipped with TCAS II, the Mode-S datalink will coordinate the evasive maneuvers between the two.

TCAS I

TCAS I is a low-power system that has a 40 nm range. It consists of a TCAS antenna, TCAS processor with a 125-watt peak-power transmitter, and a display unit. It is not programmed with collision-avoidance logic and, therefore, cannot compute evasive-maneuver courses. It does, however, locate aircraft in the immediate vicinity, display their locations within a given quadrant, and provide aural traffic advisories.

Although TCAS I is the least sophisticated of the airborne collision-avoidance systems, it does enhance a crew's situational awareness and their ability to visually detect their traffic.

TCAS II

The TCAS II, combined with a Mode-S transponder, scans a volume of airspace around the aircraft using two antennas; one on the top of the fuselage and the other beneath. Typically, the top-mounted antenna transmits interrogations on 1030 megahertz (MHz) at varying power levels in each of the four 90-degree azimuth segments. Mode-S transponder replies are received on 1090 MHz and are sent to the TCAS computer. The bottom-mounted antenna (the receiver) provides range and altitude data on targets that are below the TCAS aircraft.

The 1030/1090-MHz environment, in conjunction with the upper and lower antennas, optimizes signal strength and reduces multipath interference. The scanning area covers 14.7 miles to the front of the aircraft, 7.5 miles behind and 7000 feet above and below the aircraft. These distances are sufficient to handle an impressive 1200-knot closure rate off the nose and an altitude-closure rate of 12,000 feet per minute. The dual-antenna feature also enables the aircraft to have full and continuous coverage even in a banking turn. Additionally, TCAS II is able to track up to 45 aircraft within a 15-mile radius and can display up to 30 aircraft at one time.

NOTE: The remaining discussion and illustrations pertain to TCAS II.

TCAS II OPERATION

A TCAS II system monitors the airspace surrounding the aircraft by interrogating the transponders of intruding aircraft. The interrogation reply enables TCAS II to compute range, relative bearing, altitude and vertical speed, and closing rate.

TCAS II initial transponder acquisition

Figure 15-1 illustrates the initial phase of transponder interrogation. When an aircraft enters the *invisible envelope,* it is considered to be a potential threat. Therefore, TCAS II first determines the range of the intruder aircraft.

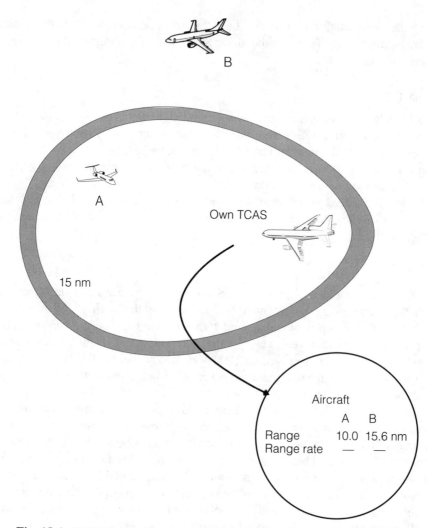

Fig. 15-1. *TCAS II initial transponder acquisition. The initial acquisition of a TCAS II's invisible envelope where own TCAS determines range of potential threats.* Adapted from Allied-Signal

TCAS II second phase of transponder interrogation

Once TCAS II measures the range of the intruder aircraft, it then analyzes its track. Figure 15-2 shows that the determination of range rate, or closing speed, is the second phase of transponder interrogation.

TCAS II final phase of transponder interrogation

The final phase of transponder interrogation is illustrated in Fig. 15-3. It relies on mathematical computations that TCAS II automatically derives in determining the level of

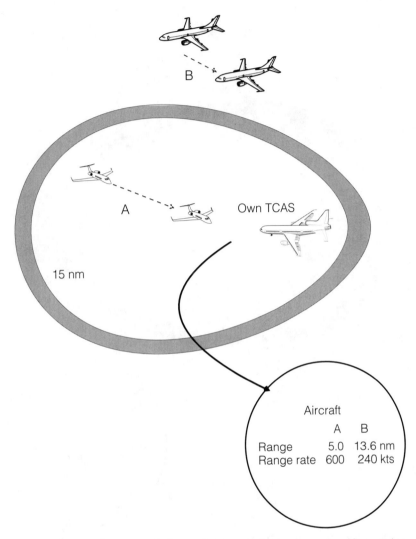

Fig. 15-2. *TCAS II second phase of transponder interrogation. Successive replies are used to establish range and closing speed.* Adapted from Allied-Signal

threat advisories and evasive maneuvers. TCAS II alerts the pilot with visual and aural warnings. The aural warnings advise the pilot what preventive or corrective actions should be taken.

Traffic advisories

TCAS II issues traffic advisories (TA) when it predicts that a transponder-equipped intruder is within 45 seconds of its closest point of approach (CPA). The TA notifies the pilot that traffic is in his vicinity and assists in visually detecting the intruder aircraft.

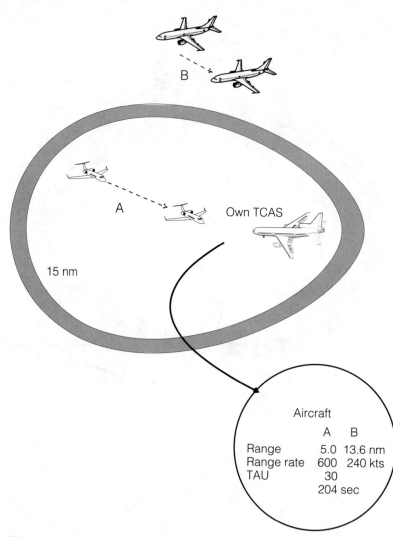

Fig. 15-3. *TCAS II final phase of transponder interrogation. The TAU for each aircraft is compared to a fixed threshold to determine the level of threat.* Adapted from Allied-Signal

An aural warning of "Traffic, Traffic," sounds in the cockpit. Because the intruder aircraft is not considered a threat, no maneuvers are required of the pilot.

Resolution advisories

TCAS II also issues resolution advisories (RA) when altitude-encoding intruders come within 30 seconds of CPA. The computer calculates the vertical rate that must be achieved to maintain safe separation from the threat aircraft. When corrective RAs are required, the aural advisories include: "Traffic, Traffic," "Climb, Climb, Climb," "Descend, Descend, Descend," "Climb Crossing Climb," "Descend Crossing Descent," "Reduce Climb," or "Reduce Descent." These are considered normal RAs. If no corrective action is needed, the alert is called a preventative RA. For preventive RAs, the aural advisory is "Monitor Vertical Speed." And after an intruder aircraft is no longer a threat, the aural message is "Clear of Conflict."

TCAS II HARDWARE

The hardware, scopes and displays of TCAS II (Fig. 15-4) are color-coded and use FAA approved symbology. The center symbol on the display screen is the "own aircraft". A 2-, 5-, 10- or 15-nautical-mile ring encompasses the center. The pilot selects the appropriate distance depending on phase of flight and volume of traffic in the area.

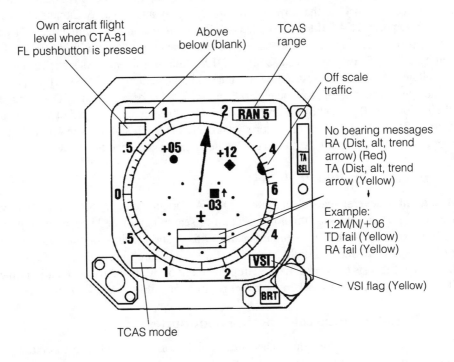

TCAS mode/failure

Standard		Optional	
TCAS STBY	(blue)	NO TCAS (blue) TCAS system in standby	
TA/RA	(blank)	TA/RA	(blank)
TEST	(yellow)	TEST	(yellow)
TA ONLY	(blue)	TA ONLY	(blue/yellow—when active TA)
TCAS	(yellow)	NO TCAS (yellow) TCAS system failure	

Display message locations

Fig. 15-4. *Display screen of a Bendix/King TCAS II.* Adapted from Allied-Signal

When an immediate threat aircraft enters the ring, a red square appears over the target to indicate an RA has been declared. A potential threat, or TA, is depicted as an amber circle. A proximity target, not a threat, is shown as an open-white diamond. If a proximity target enters the ring, and a TA or RA is already in progress, the proximity target is then displayed as a solid-white diamond.

Alphanumerics along with the symbols give the target's altitude, range and bearing. Arrows next to the symbols show whether or not the intruder is climbing or descending at a rate of at least 500 feet per minute.

TCAS LEGISLATION

In the wake of a midair collision between an Aeromexico DC-9 and a single-engine Piper Cherokee (see case study III-3) over Cerritos, California, Congress and lobbyists went into action to pass TCAS legislation. On December 30, 1987, then-President Reagan, signed an amendment to the Airport and Airway Safety and Capacity Expansion Act of 1987. It required TCAS II to be installed on all commercial aircraft with at least 31-passenger seats, operating in U.S. airspace. The amendment deadline for TCAS II was December 31, 1993. However, on December 15, 1989, then-President Bush signed a new law that allowed FAA to set new deadlines for phased implementation of TCAS II: 20 percent of the affected aircraft (approximately 1000 aircraft) had to be equipped by December 31, 1990; a total of 50 percent by the end of 1991; and 100 percent by the end of 1993. In addition, a TCAS I is required for all commuter operators using aircraft with 10- to 30-passenger seats operating in U.S. airspace. That deadline was reached February 5, 1995.

Target acquisition flight tests

As a result of several years of flight tests, the FAA concluded that the search effectiveness of the TCAS II is such that one second of search equals eight seconds of physically looking for the traffic. Remember the 12.5 second recognition and reaction rule from Chapter 12. Shaving eight seconds off a visual search is significant. Equally as important, the tests showed a marked improvement of pilots' visual acquisition times. It is easier to locate traffic when the exact position and altitude are known.

The test programs at Lincoln Laboratories

The ability of pilots to sight other airplanes in flight was evaluated during two test programs conducted by the Lincoln Laboratories of the Massachusetts Institute of Technology (MIT). Each test was part of a general research study and not part of any accident investigation. During the investigation of the Aeromexico midair, the NTSB referenced these test flights to gain a better insight on target acquisition.

The test included evaluating the subject pilot's ability to detect traffic during an unalerted search pattern. In the initial test, the subject pilot's only instruction was to fly three legs as if they were on a normal cross-country flight. Periodically, an FAA aircraft intercepted the subject, flying different collision courses. The FAA varied the aspect angle on the approach to simulated midair events to test various angles (refer to Fig. 15-5) and the visual search/acquisition.

Test results were significant: Data were obtained for 64 unalerted encounters. Visual acquisition was achieved in 36 encounters, 56 percent of the total, and the median acquisition range for these 36 encounters was .99 nm. The greatest range of visual acquisition was 2.9 nm.

According to these findings, the subject pilots saw the threat aircraft only slightly more than half of the time. When they did see them, the conflict aircraft was between only 1 and 3 miles in range, leaving little time to react.

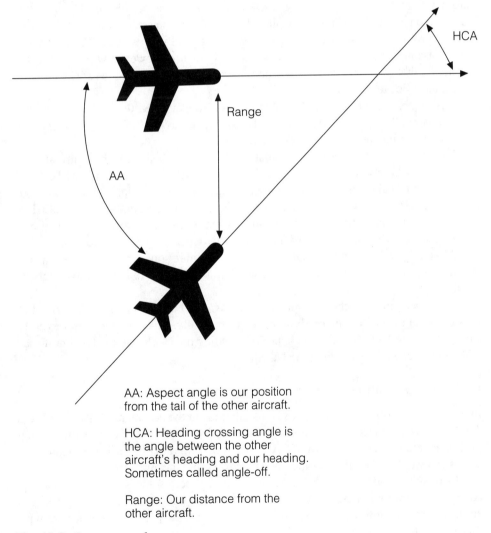

AA: Aspect angle is our position from the tail of the other aircraft.

HCA: Heading crossing angle is the angle between the other aircraft's heading and our heading. Sometimes called angle-off.

Range: Our distance from the other aircraft.

Fig. 15-5. *Intercept angles.*

With regard to the Aeromexico midair, the heading-crossing angle between the two aircraft was nearly 90 degrees. The Piper impacted the DC-9's tail section nearly head-on.

The second flight test evaluated the performance of pilots who had been alerted to the presence of an intruder aircraft. With the use of a TCAS II-equipped airplane, 66 encounters were collected. This time the subject pilots visually acquired the intruder 86 percent of the total encounters. The median range of the visual detection was 1.4 nm.

At the Safety Board's request, Lincoln Laboratories constructed two Probability of Visual Acquisition graphs founded on the extrapolation of pertinent flight data from the Aeromexico midair. The graphs were based on the closure rate between the DC-9 and the Piper, and on the results achieved by pilots having an unobstructed view of the intruder.

The probability of the Aeromexico crew seeing the Piper, without TCAS II assistance, was only 35 percent at 15 seconds prior to collision. However, with TCAS II the probability that the DC-9 flightcrew would have seen the Cherokee, was 98 percent at 15 seconds prior to collision.

Researchers at MIT's Lincoln Laboratories also examined the flight data involving a midair collision of a U.S. Army U-21 and Piper Navajo (refer to case study Fig. III-7) to determine the benefits of TCAS II.

The following variables were calculated: closing rate, bearing of both aircraft, and visual area of both aircraft. The probabilities of visual acquisition without TCAS II continued to remain consistent with previous studies. The Navajo, with one pilot onboard, had only a 27 percent chance of seeing the U-21 at 12 seconds prior to collision. Keep in mind the minimum of 12.5 seconds needed for recognition and reaction times. The U-21, with two pilots on board, had a slightly better chance at 33 percent.

Although the U-21 had a greater visual area, 48 more square feet, than the Navajo, the increase in visual acquisition for the U-21 was attributed to having two pilots on board.

The exact data were then computed with a TCAS II alert capability. The probabilities of visual acquisition naturally jumped substantially. At 12 seconds prior to collision, the Navajo had a 91 percent chance of avoiding the U-21; and the U-21 crew had a 96 percent chance of avoiding the Navajo.

TCAS transition program

In conjunction with the initial TCAS II installations in the commercial fleet, the FAA, Air Line Pilots Association (ALPA), and the airline industry are conducting a joint evaluation program, known as the TCAS Transition Program (TTP). Their agenda is to assess and validate the safe operation of TCAS II within the air traffic control network.

In monthly updates, FAA's TCAS Program Office discloses the initial findings concerning various issues, analyses, and information collected from the operational environment. The following comments are a sample of the positive feedback from pilots and controllers.

ATC task saturation

A regional-airline crew reported multiple RAs on arrival at a northeast airport. The first RA called for a reduce descent that was issued against an intruder departing the area. The second RA was issued shortly thereafter when the TCAS-equipped aircraft was at 4000 feet and was cleared to descend to 3000 feet. ATC called traffic at 11 o'clock, 3 miles, opposite direction, and at 3500 feet. As the TCAS aircraft passed through 3600 feet the crew observed the aircraft passing clear to the left.

" . . . while I was notifying ATC of the RA and that we were now climbing back to 4000 feet, we observed a second aircraft approximately 100 feet below our nose. If we had not followed the TCAS RA instruction there is no doubt in my mind that a midair collision would have occurred."

Another pilot report substantiates the capability of TCAS II in a busy ATC environment.

" . . . Ft. Worth Center was very busy, overloaded . . . but gave us a descent clearance to FL 240 at pilot's discretion. We weren't ready for the descent . . . so we maintained FL 290. TCAS was ON. Thirty seconds later the TCAS voice said "TRAFFIC" [observed] opposite direction traffic 1000 feet below. Sky conditions were clear and we watched a [Boeing] 767 pass across our nose at FL 280. After the conflict traffic passed, Ft. Worth Center asked us to maintain FL 290. TCAS definitely saved a potential midair collision."

ATC clearance coordination

A TCAS-equipped air carrier was on a missed approach when the local controller cleared it to 3000 feet and told the crew to contact departure [control]. Departure control then asked the local controller to have the air carrier maintain 2000 feet because of traffic. The local controller was unable to issue the revised altitude because the TCAS aircraft had already switched frequencies. The TCAS aircraft contacted departure stating, " . . . cleared to 3000 [feet] but descending for traffic." The controller reported that TCAS was beneficial in this situation because it perfectly coordinated the separation when ATC could not provide a quick assist.

See and avoid

All midair collisions share one common thread, the inability to see and avoid. Therefore, TCAS can be the extra third, fourth, and even fifth set of eyes that can fill the void between cockpit and radar scope.

A TCAS-equipped aircraft was on final approach at 4000 feet in VMC. During the approach, a TA was issued for an intruder at 2 o'clock, 500 feet below. As the crew of the TCAS aircraft commenced a visual search, a climb RA was declared when the intruder had closed to within 2 miles and at the same altitude. The intruder was not visually detected until it passed beneath the TCAS aircraft, at which time, ATC advised the crew of its presence. In the report the pilot stated, " . . . the controller was extremely busy at the time of the event . . . TCAS saved us from a sure midair."

A DIFFERENCE OF OPINION

Although TCAS II has proved itself as a legitimate means of collision avoidance, many in the aviation community have valid concerns over the reliability of the system. In recent years, pilots and controllers have shown skepticism and even contempt for an airborne collision-avoidance system. The reasons have been numerous, but most center on the unpredictability of man and machine.

TCAS II is not a perfect system. It has its glitches just like every other new piece of hardware and software that has come along. Since its conception, the manufacturers, ALPA, the National Air Traffic Controllers Association (NATCA) and FAA have continuously and collectively upgraded and fine-tuned the system. Although few can dispute the TCAS II success stories, some remain skeptical. The unpredictability of the human element can sometimes be just too great for the system to withstand.

According to pilots and controllers, many airlines have inadequate TCAS II training programs. Some consist of just a few hours, which can leave flight crews still scratching their heads when they leave the classroom. An ill-trained pilot behind the yoke of an active TCAS can create a very dangerous situation.

Although the FAA has gone to several ATC facilities to teach controllers about TCAS, according to NATCA, active line controllers have had very little TCAS training. Their complaint is that staff members, procedures specialists, and inactive line controllers have received the majority of the training opportunities.

NATCA's bottom line

According to NATCA, controllers support the concept of an airborne collision-avoidance system. However, because of the number of inappropriate TCAS II-induced maneuvers they do not believe the current system is the answer to the problem.

TCAS and general aviation

TCAS is not just an airline or corporate pilot's concern. TCAS affects *every* pilot— even those who fly Archers only on the weekend. But what has a TCAS II-equipped 727 got to do with a general aviation pilot? Let's look at a couple of possible scenarios.

Say you're flying that Archer into an airport that has a mix of general aviation and commercial aircraft. Approach reports your traffic as a United 727. You acknowledge with "traffic in sight," and go back to your prelanding checklist. The next thing you hear is, "United 123 climbing, RA." You quickly look out the window to see that your traffic has vanished. What started out as an uneventful approach has turned into you losing sight of your traffic during a critical phase of flight. Fortunately, you read this chapter and at least you know that he departed his flight path due to a TCAS Resolution Advisory. The bottom line is that TCAS made an impact on your flight.

You just received a clearance to descend out of 6000 feet to 4000 feet. As you're descending you hear, "United 123 descending, RA." Although you have no idea the position or altitude from where the United was starting its descent, you begin a vigilant scan of the surrounding airspace. Seconds later you see the silhouette of a 727 crossing your flight path, 2 miles off your nose. Did TCAS have an impact on your flight? Absolutely. Although the 727 was not a threat, you suddenly had unexpected traffic crossing your flight path. The positive side is that thanks to the United pilot's radio call, you were able to acquire the traffic faster and have a much keener sense of situational awareness.

TCAS affects the entire flying community. Whether or not you think TCAS is the best thing since the invention of the wheel is immaterial. It exists, and it's here to stay. The more you learn and understand how the system is integrated into the overall ATC network, the safer you're going to become.

FUTURE OF TCAS

Although TCAS II is an overall outstanding feat of technological achievement, the system is hindered by the fact that the evasive maneuvers are only in the vertical. Because there are often times when a horizontal maneuver would be best suited for the particular situation, pilots have long been requesting a TCAS II version with horizontal-maneuvering capability.

TCAS IV

For years avionics manufacturers and the FAA have been trying to develop a system that includes horizontal maneuvering. In August 1995, it was announced that a prototype TCAS IV will be delivered to the FAA by December 1999. The schedule meets the FAA's published plan for a limited installation of TCAS IV on U.S. airliners beginning in December 2000. The latest concept is based on using a Differential Global Positioning System (DGPS) to determine exact aircraft positions. The accuracy of GPS-generated positions will also allow TCAS IV to provide horizontal and vertical advisories.

TCAS TODAY

TCAS has its limitations and is certainly not the be all—end all of collision avoidance. Some of the concerns of NATCA and others in the aviation community are valid. Mix together man and machine, and the outcome has the potential for being rather unpleasant.

However, when TCAS is used properly and conscientiously it can enhance safety. The operative words here are *properly* and *conscientiously*. The system is impressive, but when a poorly-trained pilot misunderstands the advisory commands or mishandles control inputs, TCAS becomes a detriment to safety instead of a benefit. Therefore, solid pilot and controller training programs seem to have slipped through the cracks. That's a shame, because the system's true potential has yet to be seen.

TCAS has its successes, many of them. But it is just another resource. We know that no one resource ever stands alone. TCAS is meant to enhance a see-and-avoid environment, compliment the air traffic control system, and increase situational awareness. But when it's misused, the negative effects can be far more detrimental to collision avoidance than you might realize.

CHAPTER REVIEW

Traffic-alert and collision avoidance systems (TCAS)

- TCAS is an airborne collision avoidance system.
  - ~ TCAS processes transponder signals.
  - ~TCAS onboard computer determines other aircraft's position, altitude, and rate of closure.
  - ~Signals can be received and analyzed up to 40 miles away.
  - ~Information appears on cockpit weather radar scope or dedicated TCAS display.

TAU principle

- Ratio of range to range rate.

Transponders

- Onboard device that transmits a coded signal back to ATC.
  - ~Mode-A transmits position.
  - ~Mode-C transmits range, bearing, and altitude.
  - ~Mode-S transmits Mode C functions, discrete addressing, and data linking.
  - ~Mode-S transmits signals every second, known as a squitter pulse.

TCAS I

- Low-power, 40-nm range system.
  - ~No collision avoidance logic.
  - ~Displays position only.

TCAS II

- Mode-S transponder and two antennas scan volume of airspace around aircraft.
  - ~Top-mounted antenna transmits on 1030 MHz.
  - ~Bottom-mounted antenna receives on 1090 MHz.
  - ~Scanning area covers 14.7 miles to the front of aircraft, 7.5 miles behind, and 7000 feet above and below.
  - ~Can handle 1200 knot head-on closure rate and an altitude-closure rate of 12,000 fpm.

TCAS operation

- TCAS II initial transponder acquisition determines range of intruder aircraft.
- TCAS II second phase of transponder interrogation determines range rate.
- TCAS II final phase of transponder interrogation determines the level of threat.

Traffic advisories

- Predicts intruder within 45 seconds of CPA.
- Assists pilots with visual detection.

Resolution advisory

- Predicts intruder within 30 seconds of CPA.
- Provides preventive and corrective advisories.

TCAS hardware

- The displays are color-coded with FAA-approved symbology and alphanumerics.

TCAS legislation

- Midair of an Aeromexico DC-9 and a Piper Cherokee.
- TCAS I: 10- to 30-passenger seats, 5 February 1995.
- TCAS II: more than 30-passenger seats, 31 December 1993.

Target acquisition flight tests

- Unalerted visual acquisition.
- Alerted visual acquisition.
- Visual acquisition and Aeromexico/Piper midair.
- Visual acquisition and the midair collision of a U.S. Army U-21 and a Piper Navajo.

TCAS transition program

- Assess and validate the safe operation of TCAS II within the ATC network.
- ATC task saturation.
  ~ATC clearance coordination.
  ~See and avoid.

NATCA's concerns

- Man-machine interface.
  ~Pilot training.
  ~Controller training.
- Supports the concept of an airborne collision-avoidance system.
- Does not believe TCAS II is the answer.

Future of TCAS

- Horizontal maneuvers.

Final thought

- Used properly and conscientiously, TCAS can enhance see and avoid, compliment ATC, and increase situational awareness.

CHAPTER REFERENCES

Bradley, Suzanne. 1992. Simulation Test and Evaluation of TCAS II Logic Version 6.04. Mitre Corporation.

"Collins: TCAS IV on schedule to fly by 2000" Air Line Pilot. August 1995:49

Faville, Will. Statement. Federal Aviation Administration Second International TCAS Conference. Reston, Virginia, 9 September 1993.

Federal Aviation Administration Second International Conference on the Traffic Alert and Collision Avoidance System (TCAS). 8-10 September 1993. Reston, Virginia.

Krause, Shari Stamford, Ph.D. 1995. *Avoiding Mid-Air Collisions*. Blue Ridge Summit, Pa.: McGraw-Hill.

Mellone, Vincent J. 1993. TCAS Incident Reports Analysis. National Aeronautics and Space Administration.

Mellone, Vincent J. and Stephanie M. Frank. 1993. Behavioral Impact of TCAS II on the National Air Traffic Control System. National Aeronautics and Space Administration.

Radio Technical Commission for Aeronautics. Minimum Operational Performance Standards for Traffic Alert and Collision Avoidance System (TCAS) Airborne Equipment, Volume I. Washington, D.C.: 2 March 1989.

Radio Technical Commission for Aeronautics. Minimum Operational Performance Standards for Traffic Alert and Collision Avoidance System (TCAS) Airborne Equipment, Volume II. Washington, D.C.: 6 October 1989.

Steenblik, Jan. W. "TCAS Pilot Alert!" Air Line Pilot. March 1991: 33-4.

TCAS II: Collision Avoidance for Corporate Aviation, videotape, by Bendix/King.

TCAS II Pilot Training Video, videotape, by Bendix/King, December 1990.

TCAS II Proposal. Bendix-King. Allied Signal Aerospace, Inc., 1987.

TCAS Transition Program Newsletter. Federal Aviation Administration. Washington, D.C.: 29 April 1991.

TCAS Transition Program Newsletter. Federal Aviation Administration. Washington, D.C.: 3 June 1991.

TCAS Transition Program Newsletter. Federal Aviation Administration. Washington, D.C.: 12 September 1991.

Wapelhorst, Leo, Thomas Pagano, and John Van Dongen. 1992. The Effect of TCAS Interrogation on the Chicago O'Hare ATCRBS System. U.S. Department of Transportation. Federal Aviation Administration.

Williamson, Thomas and Ned Spencer. Development and Operation of the Traffic Alert and Collision Avoidance System (TCAS). 1990.

U.S. Department of Transportation. Federal Aviation Administration. Introduction to TCAS II. Washington, D.C.: March 1990.

U.S. Department of Transportation. Federal Aviation Administration. Advisory Circular: Air Carrier Operational Approval and Use of TCAS II, AC 120-TCAS. Washington, D.C.: 12 February 1991.

U.S Department of Transportation. Federal Aviation Administration. Memorandum. Immediate Distribution of a Traffic Alert and Collision Avoidance System (TCAS) Status Bulletin. Washington, D.C.: 2 May 1991.

U. S. House of Representatives. Report to the Chairman, Subcommittee on Investigations and Oversight, Committee on Science, Space, and Technology. Aviation Safety: Users Differ in Views of Collision Avoidance System and Cite Problems. Washington, D.C.: 1992.

CASE STUDY III-1: PSA Flight 182 and a Cessna 172

Safety issues: see and avoid, role of ATC, radio phraseology, cockpit discipline

On 25 September 1978, a Pacific Southwest Airlines (PSA) Boeing 727 collided with a single-engine Cessna 172 over a populated area of San Diego, California.

Probable cause

The NTSB determined that the probable cause of this accident was the failure of the flightcrew of Flight 182 to comply with the provisions of a maintain-visual-separation clearance, including the requirement to inform the controller when visual contact was lost. Furthermore, air traffic control visual separation procedures were in effect in a terminal area environment with the capability to provide lateral and vertical separation advisories to both aircraft. Contributing factors to the accident were (1) failure of the controller to advise Flight 182 of the direction of the Cessna; (2) failure of the pilot of the Cessna to maintain his assigned heading; and (3) improper resolution by the controller of the ground-radar's conflict alert.

History of flight

PSA Flight 182 operated as a regularly scheduled flight between Sacramento, California, and San Diego, California, with an intermediate stop in Los Angeles, California. The 727 departed Los Angeles at 0834 Pacific Standard Time with 128 passengers and a crew of 7 onboard.

The Cessna 172, owned by Gibbs Flite Center of San Diego, was being flown by an instrument student and his instructor.

Pilot experience

The PSA captain had 14,382 total flight hours, 10,482 in the 727. He had been with the company since 1961 and was promoted to 727 captain six years later. The PSA first officer had 10,049 total flight hours, 5800 in the 727. He had held that position since 1970. The PSA second officer had 10,800 total flight hours, 6587 in the Boeing 727. He had been qualified in that position since 1967.

The Cessna instructor pilot had 5137 total flight hours and had logged 347 hours in the last 90 days prior to the accident. The Cessna pilot had a commercial license and 407 total flight hours. He had flown 61 hours in the previous 90 days.

Weather

At the time of the collision, the San Diego weather was reported as clear with 10 miles visibility.

The accident

Around 0816, the Cessna pilots departed Montgomery Field near San Diego for an instrument-instructional flight. They proceeded to Lindbergh Field to practice ILS approaches. About 45 minutes later, and after completing their second approach, the Lindbergh tower local controller cleared them to maintain VFR and contact San Diego approach control. When the 172 pilot called San Diego approach, he reported that he was at 1500 feet and northeastbound. The controller verified that he was under radar contact. The Cessna pilot was then told to maintain VFR at or below 3500 feet and to fly a heading of 070 degrees. The pilot acknowledged and repeated the controller's instruction.

At 0853:19 Flight 182 radioed San Diego approach and reported level at 11,000 feet. They were then cleared to descend to 7000 feet. Moments later when the PSA pilot notified the controller that the ". . . airport's in sight." The flight was cleared for a visual approach. The call was acknowledged.

Shortly before 0900, the approach controller advised Flight 182 that there was ". . . traffic [at] twelve o'clock, 1 mile, northbound." Five seconds later (0859:33) the pilot answered, "We're looking." Again, in a matter of seconds (0859:39) the controller told Flight 182: "Additional traffic's twelve o'clock, 3 miles, just north of the field, northeastbound." The first officer responded, "Okay, we've got that other twelve."

Another report came only 25 seconds later, ". . . traffic's at twelve o'clock, 3 miles, out of one thousand seven hundred." This advisory was believed to have been referring to the 172. The first officer replied, "Got em," quickly followed by the captain informing ATC, "Traffic in sight."

At 0900:23, Flight 182 was then cleared to ". . . maintain visual separation . . ." and to contact Lindbergh tower. The call was acknowledged. Immediately thereafter, the controller advised the Cessna pilot that there was ". . . traffic at six o'clock, 2 miles, eastbound. A PSA jet inbound to Lindbergh, out of three thousand two hundred. Has you in sight." The pilot "rogered" the call.

About 11 seconds later, Flight 182 reported to Lindbergh tower that they were on the downwind leg for landing. The controller replied, ". . . traffic, twelve o'clock, 1 mile, a Cessna." Six seconds later the captain asked the first officer, "Is that the one [we're] looking at?" The first officer answered, "Yeah, but I don't see him now." At 0900:44, the crew informed the controller, "Okay, we had it there a minute ago." Followed shortly by, "I think he's passed off to our right." The controller responded to the call with a, "Yeah."

The crew continued to discuss the location of the traffic, and at 0900:52 the captain said, "He was right over there a minute ago." The first officer answered with a, "Yeah." Eighteen seconds later the captain told the controller they were going to extend their downwind leg 3 to 4 miles.

From the number of traffic advisories, in a relatively short period of time, it was obvious the skies near Lindbergh Field were very busy that morning. The PSA crew had been pretty successful in detecting their traffic, but there was one aircraft that kept eluding them.

At 09001:11, the first officer asked the chilling question, "Are we clear of that Cessna?" The second officer replied, "Supposed to be," followed by the captain's remark, "I guess." A deadheading PSA pilot who was riding in the jumpseat answered, "I hope." Ten seconds later, the captain remembered, "Oh yeah, before we turned downwind, I saw him about one o'clock, probably behind us now." A few seconds later the first officer said, "There's one underneath." He then added, "I was looking at that inbound there."

At 0901:47, the approach controller advised the Cessna pilot of, ". . . traffic in your vicinity, a PSA jet has you in sight. He's descending for Lindbergh." The sound of the midair collision was heard on Flight 182's cockpit voice recorder at exactly 0901:47. Eight seconds later, the captain radioed, "Tower, we're going down, this is PSA."

Impact and wreckage area

According to witnesses, Flight 182 was descending and overtaking the Cessna, which was climbing in a wing level attitude. Just before impact, the jet banked to the right slightly and the Cessna pitched noseup and collided with the right wing of the PSA. The Safety Board confirmed that scenario from the damage noted on each aircraft. Marks on the Cessna's propeller matched those on the 727's number 5 leading edge flap actuator, indicating that the impact occurred on the forward and underside portion of the jet's right wing. Witnesses also testified that the Cessna broke up immediately and exploded. Parts of Flight 182's right wing and empennage were ripped off and fell to the ground.

A bright orange fire erupted near the right wing of the jet as it began a shallow right descending turn. The blaze intensified, and it was reported that the bank and pitch angles of the airplane reached about 50 degrees at impact.

The remains of both aircraft were scattered in residential areas about 3500 feet apart and were destroyed from the force of impact and explosive fires.

Accident survivability. This accident was not survivable. The occupants of both aircraft sustained fatal injuries along with seven persons on the ground. An additional nine persons on the ground received minor injuries.

The investigation

The Safety Board focused the investigation on two significant factors: the role of ATC and the responsibilities of pilots in a collision-avoidance environment.

ATC connection. At 0901:28, which was 20 seconds before the controller notified the Cessna with the traffic advisory, the data blocks of Flight 182 and the 172 began to merge, triggering a conflict alert at San Diego Approach Control. Within a few seconds the data blocks were overlapping and the controller was unable to distinguish between either aircraft's altitude readout. The controller discussed the situation with his supervisor, and both elected not to manually offset the data blocks. The approach controller concluded that since Flight 182 said they had the "traffic in sight," and they confirmed the controller's request to "maintain visual separation," he did not believe any further action needed to be taken.

Although ATC was not required to notify a pilot if their aircraft was involved in a conflict-alert warning, the Safety Board believed that, in this case, it might have provided the PSA crew additional information that might have given them one more chance to avoid the collision. As the conflict alert progressed, however, the controller did advise the Cessna, ". . . traffic in your vicinity . . . ," albeit too late. Flight 182 was no longer on his frequency.

The Safety Board also noted their concern over the practice of San Diego approach control issuing pilots "maintain-visual-separation" clearances when the ATC system had the capability of providing that service between IFR and participating VFR traffic. Investigators believed that aircraft, especially in the high-performance categories, flying on converging, random courses should be afforded this type of separation until they are clear of each other. In the Safety Board's opinion, if this had been done, "Flight 182 and the Cessna would not have collided."

The evidence further suggested that the approach controller seemed to have relaxed his monitoring vigilance. The PSA flightcrew had reported traffic "in sight" which might have led him to believe that they had better situational awareness than he did. Even though the pilot had assumed the burden of maintaining separation, the controller should not have concluded that the pilot's ability to do so would remain unimpaired. Indeed, moments later the crew offered a few brief remarks —"Okay, we had it there a minute ago," and, "I think he's passed off to our right."— to ATC concerning losing sight of the traffic. The controller's reply was, "Yeah." The Safety Board noted that he should have been prepared to update the traffic advisory and provide additional information to a crew that was obviously looking for the threat aircraft.

The pilot connection. The acceptance of a "maintain-visual-separation" clearance requires the pilot not only to fly a safe distance from the traffic, but also to notify the controller when the traffic is no longer in sight. Although PSA's chief pilot testified that those procedures are in the company regulations, the Safety Board believed that the crew of Flight 182 might not have been aware of the requirement in its entirety. That premise was based upon the failure of the pilots to immediately advise ATC that they had lost sight of the traffic.

The investigation also found that the Cessna pilot did not remain on his assigned heading. At 0859:57, the approach controller told him to, ". . . maintain VFR conditions, at or below three thousand five hundred. Fly heading zero seven zero, vector for final approach course." This clearance was apparently for purposes of traffic separation since the Cessna was crossing and climbing toward the flight path of the descending 727. However, the Cessna turned to a downwind leg of 090 degrees prematurely and beneath the PSA. According to at least one dissenting Board member, if that had not occurred, the accident might not have happened.

Related factors. At issue in this case was whether or not the PSA crew ever saw the Cessna. There was speculation that they observed another airplane that was not known by ATC, or they misidentified the Cessna with other traffic.

The two traffic advisories concerning the Cessna placed it at 1400 to 1700 feet northeastbound, just north of Lindbergh Field and in front of Flight 182. If the crew mistook another aircraft as the Cessna, then it was logical to assume that it was flying at the same time and at a similar course and altitude. Investigators located three possible targets, however, tower and approach controllers testified that there were no primary or beacon targets near the Cessna when the traffic advisory was issued to Flight 182. In order for a third, unknown aircraft to have been misidentified as the Cessna, numerous details would have had to fall perfectly into place. All of which, according to the weight of the evidence, suggested a high improbability of such an occurrence.

Another issue to consider, although it was not causal to the accident, was the extraneous cockpit conversation between the flightcrew and an off-duty PSA pilot riding in the jumpseat. At various times during the approach, the crew engaged in non-flight essential chatter that continued at critical periods, including a time when the pilots were completing a checklist. The Safety Board commented that missed traffic advisories can easily occur when this type of cockpit conversation exists in a high-workload environment.

A final point

In its closing statement of the accident report, the Safety Board discussed the importance of the controller-pilot team. They noted that the principle of redundancy between pilot and controller has long been recognized as one of the foundations of flight safety. However, that concept can be achieved only when both parties fully exercise their individual responsibilities, regardless of who has assumed or been assigned the procedural or regulatory burden.

Lessons learned and practical applications

1. Use clear and concise communication. Avoid ambiguous—"We've got that other twelve".—and nondirective—"He was right over there a minute ago."—phraseology.

2. Immediately notify ATC when you've lost sight of your traffic. Precious seconds tick away as you hopelessly look for your traffic. Ask for help.

3. Redirect the communication. Controllers, if you don't get a clear and decisive response from a pilot, be directive. Reissue the advisory if there is any possibility the pilot has either misunderstood it, or has lost his traffic. Don't be part of the problem.

Case study reference

National Transportation Safety Board. Aircraft Accident Report: Pacific Southwest Airlines, Inc., B-727, and a Gibbs Flite Center, Inc., Cessna 172, N7711G, San Diego, California, September 25, 1978. Washington, D.C.: 20 April 1979.

CASE STUDY III-2: A Piper Aerostar and a Bell 412SP

Safety issues: see and avoid, CRM, cockpit discipline

On 4 April 1991, a Piper Aerostar and a Bell 412SP helicopter collided over Philadelphia, Pennsylvania.

Probable cause

The National Transportation Safety Board determined that the probable cause of this accident was the poor judgment by the captain of the airplane to permit the in-flight inspection after he had already concluded, to the best of his ability, that the nose landing gear was fully extended. The captain of the helicopter also exhibited poor judgment from the manner in which he conducted the inspection; and the flightcrews' failure to maintain safe separation. Contributing to the accident was the incomplete training and checking that the Aerostar flightcrew received from their company and their assigned FAA principle operations inspector (POI).

History of Aerostar (PA-60) flight

The Aerostar was owned by Lycoming Air Services and operated as an on-demand air taxi flight under FAR Part 135. The flight departed around 1022 Eastern Standard Time from the Williamsport-Lycoming County Airport, Pennsylvania, bound for Philadelphia International Airport. The captain, first officer, and one passenger were onboard.

History of Bell 412 flight

The Bell was owned by Sun Company and operated under FAR Part 91. Shortly before noon, the flight departed from the company's helicopter landing pad at Philadelphia International en route to the corporate headquarters in Radnor, Pennsylvania. The captain and first officer were the only persons onboard.

Pilot experience

The captain of the Aerostar had 1972 total flight hours, 425 in multi-engine aircraft. Approximately 115 hours were logged in the Aerostar, 72 of which were PIC time. (*NOTE:* The Safety Board recorded several discrepancies during the investigation concerning the amount of actual PIC time. This will be addressed later in the case study.) He was checked-out as a fully qualified pilot-in-command for revenue operations only nine days before the accident. He was flying his second revenue flight on the day of the midair.

The first officer had 1545 total flight hours, 194 in multi-engine aircraft. He had his second-in-command (SIC) checkride in the Aerostar about six months prior to the accident.

The captain of the Bell 412SP had about 8000 total flight hours, 2380 in Sun Company helicopters. He held helicopter and multi-engine airplane ratings. His last recurrent training in the 412 model was two months before the accident.

The first officer also had approximately 8000 total flight hours, 1629 in Sun Company helicopters. He held helicopter and multi-engine airplane ratings, including some flight experience in the Piper Aerostar. His last recurrent training in the 412 was less than two months prior to the mid-air.

Weather

The reported Philadelphia International surface weather at 1150 was as follows: Ceiling 25,000 feet scattered. Visibility 10 miles. Temperature/dewpoint 59 degrees F/40 degrees F. Wind 250 degrees at 8 knots.

Light to moderate turbulence was reported at 1157 by a pilot of a Cessna 150. The aircraft was at 3000 feet and 27 nm north of the airport.

The accident

At 1201, the captain of the Aerostar notified ATC at Philadelphia International that his nose-gear position light had not illuminated when he lowered the gear. The local controller and tower supervisor considered the situation as an emergency and called the airport fire units in position, along with closing the runway 17 approach to arriving traffic. The crew was told to maintain 1500 feet.

Meanwhile, the flightcrew of the 412 overheard the discussion between the Aerostar and ATC and offered to perform a visual inspection of the airplane's landing gear. ATC approved the maneuver. A minute later the helicopter crew reported, "that Aerostar that went past us, looks like the gear is down." The captain of the Aerostar acknowledged the reply and answered, "I can tell it's down, but I don't know if it's locked. That's the only problem." He further explained that he was able to see the reflection of the nose gear off the propeller spinner, which confirmed his belief that it was down, but he thought there was no way to tell if it was locked in position.

The local controller continued to work the flight as an emergency, and cleared the aircraft to land on runway 17. He also gave the crew the option of making a low-altitude pass by the control tower, and added, ". . . almost no traffic right now. We can do whatever you like." Although the captain agreed to a flyby, the captain of the 412 told the controller that they, "could take a real close look at that if you wanted." ATC acknowledged the comment, and at 1204:19, the 412 began to turn back towards the airport.

As the Aerostar flew by the control tower, the crew was advised that the nose gear appeared to be down. The controller gave them vectors to enter a downwind leg for runway 17, and added that the 412 was returning to provide further assistance. At 1205:30, the Aerostar captain replied, "Okay, I appreciate it."

The crew of the 412 made visual contact with the Aerostar at 1207:54. The pilots of both aircraft confirmed that they had each other in sight and that a speed of 125 knots would be maintained during the join up. Their first attempt of the maneuver was at an altitude of 1100 feet and on an extended downwind leg for runway 17. The controller advised the Aerostar crew of antenna towers 6 miles ahead, and to notify him when they wanted to turn back towards the airport.

Seconds later, the captain of the 412 told the other crew to slow down and that, "we're going to come up behind you on your left side so just hold your heading." The Aerostar captain responded that the antenna towers were straight ahead and that he might need to change the heading by 15 degrees to the left. At 1209:30, the 412 first officer said, "Aerostar, we're gonna pass around your right side now. Take a look at everything as we go by." The captain answered, "Okay." A little over a minute later, the 412 first officer stated, ". . . everything looks good from here." The captain again replied, "Okay, appreciate that. We'll start to turn in." The last transmission abruptly stopped with a tremendous noise. Shortly thereafter, the controller noticed a smoke plume to the north of the airport.

The collision. From the best available data, investigators determined that the altitude and airspeed profiles of both aircraft remained relatively constant during the join-up maneuver, except for a possible gain in altitude by the 412 just prior to the collision. Since the helicopter was behind and below the Aerostar, it would have been virtually impossible for either pilot in that airplane to maintain continuous visual contact of the 412. This situation was further complicated by the need for the Aerostar crew to monitor their flight path in relation to the antenna towers that were directly ahead of them. Another visual hindrance was from the permanent covering of the 412's cockpit overhead windows. It had been installed to eliminate flicker vertigo caused by the reflection of light off the main rotor during IFR conditions. Consequently, the pilots of the 412 would have been unable to see objects just above their helicopter.

As with most accident investigations, eyewitness accounts are usually varied, but there was a general agreement from the witnesses to this midair that confirms the helicopter was below and to the right of the airplane before the collision. Other statements seemed consistent that the aircraft were flying straight and level, and with parallel flight paths.

Impact and wreckage area

Although investigators were unable to pinpoint the exact attitude of the fuselages at impact, they did determine the position of the airplane relative to the helicopter's main-rotor blade. Based upon the impact marks on the 412's rotor blades and the marks on the Aerostar's tires, the Safety Board reached a possible collision scenario. Refer to Fig. III-A.

Parts of the Aerostar's nose gear assembly were found as far away as 750 feet from the main wreckage. The tire remained attached to the rim. However, a two-inch section below the edge of the rim had been cleanly severed, presumably by a rotor blade. The tire for the right main landing gear was located 250 feet from the main wreckage, and had been cut and torn. The wheel-hub rim had a sharp impact mark that aligned with the cut on the tire. Neither the landing gear assembly, rims, or tires showed any signs of fire damage or paint transfer.

The airplane's left engine appeared to have been operating when it hit the ground. The right engine, however, was in a feathered position which warranted a close exami-

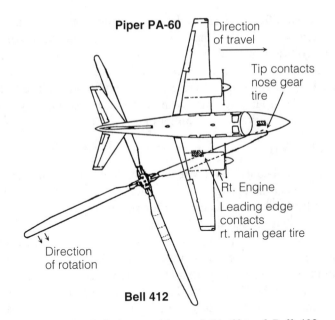

Fig. III-A. *Relative positions of PA-60 and Bell 412 just before collision.* Adapted from NTSB

nation during the investigation. The propeller governor assembly was in the full RPM stop setting, but there was no evidence of ingested metal particles or gross internal failure. It was later noted by a representative of the propeller manufacturer that if an engine lost oil pressure while it was producing power, the prop would feather in 5 to 10 seconds.

One of the four rotor blades on the helicopter had separated from the main-rotor hub assembly and was found 450 feet from the main wreckage. It was bent in two areas, and a large portion of the fiberglass trailing edge was missing. There was no fire damage to the blade, but rubber transfer marks were on the leading edge of the blade from its tip to 16 inches inboard. A section of the blade's leading edge was cut 82 inches inboard from the tip. Rubber and metal slash marks were also found 150 inches from the tip, including a 13-inch-long dent. Another blade had become detached from the rotor head, a 4-inch part of the tip was broken and missing. The blade was slightly bent and smeared with oil marks 22 inches from its tip. The mid-section of the leading edge of the blade was completely destroyed. The remaining two blades were still attached to the main-rotor hub assembly, but were heavily damaged. Evidence of slash or impact marks could not be found on those blades due to extensive fire damage.

Numerous parts of both aircraft were scattered throughout the schoolyard and a nearby residential area. A piece of the helicopter's main-rotor blade was found 1000 feet from the main wreckage. The airplane's nose-wheel fork was 875 feet from the ground impact site, and the right main landing gear had separated from its wing structure and was located 500 feet away.

Accident survivability. The Safety Board determined that the outer right wing panel from the Aerostar and one of the main rotor blades from the 412 had separated as a result of the collision. Both aircraft were, therefore, rendered uncontrollable and plummeted to the ground. Most of the Aerostar fell in the front yard of an elementary school, and the 412 came to rest just behind the school building. The five occupants and two persons on the ground were fatally injured. One person on the ground suffered severe burns, and four others sustained minor injuries.

The investigation

The Board carefully probed the training records and logbooks of the Aerostar captain. They noted a lack of certain emergency procedural training, and a few discrepancies about his actual PIC time and the method for logging that time.

The documents showed that in 1988 he successfully passed his second-in-command checkride in the PA-60 with 15.9 total hours in the airplane. However, on subsequent non-revenue flights he logged the time as PIC. Only on revenue flights did he log the flight time as SIC. The Safety Board also noted an inconsistency over the exact nature of some of the flights listed in his logbook, and whether he was acting as PIC for the whole flight, or as first officer for part of the flight. In several cases where the captain had logged PIC time, the company's records indicated that another more senior pilot was responsible for the flight.

Investigators questioned the training/check pilot for Lycoming Air Services as they reviewed the flight manual for the Aerostar. In referencing the section for a hydraulic pump failure, the manual clearly details information about lowering the gear in an emergency. There was no indication by the captain that he had lost hydraulic pressure, so let's skip over that part and get to the gear extension procedures. It states that the landing gear warning horn will sound if the throttles are set at idle and the nose gear is not locked. Therefore, a standard method to check whether the nose gear is down and locked is to simply reduce the power setting. If the horn does not sound, the gear should be locked. But if the horn does go off, the appropriate procedure is to turn off the hydraulic pump, bleed off the hydraulic pressure, and place the landing-gear handle in the down position. The gear should then free fall to the down and locked position because of gravity and a series of springs. By pulling back again on the throttles, it can then be determined if the gear is locked into place. According to the training/check pilot, he did not instruct the captain on this troubleshooting method, but he had taught him about the push-to-test function of the gear indicator lights.

Without the benefit of a CVR, the Safety Board could not conclude whether the captain took any action to isolate the problem. They believed if he had he would have informed the tower.

Three days before the accident, the captain flew his first revenue flight in the Aerostar. Coincidentally, the only passenger was a senior executive for Lycoming, the PA-60's engine manufacturer, who had also logged a few hundred hours in the airplane. Since the captain was the only crewmember on the flight, and the passenger was qualified in the Aerostar, he was allowed to sit in the first officer's seat. During the investigation, the engine company executive provided a detailed account of how he perceived the captain's piloting skills and proficiency. In his statement, he noted that the captain had problems starting the engines and that he had to give him instruction in the proper starting techniques. He described the takeoff roll as "pretty erratic" because the captain was overcontrolling the electric/hydraulic nose wheel steering to the extent that the passenger became concerned. The captain seemed to handle the aircraft quite well once they got airborne. However, shortly after they reached cruise altitude the right engine began to surge "about 200 RPM." The passenger believed that it was a problem with the fuel controller, but the captain failed to take any action. According to the passenger, he had to convince the captain to return to the airport. A subsequent maintenance inspection found that the fuel controller was defective. The total flight time was only 30 minutes.

Loggable time. Many professional pilots, especially those with minimal experience in a particular airplane, will log flight time while "deadheading" or on "positioning" trips. I've even known a few who have logged PIC cross-country time from the back seat. While the rest of us have gritted our teeth and stayed legal and watched our fellow pilots take checkrides before us, we at least have the consolation in knowing that sooner or later it *will* catch up with them. In the Board's opinion, if a pilot pads a logbook with flight time that was not based on formal training or actual hands-on flying, that pilot's ability to make sound decisions and exercise good judgment can be negatively affected. The Safety Board and the FAA are trusting that flight examiners or POIs will question a pilot's logbook when the skill level doesn't come anywhere close to reality. But, there are those who will still sign off on checkrides and allow inexperienced pilots to get in over their heads.

Pilot judgment. The Safety Board believed that the inexperience of the Aerostar captain as a PIC was a significant factor in the sequence of events that led to the collision. The evidence suggested that the captain never tried the recommended emergency nose-gear extension procedure that was previously discussed. Investigators examined the nose gear of another Aerostar and found that when the landing gear doors are closed, a very small area around the nose gear strut is left exposed. Therefore, it would have been virtually impossible for either the controller or the helicopter crew to have determined if the gear was in a locked position. The Board also believed that since the captain was able to verify from the reflection off the propeller that the nose gear was at least down, he should have proceeded to the airport for an emergency landing. The touchdown procedure was not considered extremely difficult, and according to investigators, a nose-gear collapse rarely results in a major accident or occupant injury. Also, he was aware that fire units were already standing by to assist him. Taking all those points into consideration, the Board concluded that a more prudent and experienced captain would have rejected the idea for a visual inspection, and landed.

The Safety Board went on to say that once the in-flight events were allowed to continue, both captains displayed poor judgment. First, the captain of the 412 should never have flown in the blind spot of the Aerostar crew. That was just plain dangerous. Second, there was no coordination of maneuvers agreed upon between the captains. From the accumulation of evidence, investigators noted that the Aerostar captain "relinquished the responsibility for ensuring the safety of his airplane, giving it entirely to the pilot of [the 412]." It was probable that the captain of the Aerostar did not see the helicopter, and therefore, was unaware just how close his aircraft was to those spinning rotors.

The first officer of the Aerostar had been on duty from 2100 the evening before the accident until 0600 on the day of the accident. He had logged about three hours of flight time during the night, and had reportedly slept from about 0630 to 0900. Although the airplane was legal to be flown with one pilot, Senator Heinz, the Aerostar passenger, always insisted on flying with two pilots. Since the first officer had not exceeded his duty time limit (per FAR 135.267), he might have felt rested enough to fly a few more hours. However, the Board believed that it was probably ill-advised for him to have accepted an additional flight after having been up all night. As a result of his odd work schedule the night before and lack of sleep, the Board considered his effectiveness as a SIC crewmember might have been impaired.

This case also stands as an example that low-time pilots are not the only ones who have problems with the decision-making process. The Safety Board determined that the captain of the helicopter exhibited "poor judgment," and that the flightcrew failed to maintain a safe distance from the Aerostar. Each pilot had about 8000 flight hours and held multi-engine and CFI tickets.

FAA monitoring. The Safety Board stated that because the POI's workload was so overwhelming, Lycoming Air Service did not receive an adequate level of FAA monitoring. Therefore, it was concluded that an incomplete check of the charter company's flight operations was a contributing factor to this accident.

In addition to Lycoming Air Services, the POI was responsible for 16 other certificate holders, including one scheduled commuter carrier. He told investigators that his workload was extremely heavy and that he had been unable to personally visit Lycoming until mid-January 1991. He noted that the previous POI assigned to the company became an assistant manager for them following his retirement from the FAA. Therefore, he would occasionally call the former POI and inquire about the status of the company.

In December 1990, two Lycoming pilots required recurrency checkrides from the POI. One of the pilots failed the checkride once, and the other twice. Based upon this experience, the POI decided to perform a personal inspection of the company. It was during that January visit that he discovered the training documents, pilot records, and other operational paperwork were not in compliance with the FARs. He noted that the chief pilot corrected the problems, and prior to the midair the company was in full compliance with the FARs. In the meantime, the POI gave Lycoming's check pilot a competency flight check. In the words of the POI, the flight was "pretty bad," which prompted him to suspend the pilot's Part 135 airman's privileges. However, only two days later the pilot was retested and successfully passed that checkride.

Lessons learned and practical applications

1. Understand your airplane's systems. In this case there was a simple procedure that the Aerostar pilots could have done to determine if the landing gear was down and locked. But they failed to do so.

2. Never relinquish responsibility for your own aircraft. In essence, the Aerostar pilot sat back while the helicopter pilot took over the flight operation.

3. Never fly in the blind-spot of another pilot. The helicopter pilot flew behind and below the Aerostar. As this midair proved, this is an extremely dangerous place to fly. If the Aerostar pilot had been more proactive in the situation, he would never have allowed such an occurrence.

4. Maintain cockpit discipline.

5. Avoid unusual maneuvers. Formation flying is not something that a pilot should practice without qualified instruction.

Case study reference

National Transportation Safety Board. 17 September 1991. Aircraft Accident Report: Midair Collision involving Lycoming Air Services Piper Aerostar PA-60 and Sun Company Aviation Department Bell 412. Merion, Pennsylvania. April 4, 1991. Washington, D.C.

CASE STUDY III-3: Aeromexico Flight 498 and a Piper Cherokee

Safety issues: see and avoid, airspace intrusion, role of ATC

On 31 August 1986, Flight 498, a McDonnell Douglas DC-9, collided with a single-engine Piper Cherokee over Cerritos, California, in the Greater Los Angeles Basin.

Probable cause

The NTSB determined that the probable cause of this accident was the limitations of the air traffic control system to provide collision protection. Contributing to the accident was the inadvertent and unauthorized entry of the Piper Cherokee into the Los Angeles Terminal Control Area, and the limitation of the see-and-avoid concept.

History of flights

Flight 498 was a regularly scheduled flight between Mexico City, Mexico, and Los Angeles, California, with intermediate stops in Guadalajara, and Tijuana, Mexico. The DC-9 departed Tijuana at 1120 Pacific Daylight Time with 58 passengers and 6 crewmembers onboard.

According to the flight plan of the Piper Cherokee, the pilot's proposed route was from Torrence, California, to Big Bear, California. The pilot departed Torrence at 1140 with two passengers onboard.

Pilot experience

The captain of the DC-9 had 10,641 total flight hours, 4632 in the DC-9. The first officer had 1463 total flight hours, 1245 in the DC-9. The pilot of the Piper had 231 total flight hours.

Weather

Clear skies with a visibility of 15 miles.

The accident

Flight 498 was level at 7000 feet and setting up for an approach into Los Angeles International Airport, when at 1150 the controller advised the crew that there was, "traffic, ten o'clock, 1 mile, northbound, altitude unknown." Although the crew of Flight 498 acknowledged the advisory, they never told the controller whether they had the traffic in sight. In any event, the traffic was not the Cherokee.

At 1151, Flight 498 was cleared to descend to 6000 feet. Meanwhile, a single-engine Grumman Tiger had made an unauthorized entry into the TCA and the controller was busy vectoring him out of harm's way. Less than a minute later, the controller noticed that Flight 498 had disappeared off his radar, and he proceeded to make several unsuccessful attempts to contact it. At approximately 1152, Flight 498 and the Cherokee collided over the city of Cerritos at about 6560 feet.

Impact and wreckage area

The collision occurred as Flight 498 was descending through 6560 feet on a northwesterly heading and the Piper was on an eastbound track. Based upon collision damage on the DC-9, it appeared that the Cherokee's engine struck the main support structure of the jet's horizontal stabilizer. Refer to Fig. III-B. The impact sheared off the top of the Cherokee's cabin, causing the DC-9's horizontal stabilizer to separate from the aircraft. Refer to Fig. III-C.

The best in-flight scenario that can be determined is that the Piper initially appeared 15 to 30 degrees offset to the left of the captain's windshield, and subsequently appeared in the same position through the first officer's windshield. With regard to the Piper pilot, Flight 498 was about 50 degrees to the right of the design eye reference point, and therefore, was visible out the far right-side window. Nevertheless, because the two aircraft were on a collision course, the relative motion of the Piper presumably would have been minimal, making it extremely difficult to detect.

The main wreckage sites of both airplanes were in a residential area, and within 1700 feet of each other. Except for the upper portion of the fuselage, cockpit, engine, and vertical stabilizer, the Piper remained relatively intact after the collision. The major section of wreckage fell in an open schoolyard and did not catch fire after impact.

Most of the DC-9 crashed in an area about 600 feet long by 200 feet wide. The wreckage burned to disintegration. The largest piece that was found came from the lower aft fuselage. Both engines were located near the point of impact, and were noted to have been operating at a high power.

Accident survivability. This midair collision was not survivable.

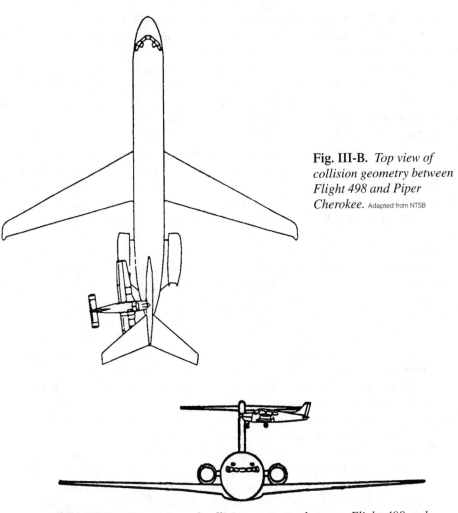

Fig. III-B. *Top view of collision geometry between Flight 498 and Piper Cherokee.* Adapted from NTSB

Fig. III-C. *Front view of collision geometry between Flight 498 and Piper Cherokee.* Adapted from NTSB

The investigation

The Safety Board quickly determined that the pilot of the Cherokee entered into the Los Angeles TCA without ATC clearance. There was no evidence that suggested the pilot had suffered a physiological disability or that he was unfamiliar with the TCA boundaries.

The recorded ground track of the airplane showed the pilot had proceeded almost directly to the collision point after he took off from Torrence. Based on a few simple calculations, the Safety Board determined that this average rate of climb was about 550 fpm. The aircraft maintained an almost constant heading and groundspeed, as if its progress was being closely monitored and managed.

The pilot had been described as methodical in his approach to flying, and had asked for advice concerning the TCA before the flight. An opened Los Angeles terminal area chart was found in the cockpit. From all accounts, the Board believed it was unlikely that he would have deliberately flown into the TCA.

Radar retrack program. Investigators sought to determine if the Piper had appeared on the controller's radar display. The recorded radar data showed that beacon returns for both airplanes were processed by the ARTS III air traffic control computer. When that data was entered into the retrack program, several "1200" VFR-coded targets, including that belonging to the Piper, were visible on controller's display screen.

ATC performance. The Safety Board believed that two required ATC procedures might have influenced the controller's monitoring of traffic that day. According to the Controller's Handbook, first priority must be traffic separation service involving IFR to IFR airplanes. ATC will provide VFR traffic advisories only if a collision is imminent or "work permitting." The Handbook also specifies that an aircraft conflict-alert advisory is limited to those airplanes when the controller, "is aware of another aircraft at an altitude which you believe places them in unsafe proximity." The Piper did not provide Mode-C information and therefore was not prioritized. As a result, the Safety Board concluded that the ATC procedures were causal to the accident in that they "set the stage for the controller to overlook or not see the Piper's target on his display."

Investigators noted that the controller's radio conversations with the various aircraft that he was working strongly suggested that his attention was directed toward the area east of Los Angeles International. At 1150, he advised Flight 498 of traffic at "ten o'clock" and then watched it pass behind the jet. He testified that after he observed that traffic clear, he "saw no traffic along its projected route of flight that would be a factor."

The Safety Board believed that a change in runways for Flight 498, coupled with the sudden appearance of another general aviation aircraft, caused the controller to direct his attention to one area of the screen. Since he had no expectation of additional traffic, the Safety Board concluded that this might have been why he did not see the Piper's target.

See and avoid. Based on cockpit visibility studies, the Safety Board determined that both airplanes were within the pilots' fields of vision for at least one minute and 13 seconds. However, there were several limiting factors that would have deteriorated the see-and-avoid environment. Theoretically, the Piper pilot should have been able to see the much larger DC-9 before the Aeromexico pilots saw him. But that was not the reality of the situation. The DC-9 was visible through the Piper pilot's right windscreen and near the outer portion of a side-to-side scanning pattern. The Piper was approaching the jet from the passenger side with less than a 30 degree offset to the left. (*NOTE: Refer to Chapter 1 if you're hazy on the following information.*) The DC-9 would have subtended to 0.2 degrees (12 minutes) of arc when it was about 6 nm away or 1 minute 23 seconds before the collision. Although some eagle-eye pilots are able to pick up traffic several miles away, the probability of this being the norm is low. In this case, the DC-9 appeared near the edge of the Cherokee pilot's scanning range and had passed through the optimum visual detection area at 6 nm.

On the other hand, the Piper was visible through the center windshield of the DC-9 but in monocular view of the first officer. It appeared in the same location for the captain but had shifted within his normal binocular vision field. The Cherokee would have subtended a visual angle of 0.2 degrees (12 minutes) of arc when it was a little more than 1 nautical mile, or 15 seconds, before the collision. Remember the 12.5 second recognition and reaction rule. The Aeromexico crew most likely failed to see the Piper due to its small size and its minimal relative motion to the jet.

Lessons learned and practical applications

1. Always contact ATC when operating near airspace boundaries. A simple radio is all it takes to notify ATC of your position. In this case, it would have prevented this accident.

2. Maintain situational awareness. This is especially true in busy traffic areas and near airspace boundaries. Be extremely diligent in monitoring your flight path if you're planning to stay out of controlled airspace. Pay particular attention to the activity on the radio. This will help in determining the location of potential threat aircraft.

3. Maintain scanning vigilance. Never let down your guard. Remember to always use effective scanning techniques.

Case study references

National Transportation Safety Board. Aircraft Accident Report: Collision of Aeronaves de Mexico, S.A., McDonnell Douglas DC-9-32, XA-JED and a Piper PA-28-131, N4891F, Cerritos, California, August 31, 1986. 7 July 1987. Washington, D.C.

CASE STUDY III-4: A Mitsubishi MU-2 and a Piper Saratoga

Safety issues: see and avoid, distraction, uncontrolled airport traffic patterns, cockpit discipline, role of ATC

On 11 September 1992, a Mitsubishi MU-2 turboprop collided with a single-engine Piper Saratoga (PA-32) approximately 2 miles northeast of the Greenwood Municipal Airport, Indiana.

Probable cause

The NTSB determined that the probable cause of this accident was the inherent limitations of the see-and-avoid concept for separation of aircraft operating under VFR. These factors precluded the pilots of both aircraft from recognizing a collision hazard and taking actions to avoid the midair collision. Contributing factors to the cause of the accident included the failure of the MU-2 pilot to use all of the available ATC services by not activating his IFR flight plan before takeoff; and the failure of both pilots to follow recommended traffic pattern procedures, as detailed in the AIM, for airport arrivals and departures.

Pilot experience

The Mitsubishi pilot had 19,743 total hours, and was also an MU-2 check pilot with 9000 hours in the airplane. The Piper pilot had a minimum of 1224 hours, 150 in the Saratoga. The passenger in the right seat of the Piper was a pilot with 412 hours.

Weather

The reported weather at the time of the midair was: Ceiling 4500 feet scattered, 25,000 feet scattered. Visibility 15 miles. Temperature 70F. Dewpoint 49F. Winds 020/10 knots.

The flight of the Saratoga

After taking off from the Eagle Creek Airport, outside Indianapolis, Indiana, the Piper pilot and his two passengers flew a short distance to Terry Airport. While there, the pilot spoke with a mechanic concerning the annual inspection that had been performed on the airplane, and by 1445 he and his passengers had departed Terry for Greenwood Airport, another airfield in the local area. The purpose of this leg was to take aerial photographs of the pilot's new office building.

Although the aircraft was operating under VFR, the most direct route between Terry Airport and Greenwood Airport was through the Indianapolis ARSA, requiring the pilot to be in contact with ATC. At 1445:17, the pilot advised the Indianapolis Departure West/Satellite controller that he had departed Terry and was en route to Greenwood. The controller issued the pilot a discreet beacon code, radar identified the airplane, and instructed the pilot to climb and maintain 2500 feet. Nearly six minutes later, the controller handed off the Saratoga to the Indianapolis Departure East/Satellite (DRE/Satellite) controller. Seconds later the pilot reported, ". . . with you at two point five [2500 feet] going to Greenwood [Airport]." The controller replied, ". . . maintain VFR, I'll have on course for you in about 5 miles." The pilot acknowledged the transmission. Approximately two minutes later the controller radioed, ". . . proceed on course to Greenwood, advise the airport in sight." That call was answered by the pilot. At 1455:51, the controller notified the Saratoga, ". . . airport twelve to one o'clock . . . three miles." The pilot replied, ". . . we have the airport." The controller immediately responded with, ". . . squawk VFR, radar service terminated, frequency change approved." At 1456:03, the pilot made his final radio call and thanked the controller.

The flight of the MU-2

On the morning of the accident, the pilot of the MU-2 departed from his home base at the Huntingburg, Indiana, Airport bound for the Greenwood Municipal Airport. The aircraft was owned and operated by a coal-mining company, and the purpose of the trip was to pick up four passengers at Greenwood and fly them to Columbus, Ohio.

The pilot had filed two IFR flight plans that day with the Terre Haute, Indiana, FSS. One was for the 30-minute flight from Huntingburg to Greenwood. The other was for the flight to Columbus, with a scheduled departure time of 1400. His passengers did not arrive at the airport until shortly after 1430. At 1456:41, the pilot contacted the DRE/Satellite controller to notify him that he was, ". . . off the ground Greenwood [Airport] standing by for [IFR] clearance to Columbus [Airport]." Seconds later, the controller gave him a discrete beacon code and told him to, ". . . maintain at or below five thousand." There was no further communication with the pilot.

The accident

According to witnesses, it was a typical day at the Greenwood Airport with little traffic. They observed the Saratoga flying southbound, while the MU-2 was climbing and turning toward the east. At approximately 1457, at an altitude of 2100 feet, the two aircraft collided. As Fig. III-D illustrates, the pilot of the Saratoga turned the airplane to the left seconds before it flew into the turboprop's empennage, shearing it away from the fuselage. Although the cockpit and cabin sections of the MU-2 remained intact, the airplane was uncontrollable and crashed in a residential area. Remarkably, the Saratoga did not break up in flight and the pilot-passenger was able to make a controlled landing before the airplane struck ground obstacles, including three houses.

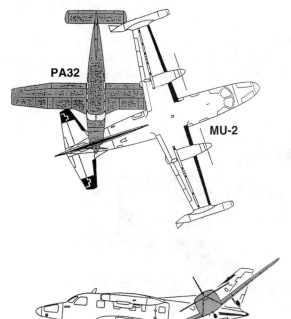

Fig. III-D. *Top and side view of collision points between MU-2 and PA-32.* Adapted from NTSB

Accident survivability. Because of the catastrophic damage caused by impact, the MU-2 was rendered uncontrollable. The pilot and four passengers onboard received fatal injuries. The Saratoga, however, was still intact and flyable after the collision. For unknown reasons, the pilot of the Saratoga became incapacitated shortly after the midair, and later died. The pilot-passenger and the passenger in the rear seat escaped the postcrash fire with serious injuries.

The investigation

The collision closure rate of the two airplanes was 234 knots. Radar data showed the Saratoga was on a track of 174 degrees, at a ground speed of 127 knots, with a rate of descent of 390 fpm. The MU-2 was on a course of 070 degrees, at a ground speed of 168 knots, and climbing at approximately 1200 fpm. The collision angle at impact was close to 90 degrees because the Saratoga pilot made a 45-degree steep bank to the left seconds before contact.

Cockpit visibility study. According to the Safety Board's analysis of the cockpit visibility study, both pilots, *theoretically,* had enough time to see and avoid each other. From the tragic outcome, however, theory and reality can be quite different. Therefore, the Board determined that the inherent limitations of the see-and-avoid concept precluded these pilots from recognizing and reacting to a collision threat.

The study showed that the Saratoga might have been visible to the MU-2 pilot for 20.5 seconds—including the 12.5 second "rule"— prior to impact. Assuming the MU-2 pilot was sitting stationary at the design eye reference point, the Saratoga could have appeared unobstructed in the lower left corner of his left windshield for four seconds. Refer to Fig. III-E. The Saratoga would then have shifted slightly, enough to be partially blocked by the MU-2 pilot's left windshield post. However, if the pilot had moved forward to adjust his radios or flight controls, or to scan outside, the Board suggested that he might have been able to see the Saratoga with both eyes.

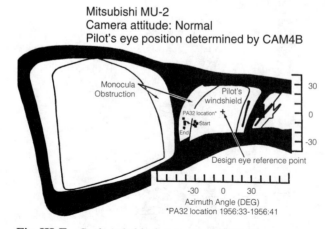

Mitsubishi MU-2
Camera attitude: Normal
Pilot's eye position determined by CAM4B

Fig. III-E. *Cockpit field of vision for MU-2 pilot.* Adapted from NTSB

The study also revealed that the MU-2 might have been in view of the Saratoga pilot for 25.5 seconds—including the 12.5 second "rule"— prior to impact. For about 13 seconds the MU-2 should have been positioned in the right windshield of the Saratoga. Refer to Fig. III-F. However, due to the obstruction from the center windshield post, this view would have provided the pilot and pilot-passenger only a monocular field of vision.

Uncontrolled airport traffic patterns. The Safety Board noted during its investigation that there is little regulation or guidance relating to arrival and departure procedures at uncontrolled airports. Since both pilots were operating under VFR, and in the vicinity of an airport, they were required to comply with FAR Part 91.127. In part, it states that: "Each person operating an aircraft to or from an airport without an operating control tower shall: (1) In the case of an airplane approaching to land, make all turns to the left, and (2) In the case of an aircraft departing the airport, comply with any traffic patterns established for that airport in Part 93."

At the time of the accident, Greenwood Municipal Airport did not have a traffic pattern established in FAR Part 93, therefore, there were no *regulatory* departure procedures. Instead, there were only *recommended* procedures published in AC 90-66 and Paragraph 4-54 of the AIM. The AC suggested a 1000 foot agl traffic pattern, whereas, the AIM described the procedure in vague terms, including a reference to general aviation traffic patterns that can extend from 600 feet to 1500 feet agl. The AIM further noted two possible departure procedures—maintain runway heading, or make a 45 degree left turn after reaching traffic pattern altitude.

Because of these discrepancies concerning uncontrolled airport traffic patterns, it was no surprise that the Board received four different answers from four local Greenwood pilots during interviews conducted as part of the investigation. Although the Airport/Facility Directory listed Greenwood Municipal as having a traffic pattern altitude

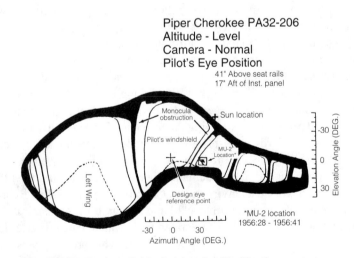

Piper Cherokee PA32-206
Altitude - Level
Camera - Normal
Pilot's Eye Position
41" Above seat rails
17" Aft of Inst. panel

Fig. III-F. *Cockpit field of vision for PA-32 pilot.* Adapted from NTSB

of 800 feet, only one of the four pilots gave the correct answer. Two chose 1000 feet and the other thought it was 2000 feet. When questioned on arrival and departure procedures at Greenwood, the four pilots came up with four different procedures, none of which resembled the recommended procedures outlined in the AIM.

According to the MU-2 backup pilot, who was one of the four local pilots interviewed, the pilot of the MU-2 had developed his own arrival and departure procedures at Greenwood. Departing on runway 36, as he did on the day of the accident, he would climb straight out 500 feet to 700 feet and then initiate a right turn. Since the airport is located only 2 miles from the southeast boundary of the Indianapolis ARSA, and because the MU-2 is a high-performance turboprop, the pilot devised that particular procedure to prevent an inadvertent penetration into the ARSA, and allow for a comfortable ride for his passengers.

Inflight IFR clearance procedures. Although the weather conditions and the MU-2 pilot's altitude request (15,000 feet) did not require an IFR flight plan, the Safety Board believed his purpose in filing one was to aid in traffic separation and to prevent inadvertent entry into the ARSA. The pilot filed the flight plan for a 1400 departure time.

According to the AIM, most centers will delete a flight plan if it has not been activated after one hour of the proposed departure time. Therefore, to ensure that a flight plan remains active, pilots should notify ATC if they encounter any delays. Otherwise, the AIM noted that due to traffic saturation, control facilities will often be unable to accept revisions to a flight plan over the radio. Revisions must then be completed by contacting the nearest FSS.

Because the MU-2 pilot intended to depart Greenwood at 1400 but was delayed until 1456, suggests that he hurried to get off the ground before 1500. The Safety Board believed that because the pilot received his IFR clearance in the air, that extra responsibility increased his cockpit workload to the point of possible distraction. It also delayed the controller's ability to identify the airplane by radar before the collision. Therefore, the Safety Board concluded that the pilot should have activated his IFR flight plan before takeoff so the controller could have provided him with traffic advisories. Because the pilot failed to take full advantage of the ATC services available to him, the Board determined that this contributed to the factors that led to the accident.

Pilot workload. In the one minute that had elapsed from the time the MU-2 pilot lifted off the ground to the point of impact, he would have performed numerous duties. These duties included: after-takeoff checklist, making radio calls to unicom and to departure control, raising the landing gear and flaps, adjusting the transponder, monitoring the engines and propellers, and flying the airplane. The Safety Board believed that in addition to those standard duties, the pilot also had concerns about inadvertently flying into the ARSA, obtaining an IFR clearance, and providing passenger comfort.

All of those physical and mental demands, jammed into one minute of flying time, produced a very high workload environment. Consequently, the Safety Board believed the pilot had a minimal amount of time available to scan for threat aircraft.

Operations near an airport. The Saratoga pilot intended to fly near his new office, so his passengers could take aerial photographs. This property was located only 3 miles from the airport, and within 1 mile of the collision point. The Safety Board believed that it was likely the pilot was looking down to facilitate the photography, thereby limiting his ability to scan for other aircraft. Likewise, the attention of both passengers might have also been distracted towards the ground.

According to the AIM, pilots flying in the vicinity of an airport should monitor and communicate on the appropriate frequency within 10 miles of that airport. The pilot and pilot-passenger were wearing headsets at the time of the collision, therefore, the Board believed that they could have monitored both the Indianapolis ARSA and the Greenwood unicom, however, the Board's report did not indicate this was asked of the surviving passengers in the Saratoga.

The Safety Board further commented on the role of the radar controller. He terminated radar services and approved a frequency change for the Saratoga pilot about 3 miles from the airport. In part, the FAA Air Traffic Controller Handbook states: "terminate ARSA service to aircraft landing at other than the primary airport at a sufficient distance from the airport to allow the pilot to change to the appropriate frequency for traffic and airport information." Since the AIM recommends pilots initiate unicom communications approximately 10 miles from the airport, the Board noted that the radar controller concluded his services too late.

The see-and-avoid concept. The Safety Board determined that the inherent limitations of the see-and-avoid concept, especially when pilots are operating under VFR and near high density traffic areas, were directly causal to this midair. As the results of the investigation proved, both pilots had an extremely short amount of time to detect the threat and to take evasive actions against it. Although the Safety Board questioned the decisions of the MU-2 pilot to airfile his IFR flight plan, and to make a VFR right turn during departure, they concluded that the deficiencies of the see-and-avoid concept were paramount to this accident.

For background information, the Safety Board studied two other notable midair collisions: an Aeromexico DC-9 and a single-engine Piper (refer to case study III-3); and an Army U-21 and a twin-engine Piper (refer to case study III-6). In each case, "the limitations of the see-and-avoid concept to ensure traffic separation," and "the deficiencies of the see-and-avoid concept as a primary means of collision avoidance," were considered causal to the accidents. The Board further analyzed the results of laboratory and in-flight studies conducted during those investigations, and concluded that there is, "great difficulty of reliably seeing other airplanes when there is no warning of an impending collision and when the opposing airplane is as small as a PA-32 or an MU-2."

Safety recommendation

Based on the commonalities between midair collisions in general, the Safety Board issued Safety Recommendation A93-127-132. It called for the FAA to assume a more active role in ensuring that instructor pilots are informed, as to the importance of emphasizing proper scanning techniques to their students during training and biennial flight reviews. It also addressed the need for better overall pilot education concerning the many factors associated with collision avoidance.

Lessons learned and practical applications

1. Take full advantage of ATC services. A simple phone call to ATC is all it takes to extend a flight plan. In this case, the Safety Board believed that the MU-2 pilot's decision to airfile his IFR flight plan prevented ATC from providing traffic advisories to him.

2. Use your dual radios. The obvious advantage of dual radios is that you can monitor two frequencies at the same time. It was unclear to the Board whether the Saratoga pilot or pilot-passenger were monitoring both the Greenwood unicom and the Indianapolis ARSA frequencies. If either one had been, the Board believed someone might have heard the MU-2 pilot's call, ". . . I'm off the ground Greenwood . . ." Likewise, the Saratoga pilot had announced his landing intentions at Greenwood Airport, immediately prior to the collision.

3. Monitor frequencies early. The AIM recommends tuning in an airport frequency when you are 10 miles away. This is also a safe practice when on a cross country. As part of your preflight planning, make a note of all the airports you will be flying near that "10-mile rule" and jot down the frequencies. Remember to always monitor and communicate.

4. Maintain a vigilant scan for traffic. This is absolutely imperative near an airport, and especially one that is uncontrolled, or in a high density traffic area.

5. Develop safe practices that reduce pilot workload during critical phases of flight. A suggestion: During your preflight planning, go down the list of every major task that must be accomplished for that flight. Then determine when it must be completed, and by whom. By taking those extra few minutes to *think* about your cockpit activities, might be all that's needed to catch potential problems before they arise. As in this case, the Safety Board noted eight physical actions that the MU-2 pilot performed during climbout. One of those was unnecessarily rushing to airfile his IFR flight plan.

6. Aggressively maintain situational awareness. Don't just look, also listen for potential traffic conflicts. If you have any doubt, ask. Whether you're near an uncontrolled airport, or cruising on a cross-country, clarify those garbled or missed radio transmissions. "The Cessna that just reported inbound, please say your position." Easy.

Case study reference

National Transportation Safety Board. September 13, 1993. Aircraft Accident Report: Midair Collision, Mitsubishi MU-2B-60, N74FB, and Piper PA-32-301, N82419. Greenwood Municipal Airport, Greenwood, Indiana. September 11, 1992.

CASE STUDY III-5: A Cessna 340 and a North American T-6

Safety issues: see and avoid, role of ATC

On 1 May 1987, a twin-engine Cessna 340 collided with a single-engine North American SNJ-4 (T-6) approximately 12 miles northwest of the Orlando, Florida, International Airport.

Probable cause

The NTSB determined that the probable cause of this accident was the failure of the Orlando-West controller to coordinate the handoff of traffic to the Orlando-North controller; and the failure of the North controller to maintain radar-target identification. Contributing to this accident was the limitation of the see-and-avoid principle to serve as a means of collision avoidance in the circumstances of this midair

Pilot experience

The Cessna 340 pilot had 2335 total flight hours, 344 in the 340. The T-6 pilot had 7118 total flight hours, 296 in the T-6.

Weather

The reported weather at the time of the collision was clear skies with a visibility of 7 miles.

The flight of the Cessna 340

The Cessna pilot and his family were about to complete a cross-country that had originated in Iowa earlier that same day. At 1538 Eastern Daylight Time, the pilot contacted Orlando approach control north sector and reported he was level at 5000 feet. The aircraft had an operating Mode-C transponder. Moments later he was cleared to ". . . descend and maintain three thousand." At 1545 the flight was handed-off to the final controller, and the pilot reported ". . . with you three thousand." The controller then advised the Cessna pilot to ". . . present heading . . . maintain three thousand . . . straight into one eight right." The call was acknowledged, which was the last transmission of the Cessna pilot. The flight had been in the Orlando area only eight minutes before the collision with the T-6.

The flight of the T-6

Earlier that afternoon, the pilot of the T-6 departed Orlando Executive Airport, approximately 7 miles north of Orlando International, for a skywriting flight over Disney

World and Sea World. Although the aircraft had a transponder, it did not have Mode-C capability. At 1542, the pilot contacted Orlando International west-sector controller and requested, ". . . like to descend [from 10,500] out to the west . . . back into Exec." The controller then cleared him to ". . . descend and maintain six thousand . . . two seven zero heading." As the pilot descended through 7700 feet the controller vectored him to 340 degrees for traffic separation.

The west controller attempted to coordinate a lower altitude for the T-6 by calling the north-sector controller, who was busy talking to other aircraft. The west controller eventually got through to the final controller, who gave the approval for the T-6 to descend to 2500 feet. The T-6's traffic was a Boeing 727 arriving from the northwest and landing at Orlando International. The T-6 pilot responded, ". . . has the traffic . . ." as the 727 passed on his right side going the opposite direction.

Seconds later, the west controller advised the pilot to ". . . maintain visual separation . . . seven twenty-seven . . . direct to the VOR. Continue descent . . . to four thousand . . . contact approach . . ." The T-6 pilot then contacted the north controller and reported, ". . . with you six thousand."

He was cleared to ". . . descend . . . one thousand five hundred." The transmission was acknowledged, followed by, ". . . proceed to the airport anytime." At 1547 the T-6 pilot "rogered" the last clearance, and seconds later collided with the Cessna.

Impact and wreckage path

Both airplanes fell into the same mobile home about 7 miles northwest of Orlando Executive Airport. The majority of the wreckage was scattered over an area about 125 feet long by about 50 feet wide. All of the major components from both aircraft were located within these boundaries, including many sizable pieces that were found inside the residence.

There was a 3-foot-long crater where the radial engine and forward fuselage of the T-6 had come to rest. One of the T-6's propeller blades had eight uniform gouge marks that matched the physical dimensions of the 340's right engine magneto drive gear.

Accident survivability

This midair collision was not survivable.

The accident

Figure III-G illustrates three views of the angle of impact. The midair occurred at 3000 feet as the C-340 was southeastbound in level flight, and while the T-6, after just completing a right turn with a bank angle of 45 degrees, was southeastbound descending wings level to 1500 feet. Allowing for a descent from 6000 feet to 3000 feet, the minimum average rate of descent of the T-6 was 2000 fpm. The ground speed of each aircraft was approximately 175 kts.

Prior to the T-6 pilot making the right turn, he was flying straight at the Cessna at a distance of a little more than 3 miles and 3000 feet vertical. At that point, however,

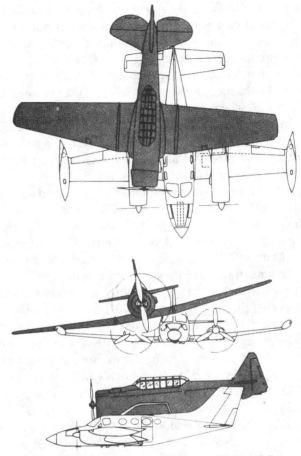

Fig. III-G. *Angle of impact: Midair collision of Cessna 340 and North American T-6.* Adapted from NTSB

he would have been watching the 727 traffic as directed by ATC, and the 340 would have been below his nose. Even during the turn and after the rollout, the 340 would still have remained well below his field of view.

The angle and altitude in which the T-6 was positioned to the 340, the Cessna pilot would have had to lean over and look up as much as 30 degrees in order to catch a glimpse of the T-6. Once the T-6 pilot had made his final turn, there was obviously no chance for the 340 pilot to see him because the T-6 was then directly behind him.

The investigation

The Safety Board centered their investigation on the role of ATC in a collision-avoidance environment.

ATC performance. Each Orlando airport is assigned a specific letter that is entered in the data tag of individual aircraft. "T" designates Orlando Executive, and "M" represents Orlando International. Provided the final controller assigns an "M", this also means that the aircraft is at or descending to 3000 feet.

When the west controller was unable to coordinate with the north controller for a lower altitude for the T-6, he subsequently received approval from the final controller for 2500 feet. Because the west controller noticed the final controller had assigned an "M" tag associated with the 340, and because of the "3000 feet" rule, he cleared the T-6 down to only 4000 feet. Although the west controller had never gotten through to the north controller for a coordinated handoff, and because he had since diverted his attention to another aircraft, the west controller, ". . . assumed that the [north controller] would see the [T-6] in the turn." The west controller described his workload as moderate.

The north controller had been working the 340 for about 20 miles when the T-6 pilot came on his frequency, ". . . with you six thousand." He remembered seeing the "T" and a "V" (representing VFR) in the data block of the T-6. He then transferred the 340 to the final controller. Moments later the north controller cleared the T-6 down to 1500 feet, since he believed the aircraft was on a northwesterly heading. The T-6 was actually in a right turn, not on a northwest course. This proved to be a key operational error on the part of the north controller. By this time, the two aircraft were about 2 miles from each other, and the 340 was at the T-6's two-o'clock position. The north controller then noticed the data block of the T-6 go into coast, but since the two aircraft were so close he was not surprised or concerned. He soon cleared the T-6 to proceed to the airport "anytime," based on a primary target he observed tracking northwest in the vicinity of the coasting T-6 data block. When the pilot "rogered" the clearance, the controller diverted his attention to another quadrant in his sector. He also described his workload as moderate.

When the final controller approved the west controller's request to descend the T-6, he stated that he did not see another aircraft to the northwest, nor was the 340 in handoff status to him. The final controller reported that his workload was light.

In the Safety Board's opinion, there were several ATC-related factors that either caused or contributed to this midair collision. First, the north controller failed to notice that the T-6 was in a steep bank, passing through a northwest (340 degree) heading. Instead, the controller had misinterpreted the aircraft's position as being on a northwest course. The T-6 was in this turn for two minutes, ample time for the controller to verify its track.

The Safety Board determined that the T-6's data tag began to coast because its antenna, which is on the bottom of the airplane, was shielded from the ARTS IIIA antenna during the turn just before the collision. In addition, because the two aircraft were so close, the system could not discriminate between each beacon code. Even the 340's data tag was intermittently coasting.

It was later discovered that the coasting data tag the north controller was tracking, was not that of the T-6. Because the primary target was heading northwest, and since he thought the T-6 was on a northwest course (not in a turn), he assumed incorrectly. This primary target ranged from 2½ miles to 5 miles away from the T-6's position. According to the Safety Board, the north controller should have been able to recognize that this was not the T-6.

The data tag of the T-6 was continuously coasting for 46 seconds. Investigators believed that the lack of proper radar identification techniques, a failure to maintain target identification, and an over-reliance on automation on the part of controllers were causal factors to this midair collision.

The Safety Board referred to the FAA Air Traffic Control Handbook that defines certain guidelines and responsibilities for controllers. It states that the controller must use more than one method of identification when the target goes either into coast status or there is doubt as to the proximity or position of targets. The handbook further states that the ". . . use of ARTS equipment does not relieve the controller of the responsibility of ensuring proper identification, maintenance of the identity, handoff of the correct target associated with the alphanumeric data, and separation of aircraft." Also, the handbook was supplemented by an Orlando International order that directs controllers: "Do not coordinate with another controller when he/she is obviously too busy to handle the distraction."

Lessons learned and practical applications

1. Say what you really mean. The T-6 pilot reported to the north controller ". . . with you at six thousand." What he actually told the controller was that he was level at 6000 feet. He was not. The T-6 pilot was descending through 6000 feet. The phraseology should have been, ". . . passing through six thousand for four thousand."

2. Don't assume anything. The west controller assumed the north controller would see the T-6 in the turn, but instead of realizing the aircraft was in a turn, the north controller thought it was on a northwest course. How you perceive a situation doesn't mean that someone else will interpret it the same way.

Case study reference

National Transportation Safety Board. 16 February 1988. Aircraft Accident Report: Midair Collision of Cessna 340A, N8716K, and North American SNJ-4N, N71SQ. Orlando, Florida. May 1, 1987.

CASE STUDY III-6: A U.S. Army U-21 and a Piper Navajo

Safety issues: see and avoid, role of ATC

On 20 January 1987, a U-21 turboprop and a twin-engine Navajo collided near Independence, Missouri.

Probable cause

The Safety Board determined that the probable cause of this accident was the failure of the radar controllers to detect the conflict and issue traffic advisories or a safety alert to the flightcrew of the U-21. The deficiencies of the see-and-avoid concept as a primary means of collision avoidance were also causal factors.

Pilot experience

The Navajo pilot had 7418 total flight hours, 4751 in multi-engine aircraft. His company records indicated that he had more than 596 hours in the Navajo, with 586 as pilot-in-command.

The U-21 pilot had 5983 total flight hours, 217 in the U-21. The U-21 copilot had 6266 hours, 1528 in the U-21.

Weather

Reported weather at the time of the collision was: Ceiling 25,000 feet thin scattered. Visibility 20 miles. Wind 230 degrees at 11 knots. Temperature/dewpoint 26F/11F.

The flight of N60SE

At 1221 Central Standard Time, the Navajo, with a pilot and two passengers onboard, departed the Kansas City Downtown Airport en route to their home base of St. Louis. The aircraft had an operating Mode-C transponder, which was squawking the 1200 VFR code. The pilot advised the local controller that he would make a left turn to the east after departure. The pilot's acknowledgment of the controller's approval of the left turn was the last known radio transmission from the Navajo.

The flight track of N60SE was reconstructed from Kansas City International Airport TRACON and Kansas City Center recorded secondary radar data from transponders. According to this evidence, the Navajo turned to an easterly heading after departing Kansas City, but remained beneath the 5000 foot base of the TCA. Its Mode-C target was detected by the TRACON at 1222:48 when the airplane was still near the Downtown Airport at 1600 feet. The target was tracked eastbound at a constant rate of climb to 7000 feet until the radar return was lost at 1227:58.

The flight of Army 18061

At 0944, the U-21 departed Calhoun County Airport, Anniston, Alabama, en route to Sherman Army Airfield, Fort Leavenworth, Kansas, with two pilots and one passenger onboard. The aircraft was equipped with an operating Mode-C transponder. The crew had filed an IFR flight plan and flew at a cruise altitude of 8000 feet.

Around 1221, the army pilots were handed-off from Kansas City Center to the Kansas City TRACON east-radar controller. The crew was advised to expect a visual approach and was given the weather as ". . . sky clear, visibility 10, wind from 260 at 7 knots . . ." At 1225, Army 18061 was notified of a traffic advisory, ". . . twin Cessna . . . southwest bound." Seconds later the crew reported, "Traffic in sight." Radar contact with the flight was lost at about 1228.

Examination of the radar data confirmed that the traffic advisory to Army 18061 did not pertain to the Navajo and the U-21 was well clear of the reported traffic when the two aircraft collided. The data further proved that the crew did not alter their aircraft's heading after the last clearance, and the airplane maintained 7000 feet until radar contact was lost.

The investigation

The Safety Board analyzed ATC and flightcrew performance with regards to this accident based on the angle of impact between the U-21 and the Navajo (Fig. III-H).

ATC performance. At the time of the accident, the east-radar position at the Kansas City TRACON was staffed by an area supervisor and a developmental ATC

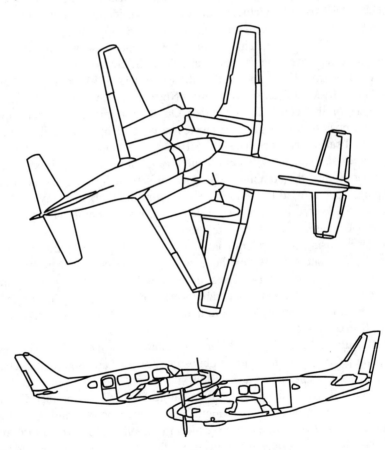

Top and horizontal views of the aircraft at impact

Fig. III-H. *Angle of impact: Midair collision of U.S. Army U-21 and Piper Navajo.* Adapted from NTSB

specialist. The supervisor had been monitoring Army 18061 for only seven minutes, but had provided a traffic advisory to the crew regarding the twin Cessna. The developmental controller had just sat down at the position, so had viewed the radar screen for only a minute before the accident. Both controllers later reported not seeing any primary or secondary radar information pertaining to the Navajo.

The day after the accident, a flight inspection of the radar system and associated TRACON radio frequencies was conducted by the FAA. No discrepancies were found. The Board requested the TRACON radar data in order to study the information on their Retrack Program Computer. Although it cannot replicate the entire radar portrayal, it duplicates the alphanumerics generated by the ARTS III program and its associated logic.

An aircraft operating under an IFR flight plan is tracked on the radarscope by a full data block (FDB). A FDB includes aircraft location, identification, altitude, ground speed, and flight plan data. A limited data block (LDB) appears on the scope to represent an untracked VFR target. The aircraft's transponder code and altitude, if Mode-C is operating, readout are the only data available to the controller.

The FDB of Army 18061 appeared on the scope at 1221:40 and remained on the display until the collision. Likewise, the LDB of N60SE came into view about 1222:45 and also remained on the scope until the midair. On the last presentation that showed both airplanes, the position tracking symbols were nearly overlapped and at the same altitude.

ARTS III safety features. The ARTS III has an automatic offset feature designed for these particular situations. To eliminate the possibility of data block information being unidentifiable, the computer will shift [offset] each block that is in danger of overlapping. According to the retrack presentation, the FDB of Army 18061 and the LDB of N60SE shifted the appropriate distances, which should have given the controllers an unobscured view of the data blocks.

Controllers using this system cannot suppress 1200 (VFR) transponder codes. They are depicted automatically on the radarscope with a computer-generated triangle over the primary and secondary targets for non-Mode-C targets. Mode-C transponder targets are shown by a computer-generated square over the primary and secondary targets. The system also displays the altitude in a three-digit code attached to the square by a ¼-inch leader line.

Conflict alert. Aural and visual alerts associated with the conflict-alert system are based on projected positional and velocity data for tracked Mode-C targets. A controller would not be alerted by the system if either of the involved aircraft was not tracked, even if it was equipped with an operating Mode-C transponder. Communication with a controller, or even operating a Mode-C transponder during a VFR flight, would provide collision avoidance protection to the pilot. However, a pilot receiving VFR flight-following services would result in the radar controller tagging the target and automatically initiating the track needed by the conflict-alert system.

During this investigation, the Safety Board chose to evaluate the usefulness of the conflict-alert system with regards to potential collisions between tracked and untracked Mode-C radar targets. The Safety Board manually tagged the LDB associated with N60SE, which automatically changed it to an FDB-tracked target. This simulated FDB remained on the radarscope until it merged with Army 18061's FDB, and ultimately vanished at the moment of collision.

The Safety Board also noted that the conflict alert visual and aural alarms activated more than 40 seconds before the actual collision, and continued until the radar targets disappeared. This would have been ample time for a controller to issue a traffic or safety advisory.

Cockpit visibility study. A cockpit visibility study was conducted to determine the location of each airplane with respect to the field of vision of the pilot(s) in the other airplane. A binocular camera was used to photograph the cockpits of two airplanes with structurally identical cockpit visibility to the accident airplanes. The camera rotated about a vertical axis that is normally 3.5 inches from the lenses, approximating the distance between the front of the eye and the pivot point about which the head rotates. As a result, photographs showed the outline of the cockpit windows as seen by a crewmember rotating his head from side to side. Monocular obstructions within the window, such as the windshield or door posts, were also defined by the photographs.

Results of the study showed that the Navajo was visible through the windshields of both U-21 pilots. The aircraft would have appeared 13 degrees left and 2 degrees below the U-21 pilot eye reference points. Since the army aircraft was in level flight, the eye reference point was the horizon. Neither pilot's view would have been obstructed by the windshield, door posts, windshield wipers, or any other airplane equipment.

The U-21 would have appeared 18 degrees to the right and 3 degrees below the Navajo pilot eye reference point. Because the center windshield post of the Navajo partially obstructed the pilot's view of the U-21, his view would have been restricted temporarily to only his left eye. The copilot's view, however, was never obstructed.

The probability of visual detection. An air-to-air visual acquisition study had been conducted previously by Lincoln Laboratory at the Massachusetts Institute of Technology. Because the circumstances surrounding the U-21 and Navajo flights closely coincided with the model produced from this study, the Safety Board used the analyses to determine the probability of visual acquisition between the army and Navajo pilots shortly before the collision. The data given were: the speeds of both airplanes, headings, the area profile at the presentation angle, the number of pilots in each airplane engaged in the traffic search, and the visual range. The outcome indicated that the probability of target acquisition would not have been high until the last few seconds before the collision. It was determined that the Navajo pilot had only a 27 percent chance of seeing the U-21 at 12 seconds before impact. Similarly, the army pilots had only a 33 percent probability of seeing the Navajo at 12 seconds before the collision. These results, however, assumed a relatively low pilot workload and unobstructed view of the opposing aircraft. If any of the three pilot's had become distracted with cockpit duties, or there were obstructions to a clear view of the other airplane, which the Navajo pilot experienced, these probabilities would have been much less.

The Safety Board's conclusion.

According to the Board, this midair could have been prevented if the Navajo pilot had requested flight-following services. He would have been assigned a discrete transponder code, giving him tracked status. If he had done so, the conflict-alert feature of the ARTS III system would have alerted the controllers of the potential conflict 40 seconds before impact. An aural alarm would have been activated and the two data blocks would have flashed.

The Safety Board believed that it was most likely that the east-radar controllers were distracted from monitoring traffic in the moments before the collision because of their position-relief briefing and associated duties. Their workload was considered light, which also might have contributed to a reduced state of vigilance on the part of both controllers. The Board believed that if the Navajo had also been tracked, the warning systems would have alerted the controllers in plenty of time to avert the accident.

Lessons learned and practical applications

1. Contact ATC when flying in a radar-controlled environment. Because the Navajo pilot was not in communication with Kansas City TRACON, there was no opportunity for ATC to provide traffic advisories to him.

2. When VFR, ask for flight-following services. Even when you're operating a Mode-C transponder, the controller sees only a 1200 VFR code that could be just one of many, particularly on a busy day. Report in and ask for a discrete transponder code. Because you are VFR, ATC will provide traffic advisories only "work permitting," but at least, they are aware of your presence and call sign.

3. Don't assume anything. The Safety Board suggested that since ATC had already notified the army pilots of traffic, and because there were very few radio calls, it seemed obvious the controllers were not busy, the U-21 pilots might have assumed that they would be alerted of any additional traffic.

Case study reference

National Transportation Safety Board. 3 February 1988. Aircraft Accident Report: Midair Collision of U.S. Army U-21A, Army 18061, and Sachs Electric Company Piper PA-31-350, N60SE. Independence, Missouri. January 20, 1987. Washington, D.C.

CASE STUDY III-7: A Falcon DA50 and a Piper Archer

Safety issues: role of ATC, controller workload, physical limitations of see and avoid, situational awareness, airspace design, high density traffic environment, night operations

On 10 November 1985, a Falcon DA50 corporate jet collided with a single-engine Piper Archer near the Teterboro, New Jersey Airport.

Probable cause

The NTSB determined that the probable cause of the accident was a breakdown in air traffic control coordination, which resulted in a traffic conflict; and the inability of the DA50 flightcrew to see and avoid the other aircraft due to (1) an erroneous and inadequate traffic advisory and (2) the physiological limitations of human vision and reaction time at night.

Air traffic control management contributed to the accident by failing to ensure that controllers were following prescribed procedures, and failing to recognize and correct operational deficiencies.

History of flight

The DA50 was owned and operated by Nabisco Brands, Inc. The two pilots were scheduled to pickup five company executives at the Teterboro Airport for a flight to Toronto, Canada.

The pilot of the Archer had departed the nearby Essex County, New Jersey, Airport and was flying under VFR. There were two passengers onboard.

Pilot experience

The captain of the DA50 had 8265 total flight hours, 817 in the Falcon. He held an ATP certificate with type ratings in Learjet, DA20 and DA50 aircraft. He had been with the company since 1980, and had completed his last proficiency check in the DA50 11 months before the accident.

The first officer of the DA50 had 4500 total flight hours, 143 in the Falcon. He held an ATP certificate with a type rating in the Learjet. He had been with the company for six months, and had completed his DA50 training five months before the accident.

The pilot of the Piper had 269 total flight hours, 92 in the Archer. He had logged 76 hours at night. He held a Private Pilot certificate. He had failed his instrument check ride about six weeks prior to the accident for VOR and ILS tracking during instrument approach procedures. He received additional instruction, but had not retaken the flight test.

Weather

The accident occurred about 40 minutes after official sunset. The 1650 surface weather observation at Teterboro was as follows: Ceiling 10,000 feet scattered, 25,000 feet thin scattered. Visibility 20 miles. Temperature/dewpoint, 65 degrees F/49 degrees F. Wind 220 degrees at six knots.

The accident

At 1654 Eastern Standard Time, the crew of the DA50 were preparing to depart the Morristown, New Jersey, Airport to meet their passengers at Teterboro. About 13 minutes later, they received their IFR clearance for the short, 16 nm flight. The crew was, "cleared to Teterboro after departure, direct to Chatham, radar vectors to intercept the VOR/DME Alpha approach into Teterboro, maintain two thousand, departure frequency . . . ." The clearance was read back correctly, and at 1709 the flight was released to contact the local controller. Meanwhile, the departure controller at the New York TRACON informed the Teterboro control tower coordinator that the DA50 was departing Morristown, en route to Teterboro. The coordinator acknowledged the call. At 1710, the flight was cleared for takeoff.

Once airborne, the crew was in contact with the departure controller and was instructed to "squawk ident," and to maintain 2000 feet. Seconds later, the controller told the crew, ". . . radar contact, two south of Morristown. Start a left turn now, heading zero eight zero." He added, ". . . you are on vectors for a VOR/DME Alpha approach Teterboro . . . good VFR visibility two zero, winds are two four zero at eight . . . overhead the airport, left traffic for runway one niner." The captain acknowledged the information.

Following the transmission, the departure controller coordinated the flight to pass through the Essex County airport traffic area, 2 miles east of the field. At 1715, the controller told the crew that they were following a twin Cessna to Teterboro, and to reduce their airspeed to 180 knots. He further advised that their turn for final approach would occur ". . . just outside of CLIFO [final approach fix]." About four minutes later, the crew reduced their speed to around 180 knots.

Meanwhile, the pilot of the Archer had departed the Essex County Airport at 1713. The airports are 10 nm apart. The Archer pilot proceeded eastbound, and three minutes later reported to the tower that he was clear of the area and requested a frequency change.

As the DA50 crew neared Teterboro, the sector departure controller provided two traffic advisories. One airplane was at 4 miles, off their nose and heading westbound at 2500 feet. The other was about 2 miles eastbound, at their 11:30 position at 1800 feet. According to the CVR, the crew sighted both aircraft. The cockpit conversation revealed that the crew was well aware of the volume of traffic flying that evening as they commented on additional airplanes. At one point the captain said, "It's the ones below the horizon that are hard to see . . . they can get you, too."

At 1716, the departure controller instructed the crew ". . . 3 miles west of CLIFO, turn right heading zero niner zero, two thousand until on the Teterboro three zero five radial, cleared VOR/DME alpha approach." The captain repeated the clearance. Almost immediately thereafter, the captain said to the first officer, "They're [traffic] everywhere . . . ," and the airplane overshot the boundary of the airport traffic area. About 12 seconds later, the controller stated ". . . that was obviously not the best turn, sir. Continue right one four zero to pick it up." The captain replied "we'll figure this one out . . . ," To which the controller answered, "Guess I need all the help I can get." The first officer instantly remarked to the captain, "Got a guy over here."

At 1717, the pilot of the Archer contacted the Teterboro control tower, said that he was 10 miles west at 1500 feet, and requested permission to proceed overhead of the airport to the Hudson River. The controller told the pilot to report 1 mile west of the airport. It was later revealed that the pilot was not 10 miles from the field, but rather 5 miles.

Seconds later, the departure controller informed the DA50 crew that "traffic as you turn back around, twelve o'clock, less than a mile, one thousand seven hundred unverified westbound." The captain acknowledged the traffic's position. At 1718:08, the departure controller gave the Teterboro tower controller a progress report on the DA50, and told him that the jet was a mile from CLIFO.

The crew was informed around 1718:12 that, "traffic you're following [a twin Cessna] just overhead Teterboro this time. Contact the tower . . . ." The captain responded to the call. As they approached the field, there were several aircraft arriving and departing from runway 19. The local controller was working six airplanes when the crew switched to the tower frequency. According to the CVR transcript, the first radio transmission that the crew would have heard was an inbound "twin Bonanza . . . 2 miles northwest of airport." The controller stated, ". . . continue overhead, report overhead the field." The airport traffic area was filled with a mix of single- and twin-engine general aviation aircraft along with high performance airplanes such as the DA50.

At 1718:39, the pilot of the Archer contacted the tower and reported that he was 1 mile west of the airport. He was actually 2½ miles to the west. The controller asked the pilot to flash his landing lights, then informed him that he was in sight and to advise when clear of the airport traffic area to the west. Less than a minute later, the DA50 crew told the controller that they were, "inside CLIFO [fix is 4.8 DME], airport's in sight." About the same time, a twin Beech Baron was cleared for takeoff from runway 19, and was instructed to "remain east of the Hackensack River." Immediately thereafter, the twin Bonanza reported overhead the field, and the controller advised the pilot that he was number two following traffic on a left base leg over Route 80. That aircraft was another twin Bonanza, but the controller did not mention the type of airplane in his traffic advisory. At this same time, the DA50 crew was in a 10-degree left bank at about 185 knots. They were heading 146 degrees when the captain told the first officer, "Better slow it up . . . we're following that guy." The first officer replied, "Okay," and called for flaps.

At 1719:54, the controller asked for the call sign of the airplane that had reported over CLIFO, and the captain answered, "Falcon . . . coming up overhead with the traffic in sight." The controller stated, ". . . understand you're overhead the field . . .?" The captain responded, "Yes, sir." The controller then told the crew to ". . . plan number three, following traffic turning downwind abeam the tower, additional traffic is at your one o'clock westbound at one point five [1500 feet]." By this time the DA50 was in level flight, on a heading of 121 degrees, at about 163 KIAS. The captain replied, "We're lookin' . . . ." At 1720:14, the jet's speed had been reduced to 151 KIAS when a right turn was initiated. The controller immediately notified the crew that ". . . you're closing on him . . . he's . . . light aircraft . . . you're one to twelve o'clock, westbound." While the crew was receiving the additional traffic advisory, the first officer called for, "flaps twenty . . . gear down before landing checklist." The captain quickly commented, "you're eatin' 'em up." The sound of the landing gear warning horn was heard, as the crew established a five-degree right bank for 41 seconds while the speed was decreasing.

The controller quickly notified the pilot of the Archer that "traffic is a Falcon jet overtaking you from your . . . six o'clock . . . ." It was unknown if he heard the advisory and tried to respond, since another pilot flying north of the airport "rogered" the call and blocked the transmission. According to the CVR, at 1720:39 the captain of the DA50 said, "Another one down low." The first officer's reply was unintelligible, but it most likely concerned the traffic because the captain's next comment was, "beneath him." The first officer answered, "I see him." Immediately thereafter, the controller asked the crew, ". . . you have the traffic, sir?" The captain said, "Affirmative," to which the controller instructed, "Okay . . . maintain visual." The crew did not acknowledge the call.

The controller then turned his attention to other traffic. At 1721:00, the first officer stated, "What kind of Cessna is that?" The captain replied, "I don't know, I'll ask him." At this point the jet was almost wings level, rolling out of the slight right turn. The captain never had a chance to contact the controller since there were five rapid transmissions made between ATC and three other airplanes, one of which was the twin Bonanza that had been cleared for a short approach and landing.

Anyone who has flown in congested airspace, especially at night, can relate to the situation in which the crew of the DA50 had found themselves. The crew's comments made in the final nine seconds before impact prove to be the most chilling. The following are those events recorded by the CVR up until the time of the collision.

1721:21	F/O:	Let's go full flaps.
1721:22	Capt:	Hey, watch out, this guy's comin' right at us.
1721:25	Archer:	Teterboro tower, Cherokee one nine seven seven hotel, clear to the east.
1721:26	Capt:	Go down.
1721:28	Capt:	Naw, go up.
1721:28	ATC:	Seven seven hotel roger, frequency change approved.
1721:30		Sound of impact.

Impact and wreckage area

The collision occurred at 1721:30 between 1500 and 1600 feet agl over the city of Fairview, New Jersey, about 4 and ½ miles east-southeast of the Teterboro Airport. The location was 400 to 500 feet above the floor of the New York TCA for LaGuardia Airport. Radar data showed that the DA50 began a slight right turn at 1720:13 and overtook the Archer passing about ½ mile to its right on a ground track of about 130 degrees and approximately 140 knots. At 1721:02, the jet initiated a level left turn that was maintained until moments before the collision. At the time of impact, ground speeds of the DA50 and Archer were 155 knots and 115 knots, respectively. Based on the radar data, the rate of closure was 100 to 110 knots with a ground track angle of 45 degrees.

The propeller of the Archer slashed through the DA50's inboard left wing at about a 75-degree angle. It also cut the front wing spar, and one blade separated 8 inches from the tip. The other blade struck but did not penetrate the upper center wing panel. Scratch marks on the lower surface wing panels and outboard flap of the DA50 indicated that the Archer traveled under the jet's left wing, tearing off a section of the outboard flap and causing the top of the single-engine's cabin to disintegrate. The pressurized fuel cell in the DA50's left wing exploded when the propeller sliced the front wing spar. The damaged spar and upper wing panels destroyed the structural integrity of the wing structure, forcing the wing to separate in an upward direction.

The main wreckage sites of the airplanes were about 700 feet apart, with debris scattered over a four-by-eight city block area. The Archer hit a populated area in a nearly vertical descent. Pieces of the wreckage caused substantial damage to property and automobiles. The DA50 crashed into three residences and several vehicles.

Accident survivability

The collision was not survivable. One resident of an apartment building was killed and two bystanders were seriously injured.

The investigation

The focus of the investigation centered on the role of ATC and the actions of the pilots involved.

ATC performance. According to the Safety Board, the accident sequence of events began when the automated radar tracking system (ARTS IIIA) computer at the New York TRACON did not pick up the DA50 after it departed Morristown. The crew's proposed departure time had been for 1730, but they took off 19 minutes earlier. As a result, the DA50's identification and flight plan were not listed in the departure controller's tabular list. Normally, this would have occurred automatically.

At 1710, seconds after the jet was cleared for takeoff, the departure controller told the coordinator at Teterboro by landline that the DA50, "is inbound just rolling at Morristown this time." The coordinator acknowledged the call. Two minutes later, the departure controller established contact with the crew, but since the computer did not automatically display the beacon target, the controller was required to initiate a manual track of the jet. In turn, this negated all subsequent automatic ARTS functions and interfacility communication, including transmission of an arrival flight strip at Teterboro.

Meanwhile, the Teterboro tower coordinator asked the clearance delivery controller to assume responsibility for his position while he took a quick break. The tower coordinator did not sign off his position in the appropriate log, as required, nor did he provide a relief briefing in accordance with the prescribed checklist. The Safety Board believed that the lack of a proper briefing was a factor that led to this accident. The coordinator did, however, tell the clearance controller that there was no inbound traffic that the local controller was not already aware of.

At 1718, the departure controller told the tower that, "Falcon . . . is a mile from CLIFO." The clearance controller took the progress report and said to the local controller that the DA50 was at CLIFO. However, the local controller was wearing a headset and apparently did not hear the information since less than two minutes later he asked for the call sign of the airplane that reported over CLIFO.

The clearance controller further stated that after giving the report to the local controller, she saw the ground controller get up, move over to the local controller, and reposition a flight strip. She assumed that the strip was for the DA50, reinforcing her belief that her coordination report had been accomplished. The ground controller, however, testified that he did not write a strip until after the jet had reported overhead. None of the controllers in the tower was able to tell investigators if a machine-generated flight strip ever existed for the DA50.

Upon learning of the DA50's position, at approximately 1720:05, the local controller notified the crew that the Archer was at one o'clock, and on a westbound heading. The Piper was actually eastbound, as it had been for its entire flight through the airport traffic area. When the controller asked the Archer pilot to "report clear of the airport traffic area to the west," he should have said "to the east." Because he was not corrected by the pilot in his acknowledgment, the controller had no reason to believe thereafter that he had erred. He repeated the mistake on two subsequent instances. The Board also noted that the controller failed to provide to the crew information about type identification for the twin Cessna and the Archer; and did not inform the crew that the Archer was an overflight. This might have hindered a timely and accurate acquisition of the threat traffic.

The local controller testified that he would have cleared the Archer to transit the area even if he had prior knowledge of the inbound DA50. He told investigators that all he needed was an inbound call at CLIFO in order to provide adequate advisories for traffic sequencing. He admitted that he was surprised to learn of the jet's position, and that he was busy with a traffic volume that was "moderate and building." He also stated that he did not know to which airplane the crew of the DA50 was referring when they flew overhead the airport at 1719:56 and called "traffic in sight." According to him, it could have been the twin Cessna, twin Bonanza, or the Archer. Despite this confusion, once the controller instructed the crew to "maintain visual," any further ATC efforts to sequence the traffic ended.

The Safety Board also noted that the local controller had the authority to make on-the-spot decisions concerning transiting aircraft. Although investigators believed the practicality of such a procedure was "questionable," they pointed out that the controller had a few options that would have provided a better flow of traffic. Two of these options were, not permitting arrival VFR aircraft to enter the traffic pattern, and holding departing aircraft on the ground. In the Safety Board's opinion, if the controller had taken this type of aggressive measure instead of trying to accommodate every pilot's request, the frequency congestion and his own workload would have been lessened.

During the investigation, the tower supervisor stated that under conditions of increasing air traffic he has the authority to establish a second control frequency. The purpose is to reduce frequency congestion and controller workload. However, he believed that the situation was "a normal Sunday night . . . [with] late evening Sunday arrivals . . . coming back." Thus, he did not implement the procedure because he was not aware of "excessive workload on the local controller."

Based on the collection of physical evidence and testimony from controllers, the Safety Board concluded that the "most significant error on the part of the local controller was not reporting that the additional traffic was an eastbound overflight." The Board members believed it was a result of a busy workload and radio frequency congestion. Investigators noted that this type of human error is indicative to a "pronounced workload condition."

Pilot performance. According to the Safety Board, the manner in which the DA50 was flown was not in accordance with company training procedures as they pertained to speed control and landing configuration. The crew did not reduce speed to 180 KIAS in a timely fashion as requested by the departure controller. Although decreasing, the speed was faster than it should have been under the circumstances, regardless of what was permitted by regulation. The DA50 was about 163 KIAS over the airport, and the speed was not down to 140 KIAS until after it passed the Archer. The increased speed resulted not only in a rapid closure rate with the Archer, but it also caused a late configuration of the jet for landing. Consequently, the crew flew outside the normal limits of the traffic pattern at 1717:41.

It was noted that the captain warned the first officer to slow down on two separate occasions, at 1719:45 and at 1720:18. Therefore, investigators believed that the first officer might have been having difficulty managing airspeed and airplane configuration during the approach. As a result, the captain might have become distracted from devoting more attention to the traffic situation.

At 1715, the crew knew they were following a twin Cessna to Teterboro. About three minutes later, they were about 5½ miles from the airport when told that their traffic was over the field. Meanwhile, the Archer was about 30 degrees to the jet's right, and about 3½ miles away. Due to the frequency congestion, the crew was unable to make a timely call to the controller when they reached CLIFO. By the time radio contact was established, the controller had already cleared the twin Baron for a downwind departure, and cleared the twin Bonanza to be number two to land. Since the Bonanza's landing sequence was not transmitted until 1719:45, the Safety Board noted that it was probable the crew did not hear those instructions because the captain was telling the first officer to slow down at that precise moment. Investigators believed that the crew might not have been aware that the controller had sequenced the twin Bonanza between them and the twin Cessna.

Cockpit visibility studies revealed that at 1720:14, the twin Bonanza was in view of the DA50 pilots at the 10 o'clock position, and slightly below the horizon approximately 1.2 miles away. The Archer was directly off the nose of the jet at less than a mile. Still another aircraft, the twin Cessna, would have appeared 30 degrees to the right of the captain's viewing angle, about 1.7 miles ahead and moving closer to a one o'clock position than the Archer. According to the Safety Board, both twin engine aircraft were turning to the downwind direction, as the DA50 made a 15-degree right bank.

Investigators believed that although told by the departure controller that they were following a twin Cessna, the DA50 crew should not have assumed that the sequencing would not change. Thinking that the Archer was the landing twin Cessna, they might have attempted to establish spacing by making the right turn. The crew would have expected the airplane to have soon turned left for the downwind leg. As a result, the Safety Board believed that the pilots became preoccupied in obtaining separation with the Archer they thought was number two to land.

This hypothesis also accounted for the first officer's question, "What kind of Cessna is that?" Since they were expecting to be following a faster twin-engine aircraft, the relatively slower speed of the Archer most likely caused the inquiry. Unfortunately, because of the frequency congestion, the captain did not get a chance to ask the controller to resolve the ambiguity. As stated earlier, if the local controller had added the type of aircraft and correct heading of the Archer, the Safety Board believed that the DA50 crew would have realized the presence of a single-engine airplane in their vicinity much sooner. Investigators concluded that if the crew had received that information, they would have planned a different course of action, which could have prevented the accident.

Given the actual circumstances, the Safety Board determined that the crew lost situational awareness due to the combination of their preoccupation in flying the jet, the assumption that they were following a landing twin Cessna, and erroneous information by ATC. Therefore, the Board believed that the crew's loss of situational awareness was a "significant factor in the accident."

The pilot of the Archer had initially reported two inaccurate position reports that might have caused the controller to plan traffic sequencing on the wrong information. Furthermore, if the Archer pilot had corrected the controller who kept referring to the Archer as westbound, the controller might have been able to update the traffic advisory. The Safety Board could not determine if the pilot ever attempted to notify ATC of the error.

Lessons learned and practical applications

1. Report accurate information. The pilot of the Archer provided incorrect position reports to ATC. One report was off by 5 miles. Safe and efficient traffic sequencing is based upon reliable information. A "ballpark" figure is never good enough, especially when operating near busy airports.

2. Fly the airplane appropriately. It took the DA50 crew four minutes to reduce their speed from 180 KIAS, which caused them to fly overhead Teterboro at 163 KIAS. Although the speeds were legal, pilots must be aware of the broad range of aircraft types that normally fly into certain airports. Turboprop or jet pilots need to remember that speeds must be adjusted accordingly when operating near other types of general aviation aircraft. It can be just as dangerous if a pilot of a single-engine airplane attempts to be sequenced into a reliever or commercial airport while puttering along at 80 knots.

3. Stay ahead of the airplane. Don't waste valuable time that could otherwise be spent configuring the aircraft or completing checklists. Once that time is gone, don't expect to get it back. In this case, ATC afforded the DA50 crew several minutes to reduce their speed from 180 KIAS. They failed to seize the opportunity and subsequently found themselves blasting overhead the airport and out of the traffic area.

4. Don't assume anything. This is especially true in a dynamic ATC environment. The Safety Board believed that the DA50 crew "assumed" that the twin Cessna would be their traffic all the way to landing. Remember, as you near the airport traffic area, aircraft might be converging from many different directions, causing the sequencing to suddenly change. Also, be careful not to fall into a hardened mindset. A preconceived notion, like an assumption that you'll be following a certain airplane, can often be difficult to erase from your brain. As a result, you might tend to hear or see what you *think* should be correct at that point in the flight, when in reality it's not.

5. Pay attention to your surroundings. It's particularly important to listen to the radio transmissions of the aircraft in your vicinity. Remember the calls signs and the airplane's positions so you can create a mental picture of potential threats. Maintain situational awareness at all times.

6. Listen to the radio. It's easy to get frustrated on a busy day and jump in the split-second you hear silence on the radio. But, if you haven't been *listening* to the flow of ATC/pilot communications, you and your rapid trigger finger most likely just blocked or cut off important clearances and advisories. Wait that extra moment to allow your fellow pilot to read back a heading change, or to acknowledge a traffic advisory. Remember, effective radio communication is one of the most important resources in collision avoidance and situational awareness.

Case study reference

National Transportation Safety Board. 4 May 1987. Aircraft Accident Report: Midair Collision of Nabisco Brands, Inc., Dassault Falcon, DA50, N784B and Air Pegasus Corporation Piper Archer, PA28-181, N1977H. Fairview, New Jersey. November 10, 1985. Washington, D.C.

Part IV

Mechanical deficiencies and maintenance oversights

THE MECHANICAL DEFICIENCES OR MAINTENANCE OVERSIGHTS discussed in the following case studies are unique to each particular accident. Therefore, rather than addressing specific subjects in individual chapters, each case study provides a comprehensive analysis of the pertinent information.

CASE STUDY IV-1: United Airlines Flight 232

Safety issues: Catastrophic engine failure in cruise flight, CRM, cockpit discipline, ADM

On 19 July 1989, a United Airlines DC-10 experienced a catastrophic engine failure at cruise altitude. The crew struggled with the airplane that no longer responded to flight control inputs. More than 30 minutes later, Flight 232 crashed while attempting to land at Sioux City, Iowa.

Probable cause

The NTSB determined that the probable cause of this accident was the inadequate consideration given to human factors limitations in the inspection and quality control procedures used by United Airlines' engine overhaul facility. This resulted in the failure to detect a fatigue crack originating from a previously undetected metallurgical defect located in the stage 1 fan disk. The subsequent catastrophic and forceful disintegration of the disk caused debris to penetrate the hydraulic systems that operated the DC-10's flight controls.

History of flight

Flight 232 operated as a regularly scheduled flight from Denver, Colorado, to Philadelphia, Pennsylvania, with an intermediate stop in Chicago, Illinois. The DC-10-10 departed Denver's Stapleton Airport at 1309 Mountain Daylight Time with 285 passengers and 11 crewmembers onboard.

Pilot experience

The captain had 29,967 total flight hours, 7190 in the DC-10. He had been with United since 1956, and was requalified as a DC-10 captain in 1987 after serving as a 727 captain for the previous two years. His most recent proficiency check in the airplane was three months before the accident.

The first officer had approximately 20,000 total flight hours, 665 in the DC-10 right seat. He was hired by United in 1985 after flying for National Airlines and Pan Am. He successfully passed his last proficiency check 11 months before the accident.

The second officer had an estimated 15,000 total flight hours, 33 in the DC-10. He had been with United since 1986 and had just completed his DC-10 transition training a little more than a month prior to the accident. His last check ride was at the same time.

An off-duty training/check captain assisted the crew during the emergency. A pilot with United since 1968, he had 23,000 total flight hours. He had logged 2987 hours in the DC-10, of which 79 were as captain.

Weather

At the time of the accident, the surface weather observation at Sioux City Airport was an estimated ceiling of 4000 feet with broken clouds and 15 miles visibility. There were towering cumulus clouds in all quadrants. The wind was shifting between 010 degrees/11 knots to 360 degrees/14 knots.

The accident

NOTE: The following account of the accident is a combination of the Safety Board report and the captain's personal recollection of the events.

About one hour and seven minutes (1516:10 Central Daylight Time) after Flight 232 departed Denver, and at a cruise altitude of FL 370, the flightcrew heard a loud bang followed by a vibration and shudder of the airframe. The first officer immediately grabbed the yoke as the captain and second officer evaluated the situation. After checking the engine instruments, the crew determined that the number 2 aft (tail-mounted) engine had failed. To the crew's surprise, they were unable to shutdown the engine. The number 2 throttle and fuel lever had frozen in position. The crew quickly actuated the firewall shutoff valve, and the fuel supply to that engine was finally cut off. It was in those few seconds that the crew realized that they weren't dealing with a simple engine failure. About that time, too, the second officer noticed that the airplane's normal systems hydraulic pressure and quantity gauges were at zero.

Seconds later, the first officer told the captain that he could not control the aircraft. According to the captain, the first officer was applying full left aileron, the control column was laying in his lap, and he was calling for full-up elevator, yet the airplane was in a descending right turn with an increasing bank angle. The captain immediately confirmed the lack of response to flight control inputs, and reduced power in the number 1 engine. The airplane began to roll back to a wings-level attitude. The crew also tried to restore hydraulic power by activating the air-driven generator, which operates the number 1 auxiliary hydraulic pump. With the pump selector "on," the hydraulic system was still dead.

At 1520, the crew radioed Minneapolis center and requested emergency assistance and vectors to the nearest airport. The Sioux City Gateway Airport was about 50 miles away. Because Flight 232 was already heading in that direction, the captain decided to accept Sioux City as the emergency airfield.

Passengers were informed of the situation shortly after the engine failure, and the senior flight attendant was called to the cockpit. She was told to prepare the cabin for an emergency landing. Another flight attendant advised the captain that an off-duty United DC-10 training/check pilot was onboard and had volunteered his services. He was immediately invited to the cockpit and briefed of the problem. Shortly thereafter, the captain directed him to try and determine the damage as seen through the aft cabin windows.

In the meantime, the crew confirmed the state of the hydraulic pressure had not changed and around 1531 the captain said, "we're not gonna make the runway, fellas." The check pilot soon returned and reported that, "both your inboard ailerons are sticking up. That's as far as I can tell." With the check pilot eager to help, the captain drafted him to operate the throttles. For about 15 minutes, the captain and first officer had struggled with handling the yoke and power adjustments. Alternating opening and closing the numbers 1 and 3 engines was the only way they could maintain some flight control. The captain told the check pilot to move the throttles in response to their commands, hoping that by one person concentrating on the necessary series of adjustments they would experience a smoother ride and a little extra control.

By that time, the crew had already dumped fuel and was about 35 miles northeast of the Sioux City airport. Concerned that the cabin was not yet prepared for landing, the captain explained to the senior flight attendant the seriousness of the emergency and that they were heading for Sioux City. He told her that it would be a difficult landing and that he had doubts as to the outcome. He concluded by telling her that when the flight attendants hear the warning, "brace, brace, brace," over the public address system, that would be their signal to prepare the passengers for landing.

As the crew continued to wrestle for control of the airplane, they had to compensate for the phugoids, each lasting between 40 and 60 seconds. A phugoid is a longitudinal oscillation caused when an aircraft is displaced along the longitudinal axis. Provided an airplane is trimmed and the power setting is constant, phugoids are suppose to dampen themselves out after a few nose-up and nose-down cycles. However, as the captain recalled, the crew was never able to eliminate the phugoids altogether because of the inherent flight characteristics of the DC-10 and the inability of the damaged aircraft to maintain constant, level flight.

The captain explained that the DC-10 is designed to stay trimmed to level flight. Because Flight 232 was trimmed at a 270-knot cruise speed before the engine failed, that is what the aircraft attempted to maintain. The crew was often adding large amounts of thrust which raised the nose, causing the airplane to retrim, and enter into another phugoid. According to the captain, the technique to stop phugoids is the opposite of what pilots might normally think. For example, when the nose pitches down and the airspeed increases, the pilot must add power in order for the nose to come up. This is due to the pitch-up characteristics created by the two underwing-mounted engines. Once the nose starts to pitch-up, and the airspeed falls off, the pilot must then close the throttles. In the case of Flight 232, this maneuvering was especially difficult because each time the crew needed to add or decrease the power for phugoid control, it was necessary for them to increase or take away power on either side. Otherwise, the airplane would have rolled over.

The crew faced additional problems of the precise timing required to arrive at the airport at the correct heading and altitude. Because they were unable to maintain a constant rate of descent, they used a basic DC-10 formula as a guide. For every one thousand feet of descent, the aircraft should travel 3 miles.

Since the crew had to contend with an airplane that wanted to turn only to the right, they compensated by making a series of right turns until they reached the vicinity of the airport. Refer to Fig. IV-A. When Flight 232 reached 21 miles north of the field, the approach controller requested they widen their turn slightly to the left in or-

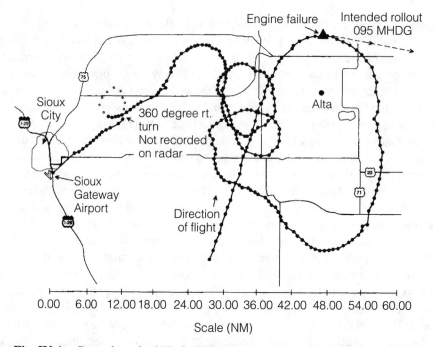

Engine failure   Intended rollout
095 MHDG

Sioux
City

360 degree rt.
turn
Not recorded
on radar

Alta

Sioux
Gateway
Airport

Direction
of flight

| 0.00 | 6.00 | 12.00 | 18.00 | 24.00 | 30.00 | 36.00 | 42.00 | 48.00 | 54.00 | 60.00 |

Scale (NM)

Fig. IV-A. *Ground track of Flight 232.* Adapted from NTSB

der to get aligned with the runway, and to stay away from the city. The captain emphasized, "whatever you do, keep us away from the city." Around 1551, the controller gave the crew a heading of 180 degrees, quickly followed by a warning that a 3400 foot tower was located 5 miles to their right.

Although the crew tried to increase the aircraft's bank angle to 30 degrees, the airplane did not respond normally, so they elected to fly straight ahead. At 1555, the controller advised the crew that if they could hold their altitude, their right turn to 180 degrees would put them 10 miles east of the airport. The captain answered, "that's what we're tryin' to do," and added that he wanted to get as close to the airport as possible.

The crew was able to maintain a heading of 180 degrees, and at 1557 the controller told them that the airport was, "twelve o'clock and one three miles." One minute later, the captain reported the runway in sight and thanked the controller for his help. The captain later noted that when they got down to 3500 feet and actually saw a runway off their nose, they were in shock. They had found the runway, but now it was time to safely land on it.

Flight 232 was aligned with runway 22, which had been closed to accommodate the fire equipment. The controller had been trying to vector the aircraft to runway 31, and that's where the focus of the rescue centered. Nevertheless, the controller quickly cleared runway 22 as the DC-10 approached the threshold. He and the captain discussed the runway length, 6600 feet compared to runway 31's 9000 feet, and the open field at the end of the runway.

At 1559:29, the "brace" call was announced in the cabin, followed 30 seconds later by a series of ground proximity warning (GPWS) alerts. From the captain's account, just as the aircraft came over the trees near the threshold, the airplane entered into yet another phugoid. They were at 300 feet, when the DC-10 started to pitch nose-down. On the CVR tape, at 1600:01, the captain said, "close the throttles." The check pilot responded, "I can't pull 'em off or we'll lose it. That's what's turnin' ya." Four seconds later, the first officer repeated "left throttle" several times. The captain remembered that as the nose lowered, the rate of descent and airspeed increased. At 1600:16 the aircraft struck the ground.

Impact and wreckage path

The airplane's right wing tip, right main landing gear, and the number 3 engine nacelle hit the ground nearly simultaneously. As the captain recalls, the nosewheel came down instantly, followed by the left main gear. The tail and right wing tip broke off, spilling fuel along the wreckage path. With no weight in the empennage, the aft section of the airplane came up causing the nose to bounce three times. For a matter of seconds the aircraft became airborne, but slammed back down, ripping the cockpit compartment away from the fuselage.

At the initial point of impact, an 18-inch hole was bored into the foot-thick concrete of the runway. The cause of such explosive damage was attributed to the final phugoid and a quartering tailwind that forced the airplane to touch down at 215 knots, 75 knots faster than a normal landing. The rate of descent was also recorded at an abnormally high 1854 fpm, instead of the usual 300 fpm.

Although most of the pieces of aircraft were found in a localized area, parts of the number 2 engine were found scattered in farm fields as long as nine months after the accident. The center fuselage was heavily damaged by the ground impact, as was the forward cabin section. The post-impact fire consumed various portions of the airplane, including both wings.

Accident survivability. Of the 296 persons aboard, 110 passengers and 1 flight attendant sustained fatal injuries. When the cockpit separated from the forward cabin, the first-class section became unprotected and exposed to the brunt of the crash. Seventeen of those passengers died, and the remaining eight received serious injuries.

The largest intact section of the airplane was the center portion of the fuselage that contained seat rows 9 to 30. This section came to rest inverted in a corn field and was destroyed by the postcrash fire. With the exception of two elderly passengers who died of asphyxia from smoke inhalation, all occupants in rows 9 to 21 were able to evacuate. Although the ceiling structure collapsed throughout the fuselage, the greatest amount of damage was found on the left side of the cabin near seats 22 to 30. Consequently, 33 passengers in that section died from either smoke inhalation or blunt trauma injuries. However, most of the passengers in the same numbered seats on the right side of the cabin were able to escape because there was less crushing damage in that area.

There were four infants and small children onboard. The flight attendants instructed the parents of each to place them on the floor and hold them there when the "brace" command was issued. The mother of a two-year-old stated that her son "flew up in the air" upon impact, but that she was able to grab and hold on to him. Two of the other children were thrown into the wreckage, but sustained only minor injuries. One infant died of smoke inhalation.

The investigation

The Safety Board determined that the accident sequence was initiated by a catastrophic separation of the stage 1 fan disk from the number 2 engine during cruise flight. The separation, fragmentation, and forceful discharge of uncontained stage 1 fan-rotor-assembly parts led to the loss of the three hydraulic systems that powered the airplane's flight controls. Refer to Fig. IV-B.

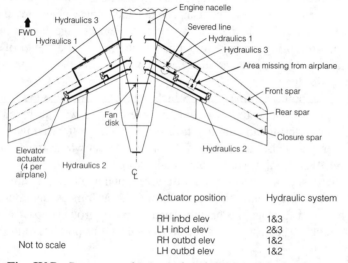

Actuator position	Hydraulic system
RH inbd elev	1&3
LH inbd elev	2&3
RH outbd elev	1&2
LH outbd elev	1&2

Fig. IV-B. *Damage to horizontal stabilizer and hydraulic systems.* Adapted from NTSB

Stage 1 fan disk. As Fig. IV-C illustrates, the General Electric CF6-6 engine fan rotor assembly consists of the large stage 1 disk and its attached fan blades and retainers. A smaller stage 2 disk and its attached blades and various mounting and balancing hardware also comprise this major assembly.

The stage 1 fan disk weighs 370 pounds and is machined titanium alloy forging about 32 inches in diameter. The rim is about 5 inches thick and is the outboard portion of the disk. The rim contains the axial "dovetail" slots that hold the fan blades. The bore is 3 inches thick and is the enlarged section of the disk adjacent to the 11-inch-diameter center hole. A disk web extends between the rim and the bore. The stage 2 fan disk is bolted to the aft face of the rim.

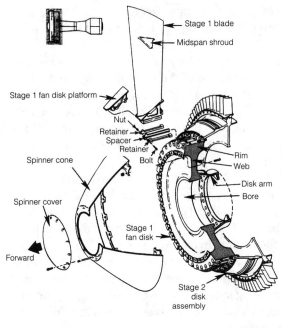

Stage 1 blade

Midspan shroud

Stage 1 fan disk platform

Nut
Retainer
Spacer
Retainer
Bolt

Spinner cone

Spinner cover

Forward

Stage 1
fan disk

Rim
Web

Disk arm
Bore

Stage 2
disk
assembly

Fig. IV-C. *DC-10 fan rotor assembly.*
Adapted from NTSB

NOTE: Stage 1 fan disk
highlighted

The primary, radial loads imposed on the stage 1 fan disk are in the dovetail slots. These loads come from the disk holding the fan blades against centrifugal forces during rotation of the assembly. The radial stress generally decreases toward the bore and are replaced by circumferential stresses. The forward most corner of the bore is exposed to the maximum level of circumferential stress.

Analysis of the fan disk. Examination of the fracture surfaces of the fan disk showed a region of fatigue on the inside diameter of the bore, which was believed to have begun in the early life of the disk. Investigators noted that the defected area cracked when the disk was exposed to the stresses associated with full engine power. Due to the particular geometry of the fan disk and the load paths within the disk, the fracture created a bending moment in the disk arm and web. This overstressed the disk, leading to the rupture of a blade segment. As soon as that occurred, the remainder of the disk was out of balance, causing other blades and fragments to blow outward. The right horizontal stabilizer and the aft lower fuselage area were subjected to the primary damage from this violent reaction.

Manufacturing defect. There are three primary steps in the manufacturing of titanium alloy fan disks: Material processing, forging, and final machining. In the first step, raw materials are melted and processed into a titanium alloy ingot (a casting mold for metal). The ingot is then shaped while in a molten furnace, and reformed into a billet (an ingot after it's mechanically elongated and reduced in diameter) for further processing. The second step involves cutting the billet into smaller pieces that are then forged into geometric shapes. The last step is the machining of the forged shape into an actual part.

The Safety Board determined that the half-inch long fatigue crack was formed in the titanium alloy material during the manufacture of the ingot from which the disk was forged. General Electric conducted many types of non-destructive inspections (NDI), including a macroetch process that indicates material-related defects. The process that was used in 1989 was similar to that used in 1971, and was performed on the machine-forged shape. According to the Board, the flaw would have been apparent if the part had been macroetched in its final part shape.

Hydraulic system. It was considered by Douglas Aircraft Company that a complete hydraulic failure on a DC-10 was virtually impossible. Because of the triple redundancy of the system, the aircraft was not designed for the pilot to manually operate the flight controls. The concept of a backup for a backup had worked so well that the airplane could still fly with only one functioning hydraulic control system. But in this case, the hydraulic lines to all three systems were severed or ruptured by the exploding shrapnel. As a result, the crew was at FL 370 with no ailerons, rudders, elevators, leading edge devices, trailing edge devices, wing spoilers, nosewheel steering, or brakes.

Two months after the accident, the FAA mandated Douglas Aircraft to enhance the design of the DC-10 hydraulic system that would preserve satisfactory flight control if a similar catastrophic failure were to occur. Although these enhancements appeared to have had the capability to protect the airplane, the Safety Board noted that it would not provide an additional margin of safety for certain failures; and the vulnerability of the DC-10 in such an event is still not fully known.

Flightcrew performance. The Safety Board believed that under the circumstances, the "flightcrew's performance was highly commendable and greatly exceeded reasonable expectation." They added that the interaction between the pilots during the emergency was "indicative of the value of cockpit [crew] resource management."

The captain is a tremendous supporter of CRM and has said: "I am firmly convinced that CRM played a very important part in our landing at Sioux City with any chance of survival. I also believe that its principles apply to no matter how many crewmembers are in the cockpit." Confirming the evidence and information in this text's chapter on CRM, the captain continued to address the subject for those who fly single-pilot aircraft: ". . . CRM does not imply just the use of . . . sources only in the cockpit—it is an 'everybody resource'—[there are] all sorts of resources available to them." He believed that although there were 103 years of cumulative flying experience in the cockpit, the crew would not have been able to get to the airport without the steady guidance of one controller from Sioux City approach. As the captain concluded: "Use them [your crew and ATC] as team members—you are not alone up there."

Lessons learned and practical applications

No better lessons can be learned from this accident than those described by the captain, himself. Use all of your available resources:

1. Work as a team. Tap into your fellow pilots' knowledge, skill, expertise—and hands. As noted in CRM research, by allowing the first officer to fly the airplane in an emergency situation, the captain then has the opportunity to evaluate the problem and make sound decisions.

2. Be open to suggestions. The captain viewed each crewmember's ideas as instrumental to the safe outcome of the flight.

3. Communicate clearly and directly. This applies to the entire flightcrew. If you get a chance, read the CVR transcripts from this accident, and those from less successful flights. You will immediately notice that every crewmember from Flight 232 communicated in a clear manner. There were no disjointed comments, confusing statements, or domineering attitudes.

4. Maintain cockpit discipline. The crew did not allow themselves to become distracted. They remained vigilant of the situation throughout the flight.

5. Keep ATC in the loop. The captain had commented that tensions were high, but hearing the steady voice of the approach controller provided a tremendous calming influence to the crew.

6. Brief flight attendants. Don't keep an emergency situation a secret. Passenger survival depends on a prepared cabin crew.

Case study references

Haynes, Alfred C. October 1991. "United 232: Coping with the Loss of All Flight Controls, Part I." *Air Line Pilot*: 10-14, 54.

Haynes, Alfred C. November 1991. "United 232: Coping with the Loss of All Flight Controls, Part II." *Air Line Pilot*.

National Transportation Safety Board. 1 November 1990. Aircraft Accident Report: United Airlines Flight 232. Sioux Gateway Airport, Sioux City, Iowa. July 19, 1989. Washington, D.C.

CASE STUDY IV-2: Air Canada Flight 797

Safety issues: In-flight fire, CRM, ADM

On 2 June 1983, an Air Canada DC-9-32 had been airborne for less than 90 minutes, and was at a cruising altitude of 33,000 feet when a fire broke out behind the wall panels in the aft lavatory. The crew made an emergency landing at the Greater Cincinnati Airport, Ohio.

Probable cause

The NTSB concluded that the probable causes of the accident were a fire of undetermined origin, an underestimation of fire severity, and misleading fire progress information provided to the captain. The time taken to evaluate the nature of the fire and to decide to initiate an emergency descent contributed to the severity of the accident.

NOTE: Additional notes and statements from the flightcrew were sent to the Safety Board that provided a thought-provoking commentary on the actual events as they unfolded. The first officer, in particular, refuted certain accusations and criticisms that the Board had imposed against the flightcrew's actions. Also, the Air Line Pilots Association (ALPA) submitted to the Safety Board a "Petition for Reconsideration of Probable Cause" that discussed the specific concerns of the captain.

Although the Board redressed their findings and made only minor revisions to the original accident report, the comments presented by the first officer and ALPA have been incorporated in this case study. The notes provided a unique perspective of the decision-making process by crewmembers who were suddenly thrown into an emergency situation.

History of flight

Flight 797 operated as a regularly scheduled international flight from Dallas, Texas, to Montreal, Canada, with an intermediate stop at Toronto, Canada. The DC-9 departed Dallas at 1625 Central Daylight Time with 41 passengers and 5 crewmembers aboard.

Pilot experience

The captain had approximately 13,000 total flight hours, 4939 in the DC-9. He had been a qualified DC-9 captain since 1974. The first officer had 5650 total flight hours, 2499 in the DC-9. He had held that position since 1979.

Weather

A warm front that extended from southern Ohio through western Kentucky caused rain showers to develop at the Cincinnati airport between 1830 and 1930. Radar echoes were spotted south and west of the airport and extended out to about 100 nm. The tops were reported at 13,000 feet and 14,000 feet.

The 1930 surface weather observations for the Cincinnati airport were as follows:

Ceiling 2500 feet scattered, measured 8000 feet overcast. Visibility 12 miles, light rain. Winds 190 degrees at 7 knots. Temperature/Dewpoint 63°F/55°F.

The accident

The investigation revealed that at 1851, the three circuit breakers connected to the aft lavatory's flush motor tripped in rapid succession. The captain immediately tried to reset them, followed eight minutes later with yet another unsuccessful attempt. At that point, the captain believed the flush motor had probably seized.

About then, a passenger seated in the last row notified a flight attendant of a strange odor that was quickly determined to be coming from the aft lavatory. The flight attendant grabbed a fire extinguisher and opened the lavatory door a few inches. She stated that a light gray smoke had filled the lavatory from the floor to the ceiling but there were no visible flames. After a cursory inspection, she closed the door and informed the other two flight attendants of the situation.

As soon as the lead flight attendant was advised of the fire, he immediately instructed the one flight attendant to move the passengers forward and to direct the air vents to the rear of the cabin. He told the other flight attendant to notify the captain, while he went back to the lavatory with the fire extinguisher. By the time he opened the door, an estimated one to two minutes from when the fire was discovered, the room was filled with thick curls of black smoke that originated from the seams of the wall just above the wash basin and at the ceiling. He then proceeded to saturate the washroom with CO_2 by spraying the paneling and the seam from which the smoke was seeping. He also sprayed the door of the trash bin, and then closed the lavatory door.

According to the CVR, at 1902:40, a flight attendant entered the cockpit and told the captain: "Excuse me, there's a fire in the washroom in the back, they're just . . . went back to go to put it out." Donning his oxygen mask, the captain immediately instructed the first officer to inspect the situation. Thick smoke had already reached the cabin and was hanging over the last two or three rows of seats when the first officer arrived to investigate the fire source. From his own account, the acridity and density of the smoke physically inhibited him from reaching the lavatory, and he quickly returned to the cockpit. He did not believe, as the Safety Board suggested, that wearing smoke goggles would have helped him get closer to the lavatory because the air was already too difficult to breathe.

At 1904:07, the first officer told the captain that "heavy" smoke had prevented him from entering the lavatory and added: "I think we'd better go down." He later testified that when he made that statement, he was not referring to an emergency descent because he "did not know enough" of the facts. He added that in an emergency descent he would have been required to take his seat, and since he thought he "could do more" by getting a closer look at the source or actually fighting the fire, he did not believe an emergency descent was necessary *at the time*.

Although the lead flight attendant had explained his firefighting efforts to the first officer, including his opinion that the fire source was not coming from the lavatory's trash bin, the captain did not receive that information. Consequently, the captain based his decisions on the assumption that the fire had originated in the trash bin, and "expected it [the fire] to be put out." In referencing ALPA's petition, between 1904:07 and 1906:54 the captain was given five separate and positive indications by three different crewmembers that the smoke was dissipating. The statements included those from flight attendants such as: "you don't have to worry, I think it's gonna be easing up," ". . . your first officer wanted me to tell you that [the lead flight attendant] has put a big discharge of CO_2 in the washroom, it seems to be subsiding . . ," and, "getting much better . . ." The first officer also informed the captain that, "it's starting to clear now." The first officer again offered to investigate the situation, so the captain gave his own smoke goggles to him because they were more easily accessible. The first officer testified that during the approximately 40 seconds between his returning to the cockpit and when he left the second time, he and the captain "did not discuss the type of fire at all."

Less than one minute later, while the first officer was out of the cockpit, the airplane experienced a series of electrical malfunctions. According to the captain's testimony, the master caution light illuminated, indicating that the aircraft's left ac and dc electrical systems had lost power. In seconds, the captain notified Indianapolis center of their "electrical problem," and to "standby." About 30 to 45 seconds later the Louisville high radar sector controller, who was working Flight 797, lost their radar beacon target but through a computer enhancement, the controller was able to regain the flight's position.

Shortly thereafter, the lead flight attendant informed the captain that the smoke was clearing. In actuality, the fire was intensifying and by the time the first officer reached the lavatory door, it was too hot to open. As he instructed the cabin crew to keep the door closed, he noticed a flight attendant towards the front of the airplane signaling him to hurry back to the cockpit. He quickly returned to the flight deck, and at 1907:11, he told the captain, "I don't like what's happening, I think we better go down, okay?" The captain told investigators that from the first officer's tone of voice, he knew that the first officer believed the fire was out of control and that he had to descend immediately.

About 30 seconds later, the master warning light illuminated along with the associated annunciator lights indicating the emergency ac and dc electrical buses had lost power. Both attitude directional indicators tumbled, and the captain ordered the first officer to activate the emergency power switch which would start battery power to the emergency buses. After doing so the attitude indicators steadied. However, because of the loss of ac power, the stabilizer trim remained inoperative for the duration of the flight.

The harrowing descent. *NOTE:* Due to the loss of electrical power, the CVR stopped taping and therefore only ATC transcripts and eyewitness testimonies were made available to the Safety Board.

At 1908:12, the high-radar-sector controller at Louisville received the "Mayday, Mayday, Mayday," call from Flight 797 and was told that there was a fire onboard and they were going down. In later testimony, the captain recalled beginning the rapid descent almost simultaneously with the emergency call. He said that the smoke continuously flowed forward and into the cockpit, as he donned smoke goggles in addition to the oxygen mask that he was already wearing. The first officer opted to use only his oxygen mask. Other than a problem with the captain's goggles steaming up from perspiration, both crewmembers suffered no difficulties in breathing. The Board did note, however, that the cockpit door was left open throughout the descent, although the captain did not remember it being open.

The captain stated that their steep descent rate was caused by the spoiler/speed brake handle being inadvertently moved to the full aft position, and the spoiler panels being deployed to the full-up position. As a result, they flew the descent at 310 KIAS and at a rate in excess of 6000 fpm.

The flight was 25 nm from Cincinnati when the controller asked the crew if they could make the airport. When they verified to ATC that they could make the field, the controller cleared them to 5000 feet. The controller later testified that because it was obvious to him that Flight 797 had to descend "immediately," he decided to issue the clearance and then coordinate it with the other ATC sectors. Due to the airplane's electrical failure, neither its transponder, which the crew had tuned it to the emergency code of 7700. or its Mode-C altitude information ever appeared on radar. Consequently, ATC could not track Flight 797's descent progress.

At 1909:29 the controller vectored the flight to 060 degrees, which placed the airport off the nose of the aircraft 20 miles. Meanwhile, Indianapolis center notified the approach controller at Cincinnati airport's TRACON of the emergency and for that facility to expect Flight 797 on their frequency momentarily. The tower's local controller then alerted the airport crash and rescue unit to be ready at the runway, and informed them of the nature and location of the fire.

The flightcrew contacted Cincinnati approach at 1910:25, declared an emergency, and said they were descending. The controller told them to plan for a runway 36 ILS approach, and cleared them to a 090° heading. Seconds later, the crew reported the cabin was filling with smoke.

At 1912:54, ATC noticed that at its present altitude of 8000 feet, Flight 797 was too high and fast to land on runway 36. The flight was then vectored to runway 27L, but because the crew did not have an operable heading indicator, the controller could only direct them to turn left or right. At 1914:03, the controller told the crew that he would assist them with a "no gyro" approach, in which ATC observes the radar track and instructs the crew with either "turn left/right," or "stop turn," as appropriate, to 27L, and cleared them down to 3500 feet. As the flight neared the airport, he continued to provide range information to the crew.

Flight 797 was about 12 nm from the airport when the crew reported being level at 2500 feet and in VFR conditions. The captain told investigators that they were in the clouds from about FL 250 to nearly 3000 feet, and did not encounter turbulence or icing. When he leveled off they were still in and out of the cloud bases so he decided to descend into better conditions. Before leaving 3000 feet, however, the captain instructed the first officer to depressurize the airplane in preparation for landing. After doing so, the first officer felt that, "because the smoke was getting bad at that point," and that he needed, "to do something," he turned off the air conditioning and pressurization packs because he believed that "they were just feeding the fire." Shortly thereafter, he opened his sliding window in an effort to clear the smoke from the cockpit, but closed it almost immediately because of the intense noise. He did, however, open and close his window several times during the final stage of flight.

At 1917:11, the controller asked the crew for passenger counts and the fuel amount, but was told, "We don't have time, it's getting worse here." About 30 seconds later, the flightcrew reported having the runway in sight, and was then cleared to land. According to the captain, because the horizontal stabilizer was inoperable, he had to extend the flaps and slats incrementally through the 0°, 5°, 15°, 25°, and 40° positions. He allowed his indicated airspeed to steady at each flap setting as he slowed to approach speed. The captain was able to maintain 140 KIAS for the final approach, and made a "maximum effort stop" by using extended spoilers and full brakes after touchdown. Due to the loss of the left and right ac buses, the antiskid system was inoperative, causing the four main tires to blow out on landing. Flight 797 finally came to a stop at approximately 1920.

Accident survivability and evacuation

According to survivor testimony, the toxic smoke in the cabin rapidly intensified during the descent. This was caused primarily from the loss of electrical power, which, in turn, closed the augmentation valve in the pressurization system eliminating fresh air from entering the cabin when the throttles were pulled back to flight idle. Once the crew leveled off after the descent and applied engine thrust, the exchange of air should have been restored. However, at least four minutes prior to landing, the first officer had turned off the air conditioning and pressurization packs. As a result, there was no air flow during those last few moments. The Safety Board believed that given the ferociousness of smoke build-up in the cabin and cockpit, his reaction was understandable. Nonetheless, had the packs been left on during those four minutes, almost two complete exchanges of cabin and cockpit air could have occurred. Without that exchange of air flow, the heat level and combustible gases increased at an accelerated rate, factors conducive to a flashfire.

During the descent, the flight attendants had distributed wet towels to the passengers and had instructed them to breathe through them or other articles of clothing in an attempt to filter the inhalation of the toxic gases. The Board believed that was a crucial step in aiding survival.

Survivors testified that by the time the airplane landed, the visibility in the cabin was virtually nonexistent from the ceiling down to about 1 foot from the floor. The massive quantities of hydrogen chloride, hydrogen fluoride, hydrogen cyanide, and carbon monoxide that the passengers and crew were exposed to also hindered their physical ability to evacuate. It had become nearly impossible for the flight attendants to yell instructions loud enough for the passengers to hear.

Figure IV-D illustrates that many of the victims were found either in the aisle or still in their seats. Some as close to an exit as in the first few rows. The Safety Board believed that a combination of three situations might have taken place to have prevented the remaining passengers from successfully evacuating. First, victims, most likely those in the aisle, might have been unable to quickly mobilize and were subsequently overcome with the toxic air or caught in the flashfire. Studies had determined that the cabin environment was not survivable within 20 and 30 seconds after the flashfire. Second, by the time the aircraft landed, victims might have already died or at the very least become incapacitated from the acrid fumes. And third, victims might have succumbed to a phenomenon known as *behavioral inaction*. Research has shown that under stressful conditions people might become catatonic or consciously wait until they are given further instructions. Because most of the flight attendants' oral commands were not heard in the cabin, some of the victims might have remained seated.

As soon as the aircraft stopped, two flight attendants immediately opened both forward doors, deployed and inflated the slides, and yelled several commands and directions to the passengers. They stayed at their posts until either the heat drove them away, or it appeared that no additional passengers were coming through the doors. The survivors who evacuated the aircraft through an overwing exit told investigators that

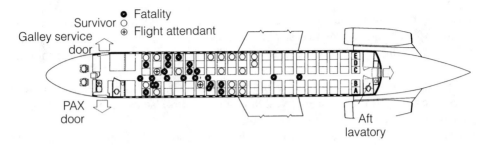

Fig. IV-D. *Diagram of location of survivors before leaving airplane and location of fatalities after the accident.* Adapted from NTSB.

they barely had the strength and presence of mind to negotiate through the dark cabin and out the window. Many survivors stated that before the landing they had memorized the number of rows between their seats and the exits. During the evacuation, they counted the rows by feeling the seatbacks as they moved aft, providing to some the only reference to an exit. Still others saw a dim glow of light through the smoke and followed it to an exit. By one survivor's account, the only way she knew she was near an exit was by a slight breeze that blew across the back of her legs.

Eighteen passengers and the three flight attendants were able to evacuate. The captain and first officer exited through their respective cockpit sliding windows. However, 60 to 90 seconds after the exits were opened, a flash fire consumed the cabin killing the remaining 23 passengers.

The investigation

The Safety Board was unable to determine the precise cause of the fire. However, they did offer a thorough analysis of cabin crew and flightcrew performance.

Cabin crew performance. In the Board's opinion, initial actions taken by the cabin crew to assess the origin and magnitude of the smoke were inadequate. Upon the lead flight attendant's investigation of the lavatory, he stated seeing smoke billowing through one of the walls, prompting him to douse the room with CO_2. Although CO_2 has no effect unless directed at the base of actual flames, it was conceivable that the lead flight attendant thought a heavy application of the chemical would at least cool down the lavatory enough to prevent the fire from spreading. In any event, the Board believed that he should have also checked the trash and supply containers in the lavatory, regardless of the assumption that the fire source was somewhere behind the wall panels.

The lead flight attendant was also aware of the procedure to use a fire ax if necessary to remove paneling and gain access to the fire. He testified that after assessing the situation, he was reluctant to use the ax, which was located in the cockpit, because he felt that he, "would have to destroy half the aircraft to get to it [the fire]." Since he did not know what critical components were behind the paneling, he was afraid of severely damaging the airplane and possibly causing a backdraft. Under the circumstances, the Board was unable to determine if any positive benefits could have been derived by actually battling the blaze. It was just as likely that the fire could have spread into the cabin.

Shortly after the fire was discovered in the lavatory, the flight attendants moved the passengers forward between rows 2 through 12. They also designated several male passengers to open the overwing exit windows in row 12.

During the descent, the cabin filled with black, acrid smoke from the ceiling down to about knee level. It was later determined by the Safety Board that the air was saturated with a variety of toxic gases including, hydrogen chloride, hydrogen fluoride, and hydrogen cyanide. It was noted that all of the survivors had covered their faces with either wet towels distributed by the flight attendants or articles of clothing. The wet towels acted as filters against the deadliest of the gases, but did not reduce the high levels of carbon monoxide. The survivors also attempted to take shallow breaths, but all reported that the smoke hurt their noses, throats, and chests, and caused their eyes to water. Regardless, the Board believed that the flight attendants' decision to hand out wet towels lessened the severity of smoke inhalation to many of the survivors.

In reference to the Air Canada Land Emergency Procedures manual, the lead flight attendant was to make numerous PA announcements advising passengers of the various safety information needed during an emergency landing and airplane evacuation. It further stated that if the PA system was inoperative, the lead flight attendant was responsible for retrieving the megaphone, located above row 2, and use it for the emergency instructions. However, the lead flight attendant in this case testified that when he realized the aft PA microphone didn't work, he thought it would have been "unwise to waste valuable time," to go forward and get the megaphone. Although all of the flight attendants attempted to brief passengers on how to assume the brace position, and other relevant emergency procedures, the smoke and toxic gases in the cabin became too overwhelming for them to adequately communicate those directives. As a result, many passengers heard only parts of the instructions, and some recalled not being able to hear anything.

The Safety Board noted that four minutes before the captain made the decision to make an emergency descent, the lead flight attendant told him not, ". . . [to] worry, I think it's gonna be easing up." Two minutes later, he again told the captain that the smoke was clearing. Although the first officer also made comments during this critical period, the lead flight attendant reinforced the captain's assumption that the fire was containable. Since the lead flight attendant was the only crewmember to have actually witnessed the rapid progression of the smoke, the Board believed that he should have made a more detailed account to the captain.

Flightcrew performance. According to the Board, when the circuit breakers tripped, there was no reason for the flightcrew to think that a serious situation was developing. The captain appropriately tried one reset, and then assumed the component was inoperative. Since the problem was in the flush motor, and its failure might ultimately cause passenger inconvenience, it could be argued that the captain should have asked a flight attendant to investigate. Although the smoke would have probably been noticed much sooner, the captain cannot be faulted for a case of 20/20 hind-sight.

However, the Safety Board believed that when it became clear to the captain that the origin and the severity of the fire could not be determined, he should have started the emergency descent. In the Board's estimation, the severity was evident at 1904:07 when the first officer returned to the cockpit after his initial inspection of the aft cabin. That was three minutes prior to the actual emergency descent. The Board noted that under their scenario, the aircraft was 14 nm from Louisville, Kentucky, which would have allowed the crew to land three to five minutes earlier than at Cincinnati. While the research data did not provide definitive answers as to whether a shortened flight time would have prevented the flashfire, the members of the Safety Board did believe that more passengers might have survived if they had not been exposed to the toxic gases for those final moments. Therefore, the Board concluded that the delayed decision to descend contributed to the severity of the accident.

In response to that statement, the ALPA petition noted that the captain was relying on the first officer to provide accurate reports on the situation. Hence, ALPA was critical of the Safety Board's determination concerning the captain's decision-making. In the opinion of the pilot association, it was incorrect to imply that the captain's decision was different from that of a more prudent captain under similar circumstances. They added, ". . . it is grossly unfair and unrealistic for the Safety Board to expect a flight crew to evaluate an incident in a dynamic situation, and reach the same conclusions in a few seconds that it took the . . . Board and all of its staff over a year to reach." Therefore, ALPA reiterated that the captain's decision was based on evidence provided by three different crewmembers on five different occasions, which indicated that the smoke was abating and that an emergency descent would not be required.

During the descent, the captain's difficulties were compounded by the condition of his flight instruments. At about 8000 feet, the emergency inverter and emergency ac bus was lost. Consequently, the airplane's ADIs, horizontal situation indicators (HSI), and radio magnetic indicators became inoperative. The only attitude-indicating instrument available to the captain was the small emergency standby ADI, which he used along with the airspeed indicator to fly the remainder of the descent, the traffic pattern, and the landing.

The cabin smoke elimination procedure required the airplane to be depressurized below 10,000 feet and both the aft pressure bulkhead door and right forward cabin door to be opened. The captain testified that he elected not to do the procedure because he needed the assistance of the first officer to fly the airplane safely. As the aircraft manufacturer noted, the procedure was intended to be used after a fire had already been extinguished, allowing any residual smoke to be sucked out the forward cabin door. However, since the fire was still burning on Flight 797, investigators believed that exposing the passengers to a rush of air would have created, "a very strong potential" that the fire would have been violently drawn into the cabin.

In the final analysis, the Board recognized the overwhelming circumstances that the captain and first officer had to endure. The captain was able to maintain control of the airplane despite the loss of engine and flight instruments, an inoperative horizontal stabilizer trim system, and unfamiliar longitudinal control forces. As a result, the Safety Board concluded that the captain exhibited outstanding airmanship without which the airplane and everyone on board would certainly have perished.

Lessons learned and practical applications

1. Be an effective crewmember. The captain relied on the first officer for accurate information on which decisions were based. Don't be the weak link of a crew.

2. Take command. When you're not receiving clear communication, or when there seems to be a conflict between verbal reports and visual observations—this also applies to a bad gut feeling—go with your own best judgment. In this case, the captain kept getting positive reports from the first officer and flight attendants, yet he could still see black smoke occasionally filling the cabin.

3. Abnormal situations might be interrelated. Eleven minutes had transpired from the time the circuit breaker tripped to the discovery of the fire. Looking at the circumstances at face value, the crew should have seen the connection between a possible problem with the flush motor, and a fire in the lavatory. A fire in the trash bin, as the captain had originally assumed, would have been quite coincidental.

4. Communicate clearly and directly. For example, when the first officer returned to the cockpit after investigating the origin of the smoke, he told the captain: "I don't like what's happening, I think we better go down, okay?" The first officer later testified that when he made the comment, he wasn't thinking of an emergency descent. Say what you really mean.

Case study reference

National Transportation Safety Board. 31 January 1986. Aircraft Accident Report: Air Canada Flight 797, McDonnell Douglas DC-9-32, C-FTLU. Greater Cincinnati International Airport, Covington, KY. June 2, 1983. Washington, D.C.

CASE STUDY IV-3: Continental Express Flight 2574

Safety issues: maintenance procedures

On 11 September 1991, a Continental Express Embraer Brasilia commuter experienced an in-flight, structural breakup near Eagle Lake, Texas.

Probable cause

The NTSB determined that the probable cause of this accident was the failure of Continental Express maintenance and inspection personnel to adhere to proper maintenance and quality assurance procedures for the airplane's horizontal stabilizer deice boots. This situation led to the sudden in-flight loss of the partially secured left horizontal stabilizer leading edge, and the immediate and severe nose-down pitchover and breakup of the airplane. Contributing to the cause of the accident was the failure of Continental Express management to ensure compliance with the approved maintenance procedures; and the failure of FAA surveillance to detect and verify compliance with approved procedures. There was one dissenting member of the Safety Board who believed that the failure of senior management to establish a corporate culture which encouraged and enforced adherence to approved maintenance and quality assurance procedures should have been a part of the probable cause.

History of flight

Flight 2574 was a regularly scheduled Part 135 commuter flight from Laredo, Texas, to Houston Intercontinental Airport, Texas. The EMB-120 Brasilia departed around 0909 Central Daylight Time.

Pilot experience

The captain had 4243 total flight hours, 2468 in the EMB-120. His last line check was a month before the accident. The first officer had 11,543 total flight hours, 1066 in the EMB-120. He was also a current captain for the airline.

Weather

About 15 minutes before the accident, the reported weather in the area included, Ceiling; 3000 feet broken, 10,000 feet broken. Visibility; 6 miles in haze. Temperature/dewpoint; 83 degrees F/74 degrees F. Wind; 070 degrees at 7 knots.

The accident

The flight from Laredo was apparently uneventful, and at 0948 the crew made their initial contact with Houston center in preparation for landing at Intercontinental. For the next several minutes, the controller issued the crew a descent clearance and a couple of heading changes. At 0959:57, the crew responded to their last radio transmission. After noticing that Flight 2574's radar beacon had disappeared, the controller began a number of unsuccessful attempts to raise the crew on the radio.

The CVR tape revealed that nothing unusual had taken place during the descent until 1003:07 when sounds of objects being tossed about the cockpit were recorded. No conversation was heard, just aural warnings from the aircraft's systems and the mechanical sounds of an airplane breaking up in flight. At 1003:13, the sound of rushing wind was picked up by the cockpit microphone, followed by the abrupt ending of the tape 27 seconds later. The airplane crashed in a cornfield, near Eagle Lake, Texas.

Impact and wreckage

The airplane was found in an upright, wings-level position, but burning and partially imbedded in the ground. Parts of the airplane, including all eight propeller blades, were scattered within a 1.5 nm radius of the main wreckage. The horizontal stabilizer had separated from the empennage before impact, and was lying about 650 feet west-southwest of the crash site. The left leading edge/deice boot was missing from the horizontal stabilizer and was located nearly ¾ of a mile west of the primary point of impact. Both wings were still attached to the fuselage, but they were severely mangled and missing small sections.

Accident survivability. This accident was not survivable due to the high impact forces. The passengers and crewmembers all sustained fatal injuries.

The investigation

The readout of the airplane's FDR showed Flight 2574 was in a normal descent until passing through 11,500 feet msl. Two cycles of a rapid decrease and increase in propeller speed were recorded. The aircraft suddenly pitched-over with a force of at least three-and-a-half negative Gs, along with severe roll and yaw moments, and wide variances of engine thrust.

Following a series of engine and propeller tests, the Board determined that the right engine had oversped well beyond its 100 percent limit, and overtorqued before impact. The left engine, however, appeared to have been operating normally. Investigators found no evidence of defects in either engine or propeller assemblies prior to the crash. The damage noted in both systems was consistent with extreme changes in the airplane's attitude, and in the case of the left engine, separation from the wing before ground impact.

Investigators also discovered the 47 screws that attached the upper surface of the leading edge assembly for the left side of the horizontal stabilizer were missing. After further examination, they found no evidence to suggest that the screws had been ripped away during the in-flight breakup. The Board noted that a "lip" had formed on the forwardmost frame on the left lower side of the horizontal stabilizer spar cap. This frame had lower attachment screws which mounted the underside portion of the left side, leading edge assembly to the stabilizer. Those screws remained in place, but the leading edge assembly had separated from the stabilizer. The spar cap showed signs of being pulled down to the point of protruding into the wind stream. According to the Safety Board, that type of pulling action was consistent with the left side, leading edge assembly having been sheared down and away from the lower attachment screws. Therefore, the physical evidence proved that the top side of the leading edge assembly had not been secured, but by the lower screws still holding the structure, caused the frame to pull down and form the "lip."

Maintenance procedures. A review of the airplane's maintenance records showed that a couple of weeks prior to the accident, a quality control inspector had noted dry-rotted pin holes in the aircraft's deice boots. They were scheduled for replacement the night before the crash.

Investigators conducted numerous interviews with the airline maintenance personnel, inspectors, and supervisors who were working the night before the accident. Those interviews revealed that the aircraft had been pulled into the Continental Express hangar at Houston-Intercontinental around 2130, and the repair took place through two overlapping shifts. Normal procedures called for the old deice boot to be stripped from the composite structure of the leading edge while it was still attached to the stabilizer. Then, the deice fluid lines would be disconnected, the leading edge removed and a new deice boot bonded onto the surface of the leading edge. The final step was to reinstall the deice boot assembly on the horizontal stabilizer with two sets of 47 attachment screws.

According to the Board, the two mechanics and inspector who initially worked on the airplane used a hydraulic lift to gain access to the tail section. The inspector removed the top side attachment screws from both sides of the stabilizer, and put them in a bag, which placed on the lift. The bottom screws were left untouched. Shortly thereafter, another shift came on duty and those mechanics removed the right leading edge assembly from the stabilizer. A new deice boot was then bonded to the assembly.

The aircraft was moved outside of the hangar to make room for a repair on another airplane. The remainder of work on the stabilizer was completed outside, with no direct light. A mechanic reinstalled the right side leading edge assembly by using old and new screws to attach the top and bottom of the assembly to the right horizontal stabilizer.

The mechanic told investigators that the second shift had not been thoroughly briefed by the previous supervisor on what work had already been done on the airplane. He added, that the supervisor explained to him that the other mechanics found a few stripped screws which prevented the right leading edge assembly from being removed before the shift change. The Board noted that there were no entries in the maintenance log by the first-shift supervisor.

Although several errors were made by individuals responsible for the airworthiness of the airplane, the Safety Board believed that the reasons for those mistakes were complex and not simply related to a single factor. Rather, they questioned the effectiveness of the airline's quality assurance program which allowed mechanics not to follow established company procedures. According to the Board, the inspectors were among the "worst offenders" in that they created a lax work environment with inadequate cross-checks. In general, the airline's maintenance department personnel were aware of the correct procedures, but management did not pursue an effective safety orientation for its employees.

In a related issue, the FAA's Principle Maintenance Inspector (PMI) responsible for the Continental Express operation at Houston-Intercontinental was only one week on the job prior to the accident. Both he and his predecessor told investigators that they had a heavy workload that sometimes required them to work on weekends and evenings. As a result, on-site inspections of the airline's maintenance facility were limited. It was also known by Continental Express mechanics that FAA visits to the hangar were infrequent, and were always announced in advance.

Aerodynamic and structural failures. According to the Board, Flight 2574 was descending at a normal 10 degrees of pitch down and at 260 KIAS. As the aircraft passed through 11,500 feet, the leading edge of the left horizontal stabilizer separated from the airframe. The loss of the leading edge exposed the front spar of the left side of the stabilizer to the wind stream, causing an aerodynamic stall. The downforce produced by the horizontal stabilizer created a sizable and violent nose-down pitching moment. A peak load factor of approximately five negative Gs was recorded in only one second.

The airframe's pitch attitude dropped down to 68 degrees as the aircraft rolled 15 degrees to the right. By the time Flight 2574 reached 9500 feet, its speed had increased to 280 KIAS. The airplane experienced another severe negative load factor, causing the left wing to fail and the right wing tip to separate.

The Safety Board believed that the lift produced by the right wing induced a roll rate that exceeded 160 degrees per second. The combination of the high airspeed and excessive roll rate caused the horizontal stabilizer and left engine to rip away in-flight. The airplane then entered into an uncontrollable and unrecoverable right spin until impact.

Crew preflight performance. The aircraft was cleared for an 0700 flight from Houston to Laredo. Investigators found no evidence that suggested the crew was aware of any overnight work to the horizontal stabilizer, nor were they required to know that information from standing FARs and airline policy.

In the Safety Board's opinion, if the pilots had known the critical nature of the repair, they might have discussed it with maintenance, or had requested a visual inspection of the stabilizer's upper surface.

Lessons learned and practical applications

Without the benefit of a CVR, specific analysis of flightcrew performance was unobtainable. However, the Board made three points that prove to be good reminders for pilots.

1. Seemingly minor distractions can lead to disaster. Maintenance personnel had become distracted with the discovery of stripped screws, the shift change, and moving the airplane. Therefore, the work was not completed.

2. Finish what you've started. With regards to flying, complete all of your checklists and procedures before you move on to another task. No matter how familiar you are with the airplane or route of flight, you can easily forget to finish important duties, if you stop half-way through.

3. Make maintenance inquiries. You are pilot-in-command, and responsible for the safety of your passengers and crew. Therefore, you have the right to know what type of recent maintenance has been performed on the aircraft before you accept it.

Case study reference

National Transportation Safety Board. 21 July 1992. Aircraft Accident Report: Britt Airways, d/b/a Continental Express Flight 2574, In-flight Structural Breakup. EMB-120RT, N33701. Eagle Lake, Texas. September 11, 1991. Washington, D.C.

CASE STUDY IV-4: Aloha Airlines Flight 243

Safety issues: airline maintenance program, aircraft flight cycles, CRM, role of ATC, FAA oversight

On 28 April 1988, an Aloha 737 experienced an explosive decompression and structural failure over the Pacific Ocean.

Probable cause

The NTSB determined that the probable cause of this accident was the failure of the Aloha Airlines maintenance program to detect the presence of significant disbonding and fatigue damage. This situation ultimately led to the failure of the lap joint at S-10L and the separation of the fuselage upper lobe.

Contributing factors were: (1) failure of Aloha Airlines management to properly supervise its maintenance force; (2) failure of the FAA to properly evaluate the Aloha maintenance program; (3) failure of the FAA to require an Airworthiness Directive inspection of all the lap joints proposed by a Boeing Alert Service Bulletin; and (4) the lack of a complete terminating action after the discovery of early production difficulties in the 737 cold-bond lap joint, which resulted in low-bond durability, corrosion, and premature fatigue cracking.

History of flight

Flight 243 was a regularly scheduled passenger flight from Hilo, Hawaii to Honolulu, Hawaii. The 737 departed Hilo with 89 passengers and five crewmembers onboard.

Pilot experience

The captain had 8500 total flight hours, 6700 in the 737. He had been with Aloha since 1977 and had logged 400 hours as a 737 captain. The first officer had 8000 total flight hours, 3500 in the 737. She had held that position since 1979.

Weather

The weather conditions were VMC, and not a factor in the accident.

The accident

On the day of the accident, the aircraft had flown on six inter-island legs prior to being scheduled for Flight 243. The pilots on the morning flights stated that the trips were uneventful, and there had been no maintenance write-ups. Likewise, the crew of Flight 243 told investigators that the climbout and cruise portions of the flight were routine. All of that changed, however, when at FL 240, the crew heard a loud "clap" or "whooshing" sound followed by a wind noise behind them. The first officer's head was jerked backward and she reported seeing debris, including pieces of gray insulation, floating in the cockpit. The captain noticed that the cockpit door was missing, and that, "there was blue sky where the first-class ceiling had been." The captain immediately took over the controls of the airplane. He described the airplane attitude as rolling slightly left and right, and that the flight controls felt "loose."

Because of the decompression, both pilots donned oxygen masks. The captain began an emergency descent, extended the speed brakes, and continued to descend between 280 and 290 KIAS. The wind noise was deafening, so the crew had to use hand signals to communicate. The first officer tuned the transponder to the emergency code 7700, and tried to notify Honolulu Center that the flight was diverting to Maui. Because of the cockpit noise level, she was unable to hear any radio transmissions.

The controller did not receive the first officer's initial communication, but saw the 7700 transponder return when the aircraft was about 23 nm south-southwest of Maui's Kahalui Airport. He made several unsuccessful attempts to contact the flight. Shortly thereafter, the first officer switched to the Maui tower frequency as the aircraft passed through 14,000 feet. She informed the Maui controller of the rapid decompression, declared an emergency, and requested rescue equipment. The controller acknowledged the call and notified the airport's fire department of the situation.

The aircraft was flying beyond the local controller's frequency range, so the crew was told to tune in Maui approach control. The flight was never heard on that frequency, so the crew continued to communicate with the local controller. The first officer informed him that: "We're going to need assistance. We cannot communicate with the flight attendants. We'll need assistance for the passengers when we land."

The captain stated that he began slowing the airplane as the flight approached 10,000 feet. He retracted the speed brakes, removed his oxygen mask, and began a gradual turn toward Maui's runway 02. At 210 KIAS, the wind noise subsided enough that the crew was able to orally communicate. Flaps were lowered to five degrees and the crew noticed that the aircraft became hard to control if the flap setting was increased. The captain also found that the jet was less controllable below 170 KIAS, so he elected to fly that speed for the approach and landing.

At the normal point in the approach path, the first officer lowered the landing gear but did not get a green light for the nose gear. After a couple of attempts to manually lower the nose gear, the light still did not illuminate. She notified the local controller that, "We won't have a nose gear . . . we'll need all the equipment you've got."

As the captain advanced the power levels for the approach, he sensed a yawing motion, and determined that the number 1 engine had failed. At 170 to 200 KIAS, he placed the engine start switch to the "flight" position, but there was no response. At 4 miles out, the captain managed to establish a normal descent profile on the final approach. He noted that the airplane was "shaking a little, rocking slightly, and felt springy."

About 13 minutes after the rapid decompression, the captain successfully touched down at Maui. The nose gear held and the pilot used the Number 2 engine thrust reverser and brakes to stop the airplane. He then lowered the flaps to 40 degrees to aid in the passenger evacuation.

Damage to airplane

Refer to Fig. IV-E. A major portion of the upper-crown skin and structure of section 43 separated in-flight. The damaged area extended from slightly aft of the main cabin entrance door rearward about 18 feet to the area just forward of the wings, and from the left side of the cabin at the floor level to the right side window level.

Accident survivability. All passengers were seated when the decompression occurred. However, the lead flight attendant was swept out of the cabin through a hole in the left side of the fuselage, near row five. One of the other flight attendants was thrown to the floor and sustained minor bruises. She subsequently was able to crawl up and down the aisle to render assistance and calm the passengers. The third flight attendant was struck in the head by debris and also thrown to the floor. She suffered serious injuries, including a concussion and severe head lacerations.

Sixty-five passengers received injuries including electrical shock burns, a skull fracture, broken limbs, and multiple face and head lacerations. Passengers seated in the forward section of the cabin received the most serious injuries. Those seated in rows 8 through 21 sustained mostly minor injuries. Twenty-five passengers reported no injuries and continued to their destinations that same evening.

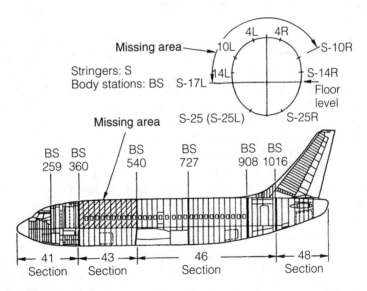

Fig. IV-E. *Boeing 737-200 Body Stations, Stringers, and Section Locations.* Adapted from NTSB

The investigation

A post-accident examination of the aircraft revealed that the remaining structure did not contain the origin of the failure. Since the missing section was not recovered, the Safety Board analyzed the undamaged airframe and the airworthiness history of the airplane. Based on the physical evidence, investigators concluded that the primary failure was located between Body Stations (BS) 360 and 540, and that butt joints were pulled away in a tension overload caused by the failure. As a result, the areas along the longitudinal separation of section 43 were examined as a likely spot for the origin of the failure. Since very little of this structure was found from the left side of the fuselage, investigators noted the collateral damage on the aircraft to make a determination. Pieces were believed to have been ingested in the Number 1 engine, and the leading edges of the left wing and horizontal stabilizer were also damaged. The remaining right side of the fuselage portion of section 43 was heavily distorted and bent outward more than 90 degrees. The floor beams at BS 420, 440, 460, 480, and 500 were broken all the way through. In addition, adjacent floor beams at BS 400 and 500A were severely cracked, and most of the center floor panels on the left side from BS 360 to 947 had lifted.

The right side of the cabin floor panels were not displaced. The comparison of the damage between the left side and right side suggested that the initial fracture was on the left side of the fuselage. Further evidence indicated that the defect ran longitudinally along the fuselage. As the cabin pressure in the upper lobe was released, the pressure in the lower lobe was contained by the floor. However, since the cabin floor was not designed to withstand a large pressure differential, it subsequently deflected upward during the decompression.

In the final analysis, investigators believed that the point of maximum floor deflection occurred at seat row 3, probably near BS 440. Based on the aircraft's damage pattern and the service history of the lap joints on earlier 737s, the Board determined that the most likely sites of the failure were at S-4, S-10, and S-14. Had the fuselage first separated along S-14L, below the window line and above the floor line, the force and internal pressure would have been insufficient to bend the structure and break the frames. But because the fuselage on the left side was torn extensively into the lower lobe, and the fuselage frames had separated above and below the floor line, the Safety Board concluded that the failure was probably at the S-10L lap joint.

Fatigue cracking. In the remaining fuselage, investigators noted several fatigue cracks running longitudinally at BS 520. The Safety Board believed that such cracking was indicative of the type of preexistent structural condition that was present along random areas of the lap joint at S-10L. A passenger later reported having seen a skin crack aft of the forward entry door while boarding the airplane. The location matched the top rivet row at the S-10L lap joint aft of BS 360. A 1974 Boeing Service Bulletin had explicitly defined the disbonding problems that 737s were experiencing with lap joints. This resulted in corrosion and probable fatigue cracking.

The Board determined that it was likely that numerous small fatigue cracks along S-10L had joined to form a large crack similar to the one observed by the passenger. It was also believed these cracks formed rapidly to cause the catastrophic failure of a major section of the fuselage.

Maintenance program. As a result of the cumulative evidence, the Safety Board recommended that sufficient information regarding lap-joint problems be implemented in the airline's maintenance program. In addition, because Aloha jets routinely flew short distances between the Hawaiian islands, the aircraft rapidly accumulated flight cycles (takeoffs and landings). The Board further noted that the program did not adequately recognize and consider the effect of this high volume of cycles. Investigators pointed out that the dominant concern in the development of fatigue cracks in pressurized fuselages was the result of flight and landing loads. Aloha's mechanics allowed one-and-a-half times the number of flight cycles to accumulate on an airplane before the appropriate inspections. Investigators also noted that the airline permitted corrosion inspections to be performed in lengthy eight-year intervals.

In the final accident report, the Safety Board stated that Aloha Airlines had sufficient information available to them that would have alerted maintenance personnel to the cracking problems associated with the deterioration of lap-joint bonds. The Board believed that the airline should have followed a maintenance program to detect and repair this cracking before it reached a critical condition. Overall, the Board determined that Aloha's maintenance program was inadequate for the unique flight regime and environment in which its aircraft normally operated.

FAA oversight. In the wake of this accident, the Safety Board issued numerous safety recommendations pertaining to the surveillance of air carrier maintenance by the FAA. They noted that a manpower shortage and insufficient training for inspectors hindered a satisfactory level of FAA guidance to airlines, such as Aloha. The Board also expressed a concern over a practice of rubber stamping approvals and endorsements of an air carrier's operations and maintenance programs.

Lessons learned and practical applications

Although the NTSB concentrated its investigation on the structural failures and maintenance oversight, there are a few valuable insights that can still be picked out of the final analysis.

1. The importance of CRM. Many pilots have a tough enough time communicating in a normal manner. Imagine the coordination needed between members of the crew of Flight 243 when they could communicate only through hand signals. If you're in a poor CRM situation during a routine flight, then you're going to be even worse off when an emergency suddenly strikes.

2. It's a team effort. When the first officer finally made initial contact with the tower controller, she told him that they were declaring an emergency. Although the radio transmission was difficult to understand, the controller did hear the words "Aloha" and "emergency." Rather than clarifying the nature of the emergency and location of the aircraft, his first inquiry was to verify the flight number. There were six separate exchanges between the controller and first officer concerning this relatively minor point. The crew needed assistance, not a debate over their flight number. Just as pilots need to prioritize their tasks during an emergency, so, too, do controllers.

Case study reference

National Transportation Safety Board. 14 June 1989. Aircraft Accident Report: Aloha Airlines, Flight 243, Boeing 737-200, N73711. Near Maui, Hawaii. April 28, 1988. Washington, D.C.

Index

A

About the author

Shari Stamford Krause, Ph.D., operates her own aviation research company. She is a faculty member at Embry-Riddle Aeronautical University and has assisted with graduate research at Pacific Western University. Dr. Krause holds a bachelor of science degree in aviation technology from Metropolitan State College, a masters of aeronautical science degree from Embry-Riddle, and a doctor of philosophy in management from Pacific Western. She has been a licensed pilot since 1980.

Other Bestsellers of Related Interest

Avoiding Mid-Air Collisions
—Shari Stamford Krause, Ph.D.
Concise, easy-to-understand information on how to steer clear of other aircraft during all phases of flight. A virtual training course—and a unique, integrated approach to a serious safety issue.
0-07-035945-8 $16.95paper
0-07-035944-X $27.95 hard

Weather Patterns and Phenomena: A Pilot's Guide
—Thomas P. Turner
A volume in the TAB Practical Flying Series, this book tells pilots how to assess aviation weather hazards correctly and confidently. Chapters cover weather theory in depth, including specific hazards such as thunderstorms, turbulence, reduced visibility, ice, and distinct regional weather patterns.
0-07-065602-9   16.95 paper

Business & General Aviation Aircraft Pilot Reports
—Aviation Week & Space Technology Magazine
Read all about the most interesting general aviation aircraft today from these first-hand accounts from America's premier aviation writers.
0-07-003092-8 $19.95

Military Aircraft Pilot Report
—Aviation Week & Space Technology Magazine
Read all about the most interesting military aircraft today from these first-hand accounts from America's premier aviation writers.
0-07-003089-8 $19.95

Flying Jets
—Linda D. Pendleton
A basic but solid overview of jet systems, theory, and operations for pilots positioning their flying careers for the big time.
0-07-049296-4 $40.00i hard

Commercial & Regional Transport Aircraft Pilot Reports
—*Aviation Week & Space Technology Magazine*
Read all about the most interesting air carrier aircraft today from these first-hand accounts from America's premiere aviation writers.
0-07-003167-3 $21.95 paper

Helicopter Pilot Reports
— *Aviation Week & Space Technology Magazine*
Read all about the most interesting helicopters from these first-hand accounts from America's premier aviation writers.
0-07-003168-1 $21.95 paper

Handling In-Flight Emergencies
—*Jerry A. Eichenberger*
Learn how to prepare for in-flight emergencies before becoming involved in one.
0-07-015093-1 $21.95 paper
0-07-015092-3 $32.95 hard

EASY ORDER FORM— SATISFACTION GUARANTEED

How to Order

Call 1-800-822-8158
24 hours a day,
7 days a week
in U.S. and Canada

Mail this coupon to:
McGraw-Hill, Inc.
P.O. Box 182067
Columbus, OH 43218-2607

Fax your order to:
614-759-3644

EMAIL
70007.1531@COMPUSERVE.COM
COMPUSERVE: GO MH

Ship to:
Name _____
Address _____
City/State/Zip _____
Daytime Telephone No. _____

Thank you for your order!

ITEM NO.	QUANTITY	AMT.

Method of Payment:
☐ Check or money order enclosed (payable to McGraw-Hill)
☐ DISCOVER
☐ AMERICAN EXPRESS Cards
☐ VISA
☐ MasterCard

Shipping & Handling charge from chart below	
Subtotal	
Please add applicable state & local sales tax	
TOTAL	

Account No. ☐☐☐☐☐☐☐☐☐☐☐☐☐☐☐☐

Signature _____ Exp. Date _____
Order invalid without signature

In a hurry? Call 1-800-822-8158 anytime, day or night, or visit your local bookstore.

Key = BC95ZZA

Shipping and Handling Charges

Order Amount	Within U.S.	Outside U.S.
Less than $15	$3.50	$5.50
$15.00 - $24.99	$4.00	$6.00
$25.00 - $49.99	$5.00	$7.00
$50.00 - $74.49	$6.00	$8.00
$75.00 - and up	$7.00	$9.00